I0715342

Around the World in 80 Pots

Text © Ashmolean Museum, University of Oxford
Design and layout © Carlton Books Limited 2023

First published by Carlton Books Limited in 2023.
This edition published by Welbeck.
An Imprint of HEADLINE PUBLISHING GROUP LIMITED

2

Cataloguing in Publication Data is available from the British Library

ISBN 9781802792393

Printed in China

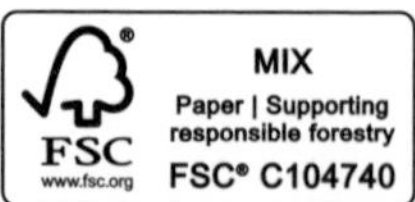

Headline's policy is to use papers that are natural, renewable and recyclable products and made from wood grown in well-managed forests and other controlled sources. The logging and manufacturing processes are expected to conform to the environmental regulations of the country of origin.

HEADLINE PUBLISHING GROUP LIMITED

An Hachette UK Company
Carmelite House
50 Victoria Embankment
London EC4Y 0DZ

The authorised representative in the EEA is Hachette Ireland,
8 Castlecourt Centre, Dublin 15, D15 XTP3, Ireland (email: info@hbgi.ie)

www.headline.co.uk
www.hachette.co.uk

Around the World in 80 Pots

Foreword by
Keith Brymer Jones

The story of humanity told through beautiful ceramics

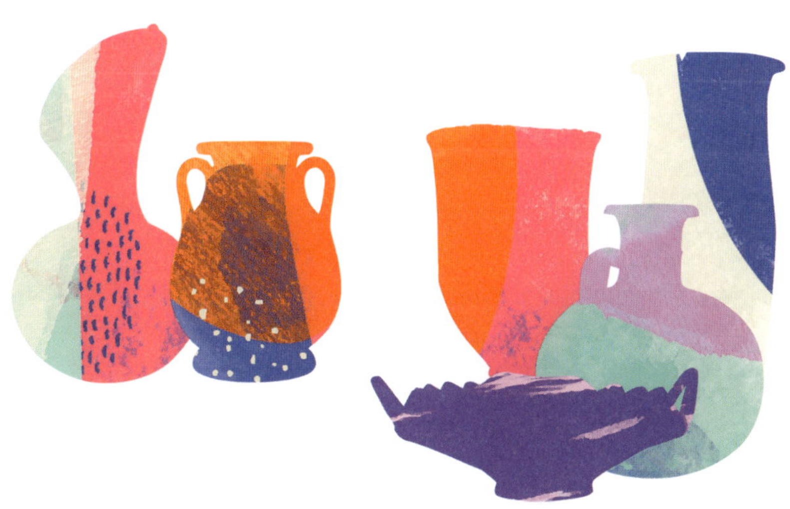

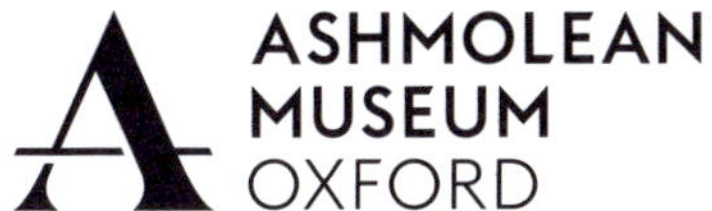

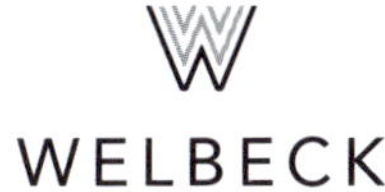

Timeline

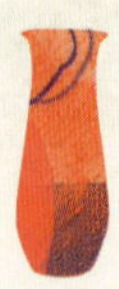

Decorated pottery jar
Egypt, predynastic period,
*c.*3800–3450 BCE

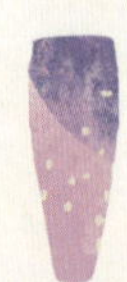

Black-topped pottery beaker
Egypt, *c.*3600 BCE

"Skidding goat" pottery vessel
Iran, *c.*3500 BCE

Bowl
Iraq, *c.*3100 BCE

Polychrome Nal funerary pottery
Pakistan, 3000–2001 BCE

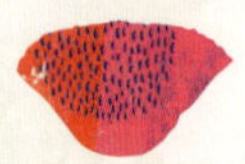

Model chariot with four-wheeler battle car
Iraq, *c.*2950– *c.*2575 BCE

Terracotta figure of a bull or ox
Pakistan, 2500–2300 BCE

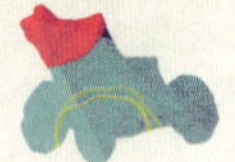

Indus Valley pots
India, 2500–1900 BCE

Pottery lion
Hierakonpolis, Egypt
Old Kingdom, Sixth Dynasty,
c. 2325–2175 BCE

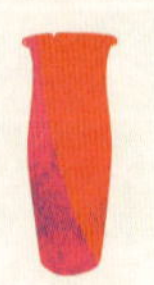

Terracotta votive figurines
Crete, *c.*2100–*c.*1700 BCE

Red polished stag-shaped figurine with three legs
Cyprus, *c.*2100–*c.*1850 BCE

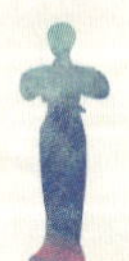

***Kernos* with two rows of containers and central bowl**
Greece, *c.*2300–*c.*2100 BCE

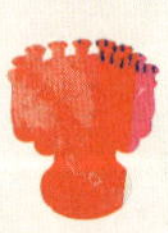

Sumerian king list
Iraq, *c.*1800 BCE

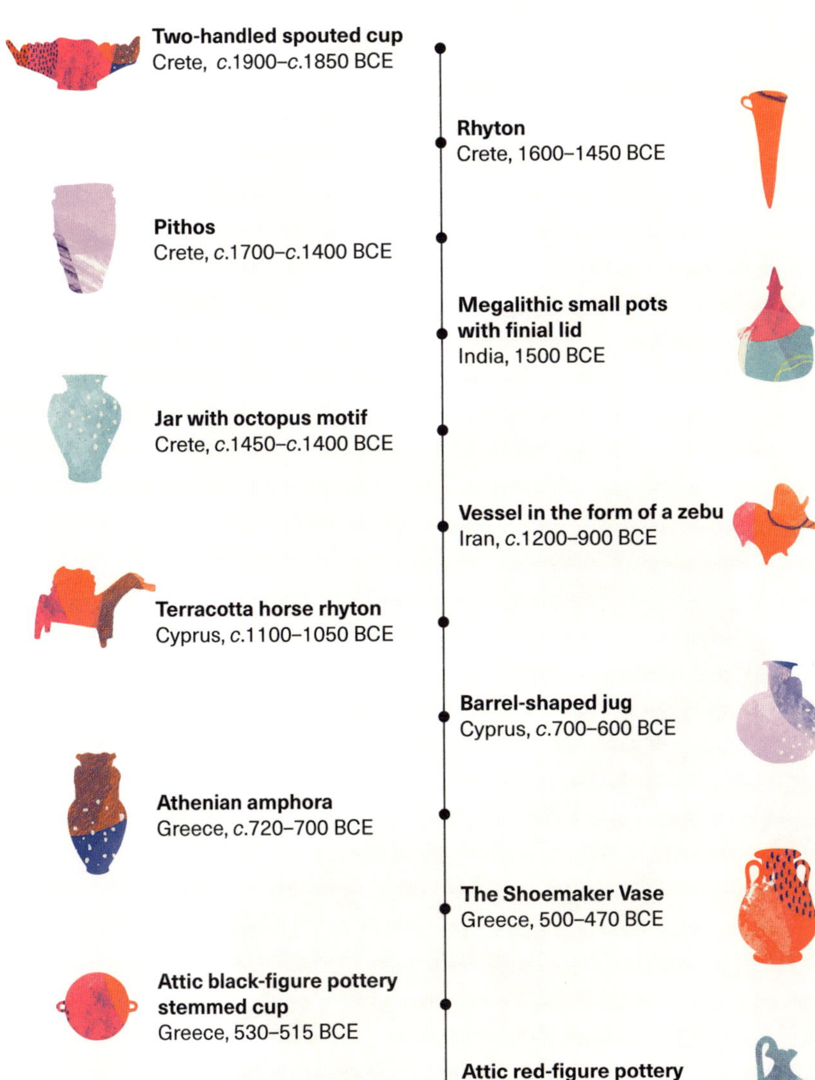

Two-handled spouted cup
Crete, *c.*1900–*c.*1850 BCE

Rhyton
Crete, 1600–1450 BCE

Pithos
Crete, *c.*1700–*c.*1400 BCE

Megalithic small pots with finial lid
India, 1500 BCE

Jar with octopus motif
Crete, *c.*1450–*c.*1400 BCE

Vessel in the form of a zebu
Iran, *c.*1200–900 BCE

Terracotta horse rhyton
Cyprus, *c.*1100–1050 BCE

Barrel-shaped jug
Cyprus, *c.*700–600 BCE

Athenian amphora
Greece, *c.*720–700 BCE

The Shoemaker Vase
Greece, 500–470 BCE

Attic black-figure pottery stemmed cup
Greece, 530–515 BCE

Attic red-figure pottery head vases
Greece, *c.*500 BCE

Boeotian black-figure *skyphos* depicting Odysseus at sea and with Circe
Greece, Fourth century BCE (400–301 BCE)

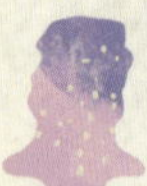

Terracotta female figurines
Pakistan, *c.*300 BCE–100 CE

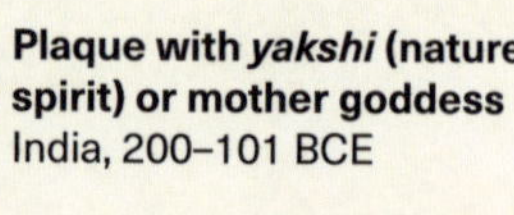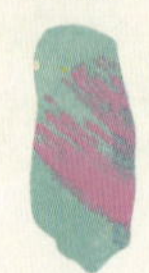

Plaque with *yakshi* (nature spirit) or mother goddess
India, 200–101 BCE

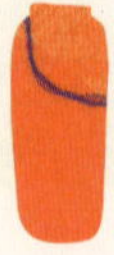

Reconstructed Dead Sea scroll jar
Jordan, 1–70 CE

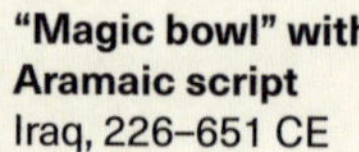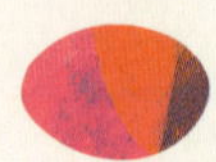

"Magic bowl" with Aramaic script
Iraq, 226–651 CE

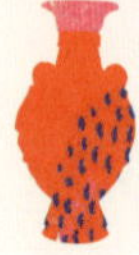

Jar with heavily moulded decoration in Central Asian style
China, 500–600 CE

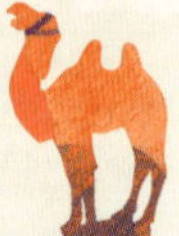

Earthenware figure of a camel
China, 618–907 CE

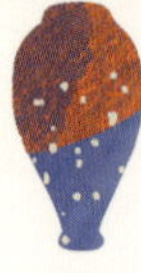

Storage jar
Iraq, *c.*700–800 CE

Two tin-glazed bowls
Iraq, 801–900 CE

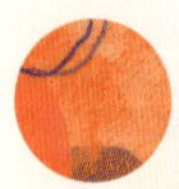

Bowl with animals and plants
Iran, 10th century CE

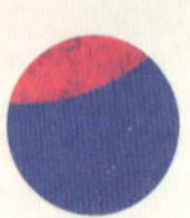

Bowl with epigraphic decoration
Nishapur (Iran) or Samarqand (Uzbekistan), *c.*10th–11th centuries CE

Black ware jar with white stripes
China, 960–1279 CE

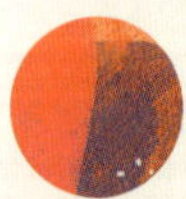

Dish with purple splash
China, 960–1279 CE

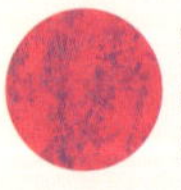

White ware dish with incised lotus decoration
China, 1100–1200 CE

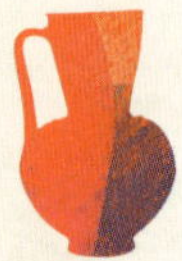

Jug with epigraphic decoration
Iran, 1151–1220 CE

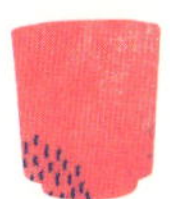

Nabeshima porcelain cup
Japan, *c.*1660 CE

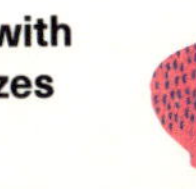

***Kintsugi* bulb bowl with purple and blue glazes**
China, 13th–14th centuries CE

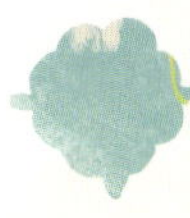

Bowl with seated figures by a stream
Iran, 1211–1212 CE

White ware vase with floral decoration
China, 1279–1368 CE

Spouted bowl with two queens flanking a pillar
Italy, *c.*1275–1375 CE

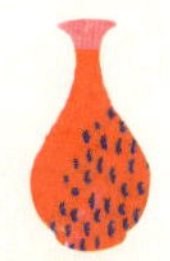

***Albarello*, or storage jar**
Syria, 1301–1400 CE

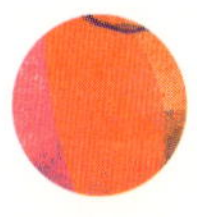

Piggy bank
Java, 15th century CE

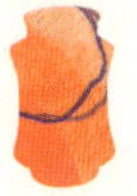

Plat de la Passion dish
France, 1511 CE

Dish with a composite head of penises
Italy, 1536 CE

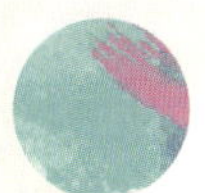

Childbirth bowl and cover
Italy, *c.*1570 CE

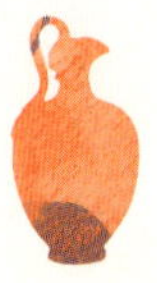

Porcelain ewer
Italy, *c.*1575–87 CE

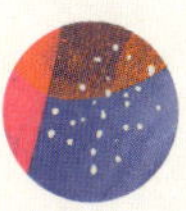

Shino-style serving dish for the Japanese tea ceremony
Japan, *c.*1590 CE

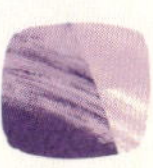

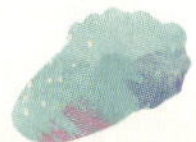

Palissy ware dish
Europe, 1601–50 CE

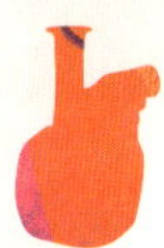

Pouring vessel, or *kendi*, in the form of an elephant
Iran, 1601–1700 CE

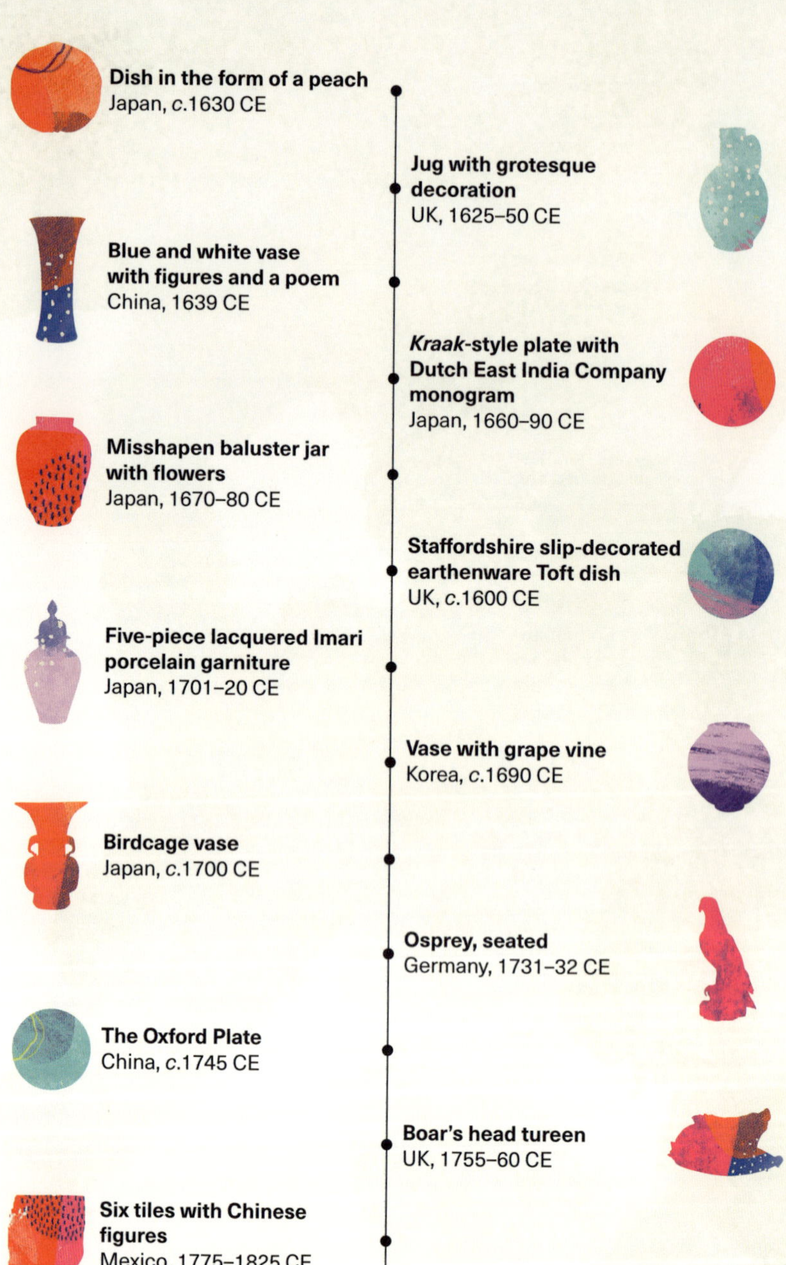

Dish in the form of a peach
Japan, *c.*1630 CE

Jug with grotesque decoration
UK, 1625–50 CE

Blue and white vase with figures and a poem
China, 1639 CE

***Kraak*-style plate with Dutch East India Company monogram**
Japan, 1660–90 CE

Misshapen baluster jar with flowers
Japan, 1670–80 CE

Staffordshire slip-decorated earthenware Toft dish
UK, *c.*1600 CE

Five-piece lacquered Imari porcelain garniture
Japan, 1701–20 CE

Vase with grape vine
Korea, *c.*1690 CE

Birdcage vase
Japan, *c.*1700 CE

Osprey, seated
Germany, 1731–32 CE

The Oxford Plate
China, *c.*1745 CE

Boar's head tureen
UK, 1755–60 CE

Six tiles with Chinese figures
Mexico, 1775–1825 CE

Teapot and lid
UK, *c.*1772 CE

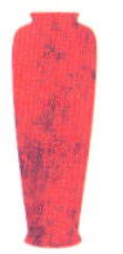

Vase with a design of a bird
Japan, *c.*1910

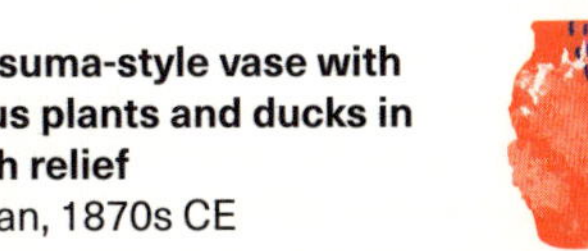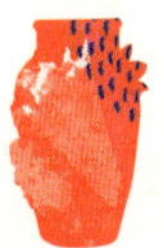

Satsuma-style vase with lotus plants and ducks in high relief
Japan, 1870s CE

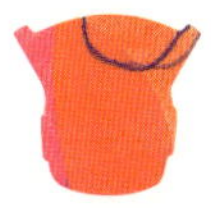

"Peruvian face" bridge-spouted vessel
UK, *c.*1880 CE

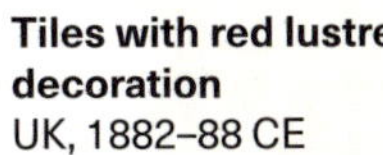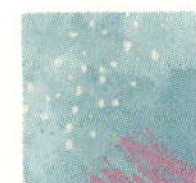

Tiles with red lustre decoration
UK, 1882–88 CE

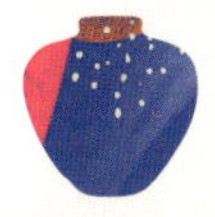

Baluster vase with flattened shoulders
Japan, *c.*1885 CE

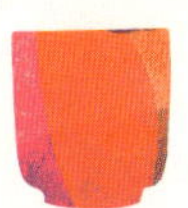

Satsuma cup with chrysanthemums and key-pattern border
Japan, *c.*1900 CE

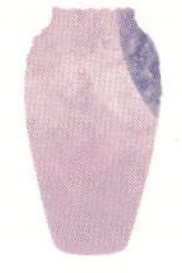

Art nouveau-style vase with a design of chrysanthemums
Japan, 1900–05

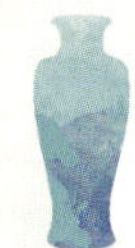

Vase with winter landscape
Japan, *c.*1910 CE

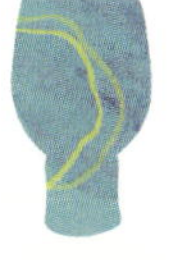

Pot
UK, *c.*1975 CE

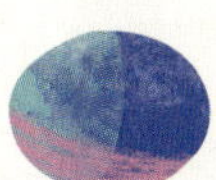

Lustred bowl
UK, 2005 CE

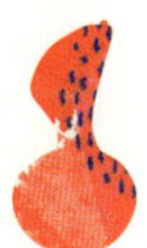

Asymmetrical "Betu" Series I
UK, 2009 CE

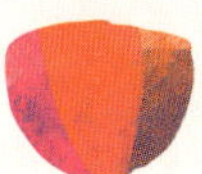

Tea bowl
Japan, 2019 CE

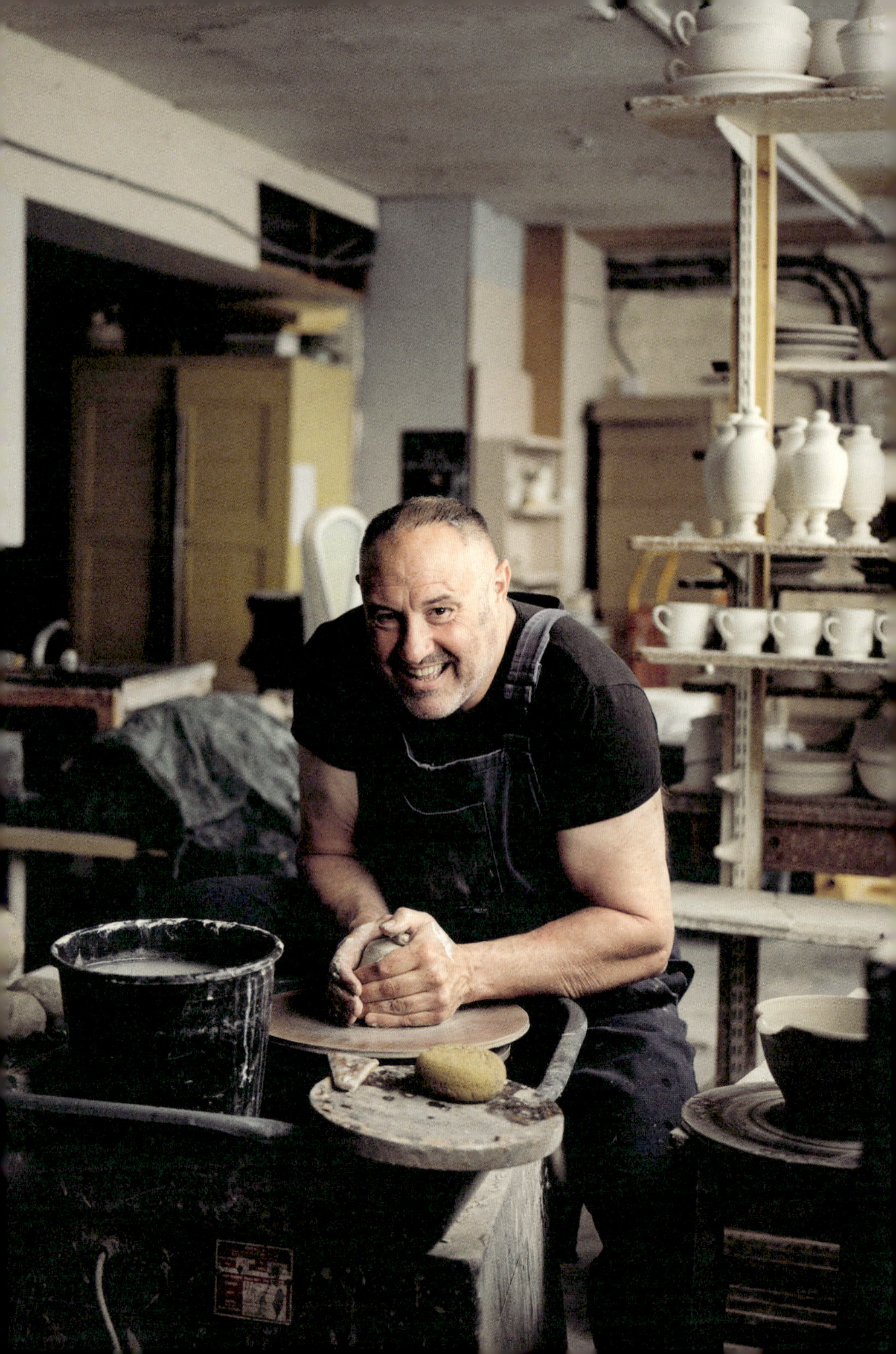

Foreword

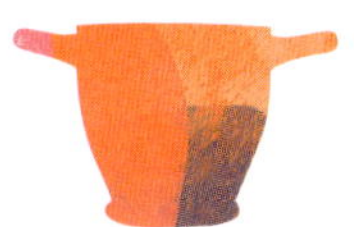

Keith Brymer Jones

I sit in my studio working with clay, as I have done for 40 years, and a notification pops up on my phone. It's an email that momentarily interrupts the programme I'm listening to on Radio 4. You see, clay and computers don't really mix, for obvious practical reasons, but also because one of the technologies I am using at this present moment in time has been with us almost since humankind started to express themselves in a creative way, while the other (certainly in mainstream form) has barely been with us for 50 years. But the email that arrives sees a meeting of these ancient and modern technologies. It is an invitation to write a foreword for a new book by the Ashmolean Museum. Yes, this book you have in your hands.

The Ashmolean was the first ever public museum – and when I say the first I mean the first in the world – and the second museum to be affiliated to a university, which is Oxford. Conceived and created in 1683, the Museum has collected artefacts from all four corners of the globe and has established itself as one of the most important collections of art and archaeological items on this planet of ours. Its "eclectic spectrum of curiosities", a description that was first used when the Museum was established, is truly incredible. To be asked to write a foreword for such an institution is indeed an honour and a privilege.

It's no surprise to me that the Ashmolean would wish to produce a book to showcase its impressive collection of ceramics. However, what is astonishing is the sheer beauty and variety of the pots that lie between its pages.

Keith Brymer Jones at the
pottery wheel in his studio.

Clay is one of the oldest materials used by the human race. From the early practice of forming pots to carry water to later, highly decorative creations, clay has always been a significant and indeed crucial part of our advancement on this earth. It has been an extremely practical medium in which to express an opinion, share an emotion, or communicate a message, and if the maker chooses, the result of their work can be not only a symbolic but also a practical object – form and function combined in one creative endeavour. Depending on what the maker intended, a piece of pottery can transcend cultures to speak in a unique and often powerful way to the onlooker.

Some of the pieces within this book date back to 3800 BCE. Not only is this an incredible achievement for the original makers, whose works have survived for thousands of years, but it's also a testament to the Ashmolean Museum, whose curators recognised their significance and are still the custodians of these wonderful objects.

As I sit at my wheel in my studio, I have often thought about the first pot ever made, by which I mean the first pot ever fired. It's this process of introducing clay to extreme heat which is why we are blessed to have these ancient pots today. When it's fired, clay goes through a complete transformation. The material itself changes its molecular structure, never to be reversed, from an organic material to a ceramic. The pots in this book are snapshots of time, thought and emotion that have endured through thousands of years of human progress. It's mind-blowing, don't you think?

Any pot can tell us about so much more than just its cultural origin or the identity of its maker. It also tells us about the technical ability that brought it into existence. Anthropologists have concluded that the development of pottery was a pivotal moment when humans developed the ability to store food in a more practical way, thus enabling them to form more permanent settlements. Before the introduction of pottery, we were just hunters,

Detail of red polished stag-shaped figurine with three legs. Cyprus, *c.*2100–*c.*1850 BCE (see page 53).

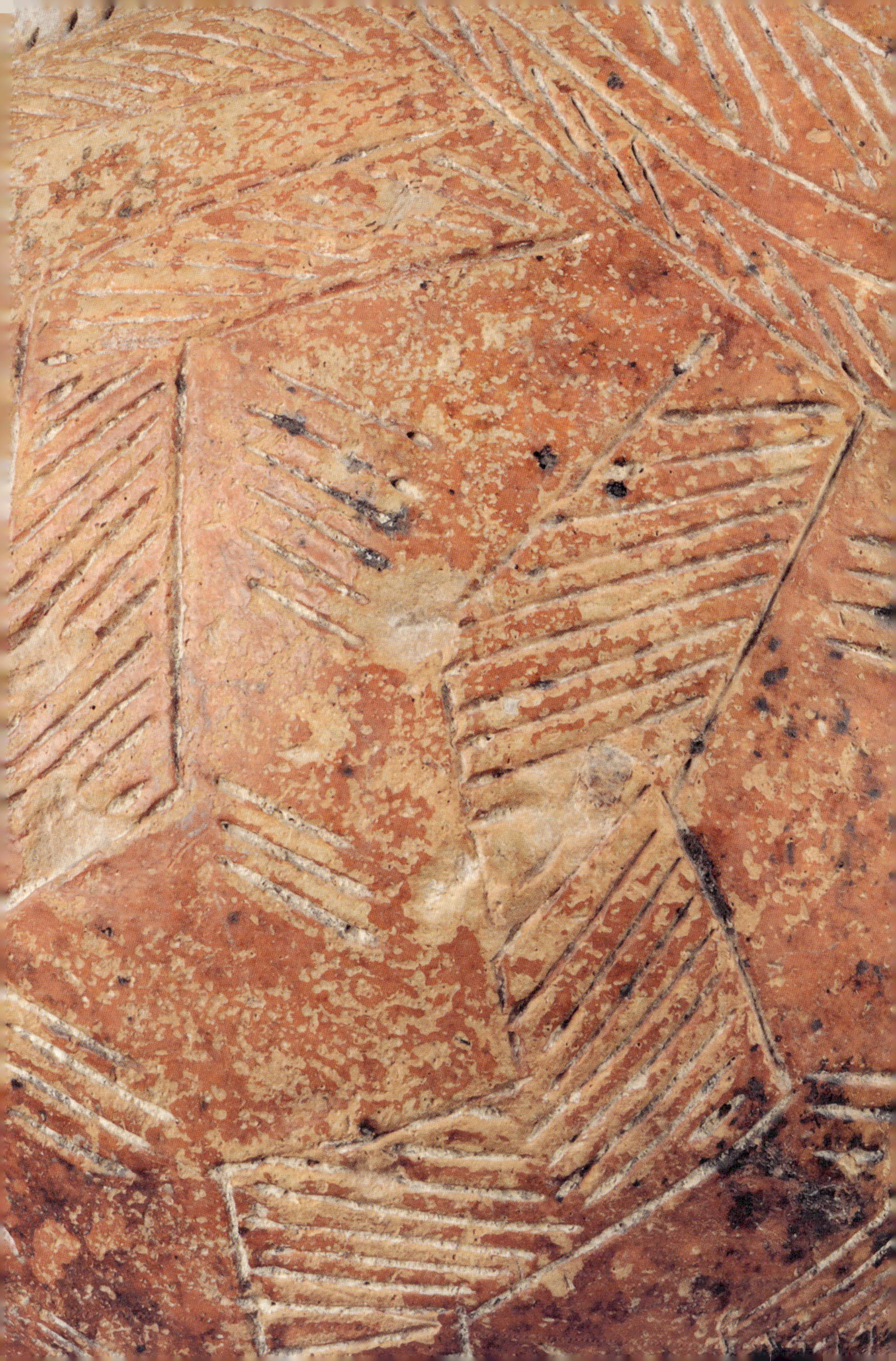

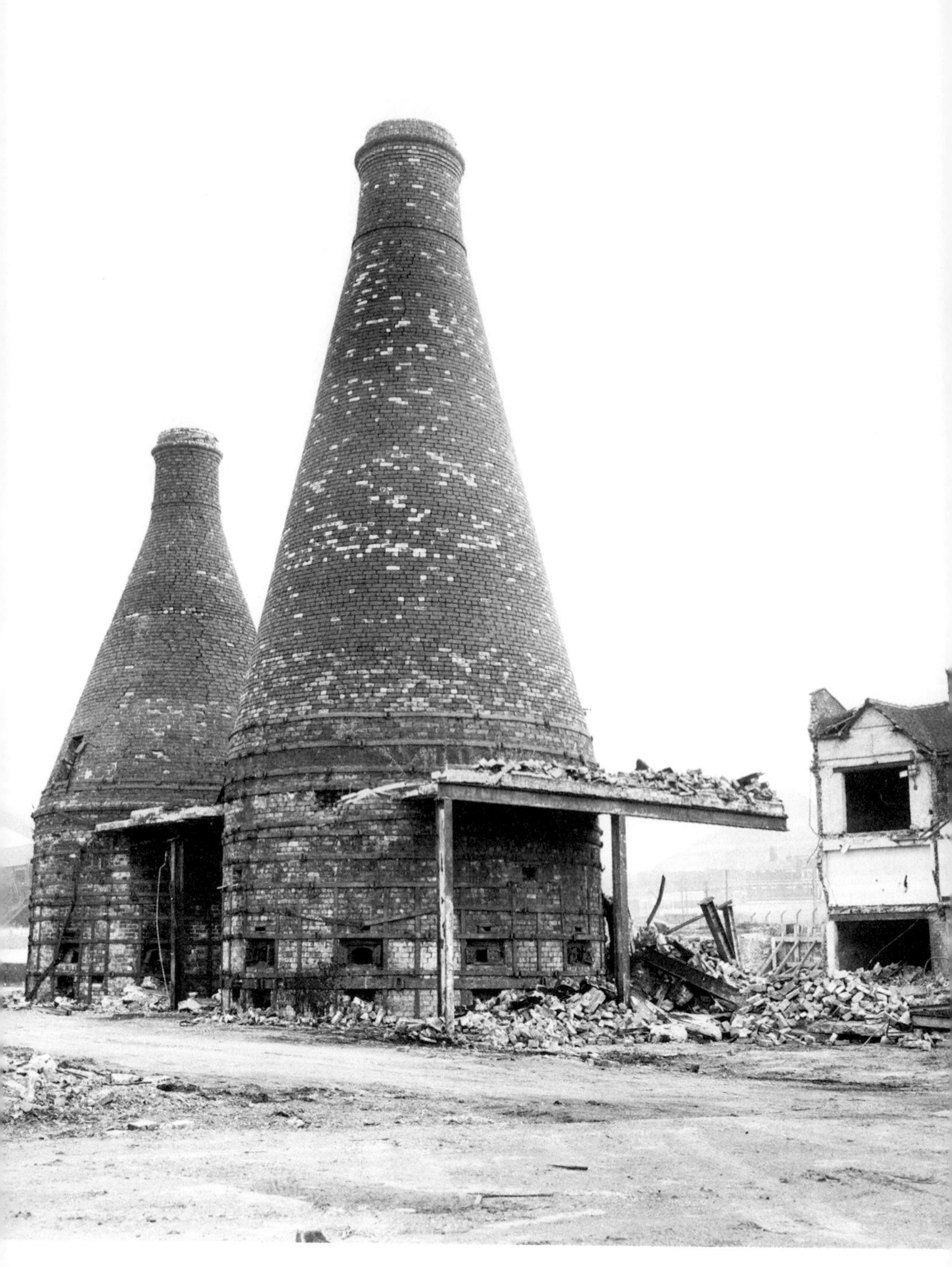

surviving on what we could acquire at any given moment. To be able to store seeds, berries and other food in permanent vessels was revolutionary in terms of our development. It gave us the time to think, create and farm our way out of an existence focused only on survival. Pottery gave us the opportunity to advance in ways that, previously, we could not have imagined.

Some of the designs and imagery that appear on the pots within these pages are unique to a particular time and place. They give us a pictorial insight into ancient civilisations and how they were connected with each other, the beliefs of their citizens, their values and the tasks that made up their daily lives.
As you turn the pages of this catalogue of creativity, you will notice that as each century passes, the techniques and decorative processes used to produce the pots increase in their visual complexity because of the colours and oxides employed, and in some cases the variety of forms. It is awe-inspiring to even imagine the years of experimentation it took to achieve these decorative effects, using the natural ingredients of oxides and the minerals offered up by the earth, to create glazes of such beauty. Is it any wonder that, at the time when these efforts were achieving ever more refined methods of production, the individuals tasked with these challenges were, in Europe, called "alchemists"? They used seemingly magical processes to change minerals into glass or glaze and created almost immortal objects in the form of pots.

"If pots could talk, what would they teach us of the things they've seen?"

With every brushstroke of decoration, with every mark of the maker's hands, their touch was immortalised in clay. This is what is so enchanting not only to the observer, but also to the creator of such pots. Maybe this is what so inspires the potter in any century to doggedly pursue the art of pottery: the desire to live on through what one has created long after the physical presence of the potter has gone. In some of the examples in this book you can actually see the potters' fingermarks, and because of this, the pot is living proof of the existence of its maker, whether 100, or, indeed 5,000 years ago.

Two bottle kilns at Josiah Wedgwood's Etruria Pottery Works, Stoke-On-Trent.

Take just one pot, the oldest within this book: an ancient Egyptian beaker from about 3800 to 3450 BCE (see page 22). The workmanship is sublime, and remarkably contemporary in its aesthetic. The fact that the beaker has survived for this length of time is a testament to the process of pottery itself. Furthermore, the way it was made and the depiction of the animals incorporated in the design tell us a lot about life in ancient Egypt. The maker knew that burnishing a pot like this one would seal the surface and make the piece far more durable and thus more practical for the user. One begins to realise that this form of creativity has been established over years of understanding the very chemistry of the earth. The maker intrinsically knew that the simple colour choice would be highly effective, and this is inspirational and incredible.

These pots have transcended time and survived wars, famines, earthquakes and other natural disasters to become a permanent reminder of the distant human past. If pots could talk, what would they teach us of the things they've seen? I would say that they do talk, telling us so much just by means of their existence.

So, as I mentioned at the beginning of this introduction, clay and computers don't really mix. But just as computers use algorithms to communicate with us, I suggest to you that pots are the algorithms of lost civilisations and are still speaking to us now. The computer is our modern-day means of communication, but can it hold water at the same time as conveying a message, whether that message is emotional or factual? Can it depict a revolution while being used as a platter? And, ultimately, will it still be around in 5,000 years' time? A pot, in all its simplicity or complexity of form, will be with us forever. When I was discussing this subject with an eminent museum curator some years ago, he mentioned that a bowl, with its simple form, is one of very few objects that we employ today that humans some 5,000 years ago would recognise, both in terms of use and meaning. Just think about that for a moment. A pottery bowl, and the art of pottery itself, are design classics that will quite literally outlive us all. Pottery is one of the very few threads in human existence that has transcended all other progress. It is the art of depicting life itself – but then, I would say that, as I am a potter.

Detail of *Kintsugi* bulb bowl
with purple and blue glazes.
China, Song or Yuan dynasty,
13th–14th centuries (see
page 140).

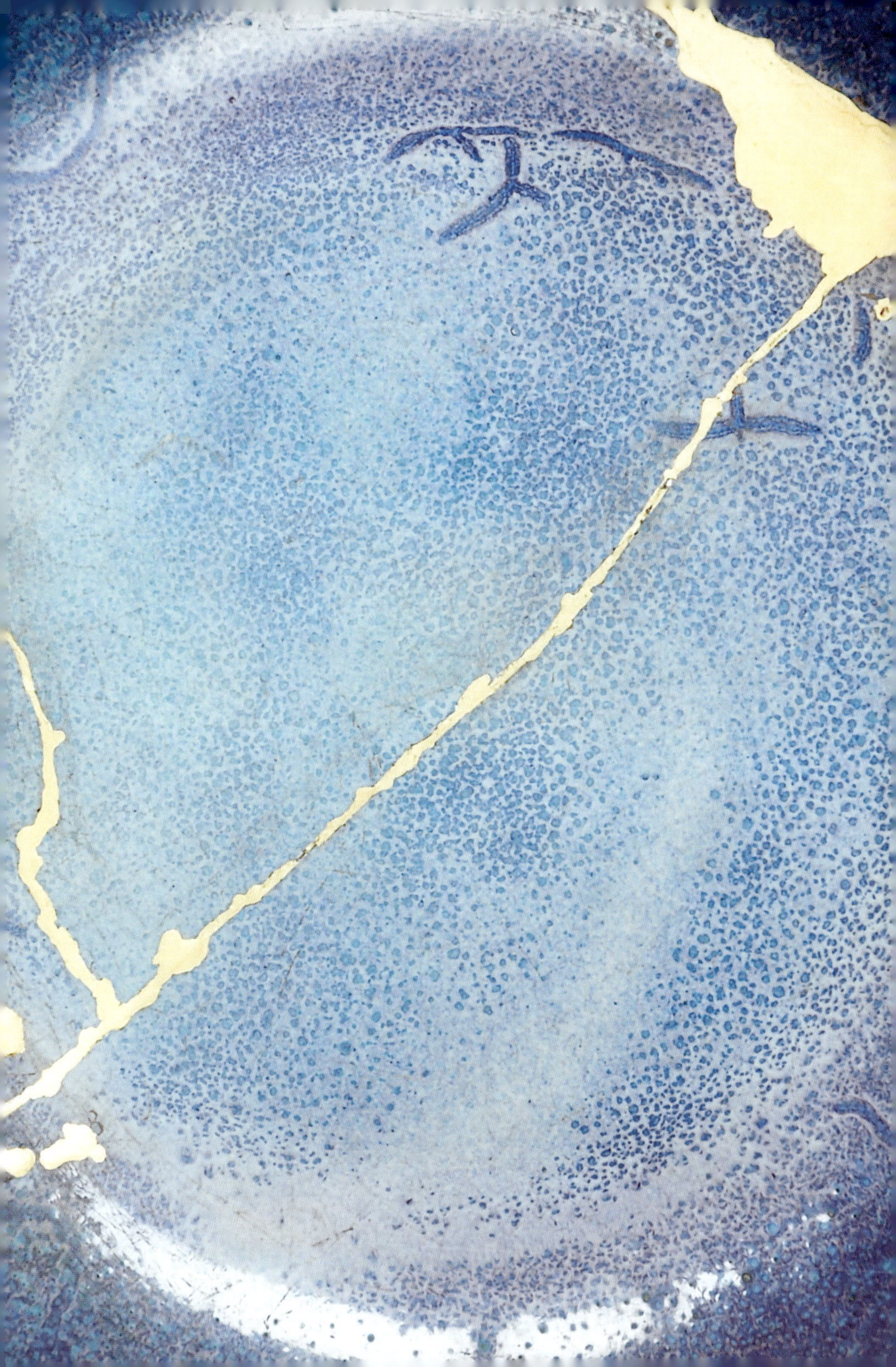

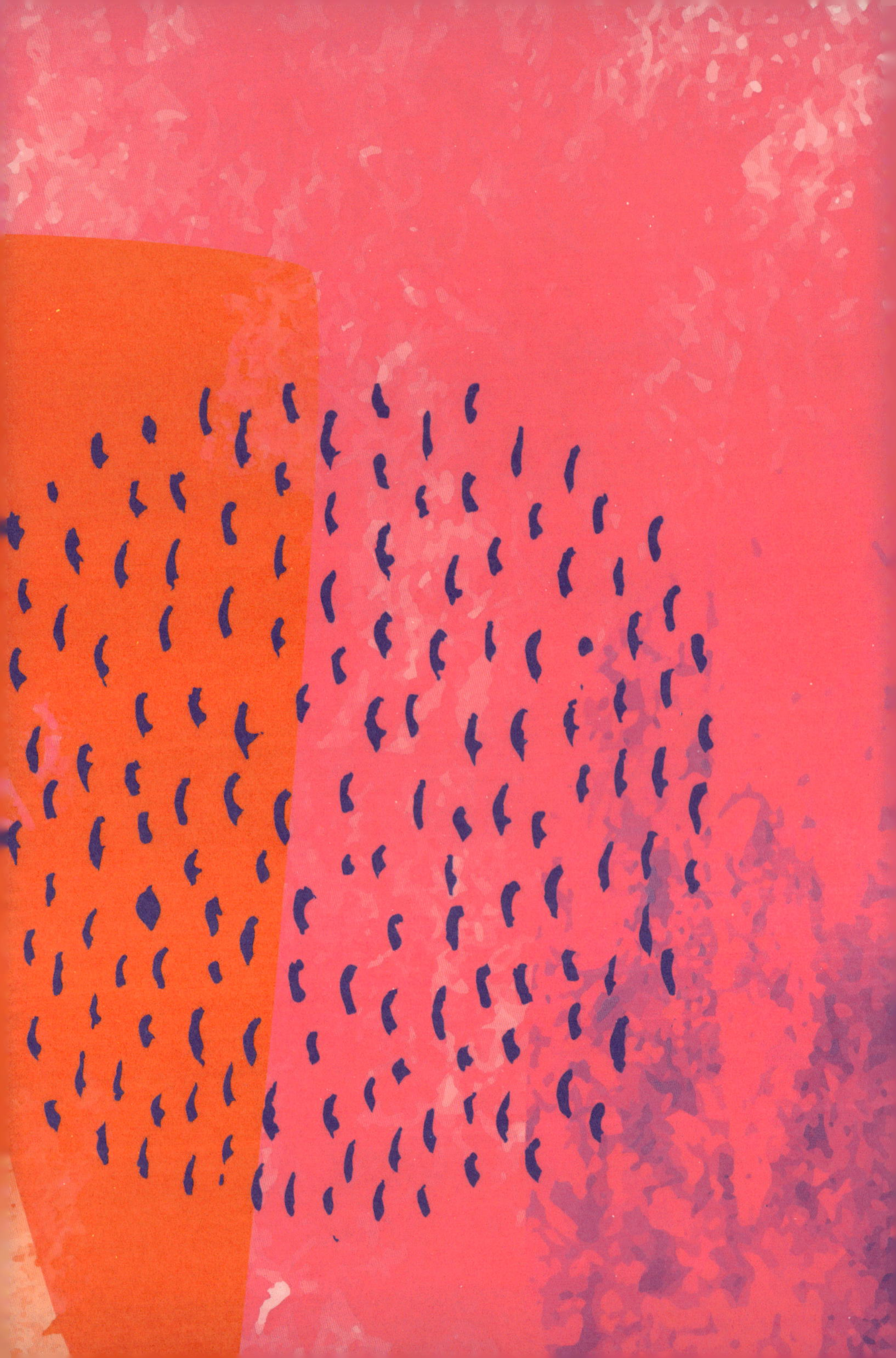

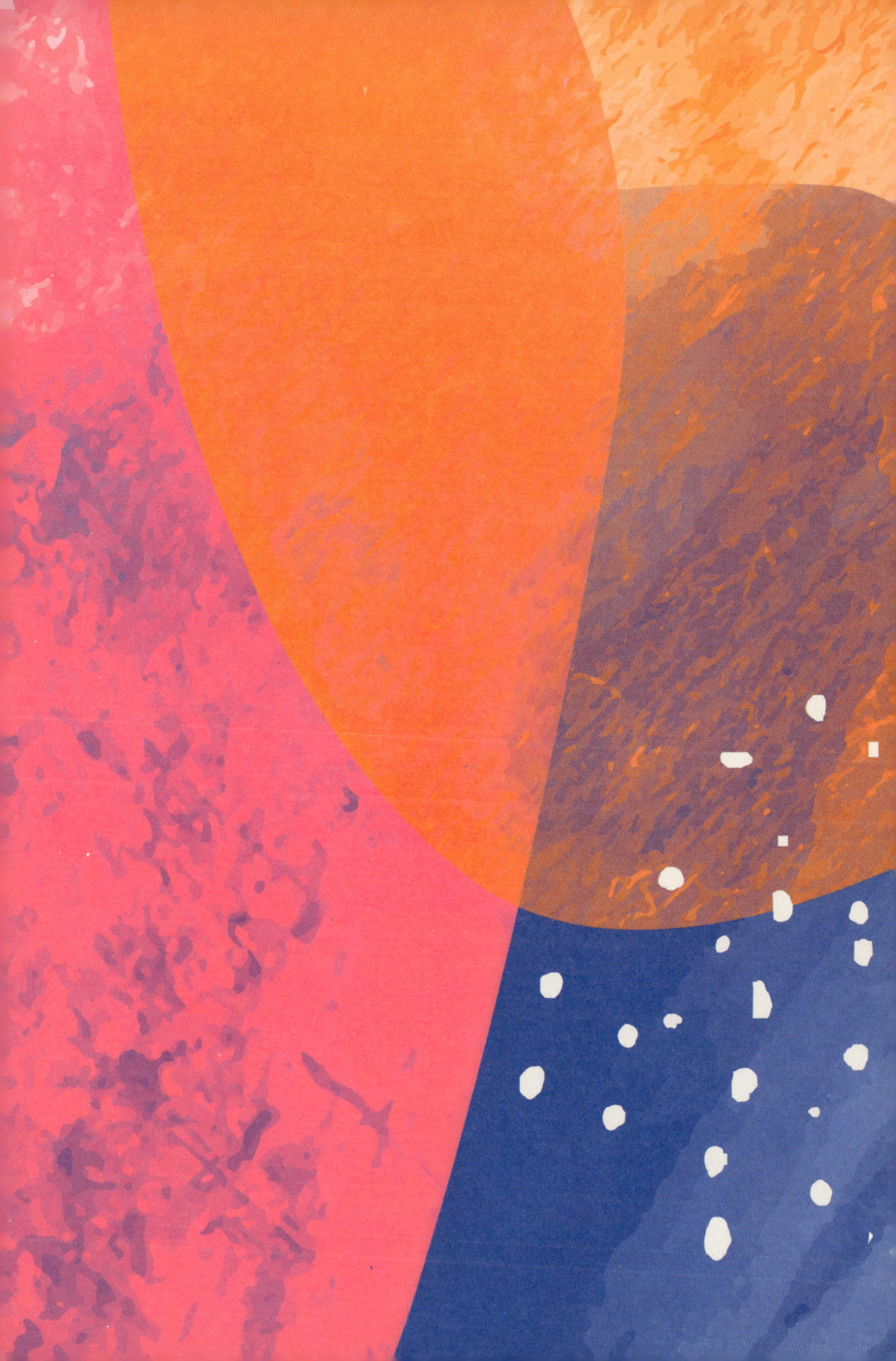

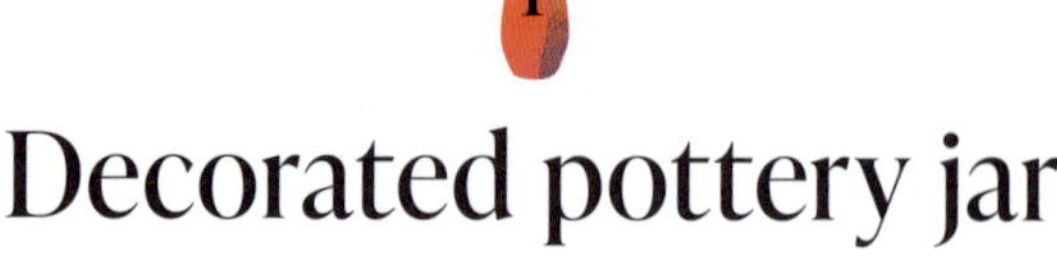

Decorated pottery jar

Egypt
*c.*3800–3450 BCE
Fired clay, 25.5 x 9.9 cm

On this jar, you can see one of the earliest depictions of figures and landscapes that survive from ancient Egypt, apart from those found in rock art. The jar is an example of "cross-lined ware", the first pottery with painted decoration made in Egypt, more than 5,000 years ago.

The man who came up with the term was the archaeologist Flinders Petrie, who created a typology of pottery to shed light on the chronology of the predynastic period. Most of the "C-ware" vessels that have been discovered are decorated with linear designs, such as images of vegetation and basketwork. But a few of them, including this one, depict animals or people. In fact, this jar is one of the finest examples of "C-ware" that exists.

The vessels were decorated in a light-coloured paint which fired in colours from creamy white to pink, standing out against the glossy red surfaces of the pots. Some examples have been analysed, revealing that they contained calciferous clay or pale iron oxide. We can also see that the bodies of the pots were handmade from Nile silt, "washed" with haematite (iron ore) and burnished using a pebble before the decoration was applied.

This tall jar features simplified but naturalistic depictions of a flock of animals, probably sheep and goats, within a border of double triangles. The largest is definitely a male Barbary sheep, its characteristic curved horns and chest mane clearly depicted. However, the others are more difficult to identify.

The animals are wild creatures, part of the desert landscape of predynastic Egypt, and unrelated to the domesticated breeds we know from later times. The rows of triangles at the top and bottom of the jar probably represent the cliffs along the edges of the Nile that would have been their natural habitat. Egyptians in this period hunted sheep and goats (the animals' bones have been found in early settlements and cemeteries) and may have occasionally eaten them. Later, hunting was one of the activities depicted in tomb scenes. The artists would have imagined the deceased would repeat these activities in the afterlife for eternity. It's likely that similar images on vessels found in predynastic graves served a similar purpose.

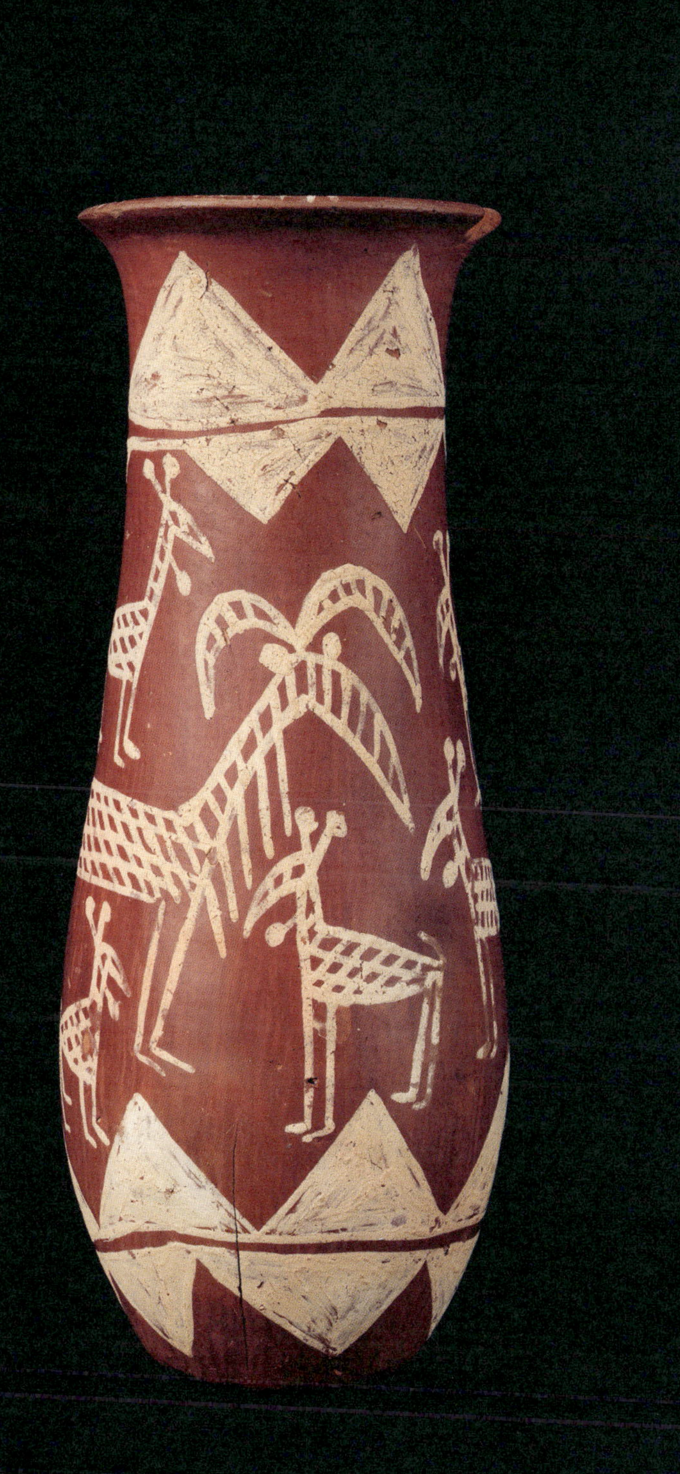

2

Black-topped pottery beaker

Egypt
*c.*3600 BCE
Clay, 42.8 cm

Black-topped ware is the most important type of pottery made during the predynastic period, which took place from around 6000 to 3150 BCE. Experts call it "B-ware", a name given to it by the pioneering Egyptologist Flinders Petrie. Potters were making B-ware nearly 7,000 years ago, but most of what archaeologists have found dates from the periods Naqada I (*c.*3900–3650 BCE) and early Naqada II (*c.*3650–3300 BCE).

B-ware finally disappeared from the potter's repertoire in Egypt early in dynastic times, about 5,000 years ago, when the fine, mostly handmade, vessels of prehistory gave way to wheel-made pottery.

These pots were named after the lustrous, almost metallic black band on their upper surface, seen inside the vessel and through the thickness of the clay. This blackening occurred because they were fired in a reduced atmosphere and so absorbed carbon. The potter would set the vessels upside down in a bonfire, and the areas that were deprived of oxygen and smoked by the burning fuel would turn black when the object was fired, while the rest of the surface would become a glossy red. The colour and sheen were the result of a coat of ground haematite, applied as a liquid to the leather-hard clay and burnished before the pots were fired. The potter could also treat the surface of the clay, and vary the firing method, to produce all-red or all-black vessels. When we try to replicate this process today, we get the best results when the bonfire containing the pots is covered with clay.

Black-topped vessels were made in lots of different shapes and occasionally decorated in relief. The decoration was modelled separately, then applied to the surface before it was coated and burnished. On the beaker illustrated here, a human face on a pole is flanked by bovid horns. Below it, a pair of arms stretch around the vessel to hold small breasts. We don't know what this decoration signifies, but it's echoed in jars of much rougher ware produced in the times of the pharaohs – over 600 years later – which were decorated with miniature breasts and the head of the goddess Hathor, who was associated with the cow.

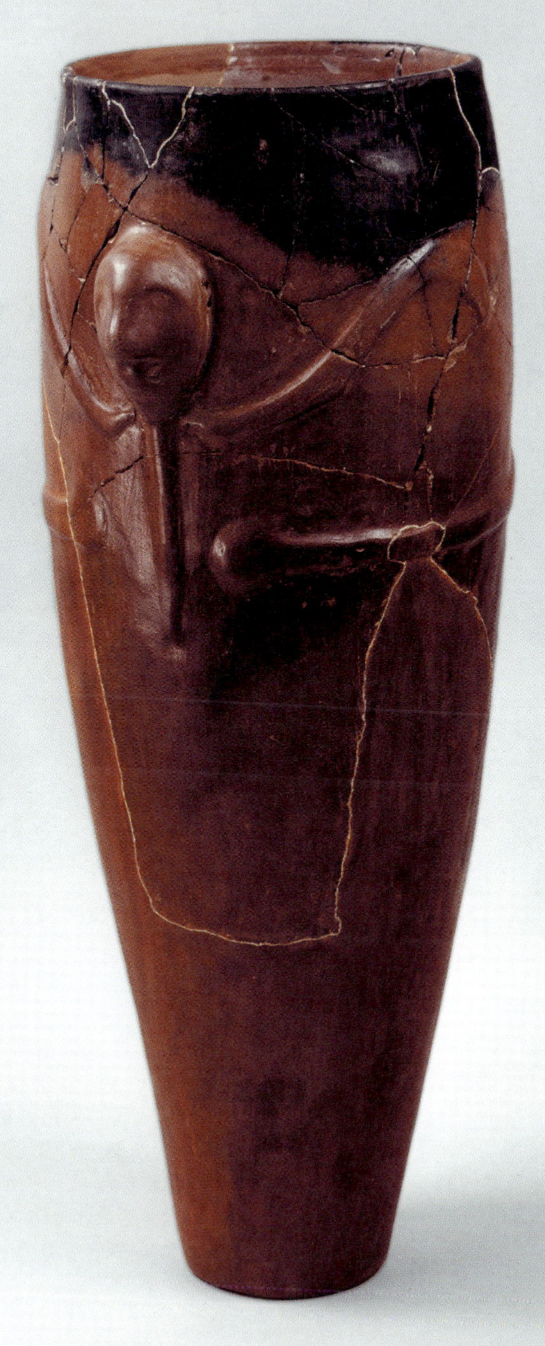

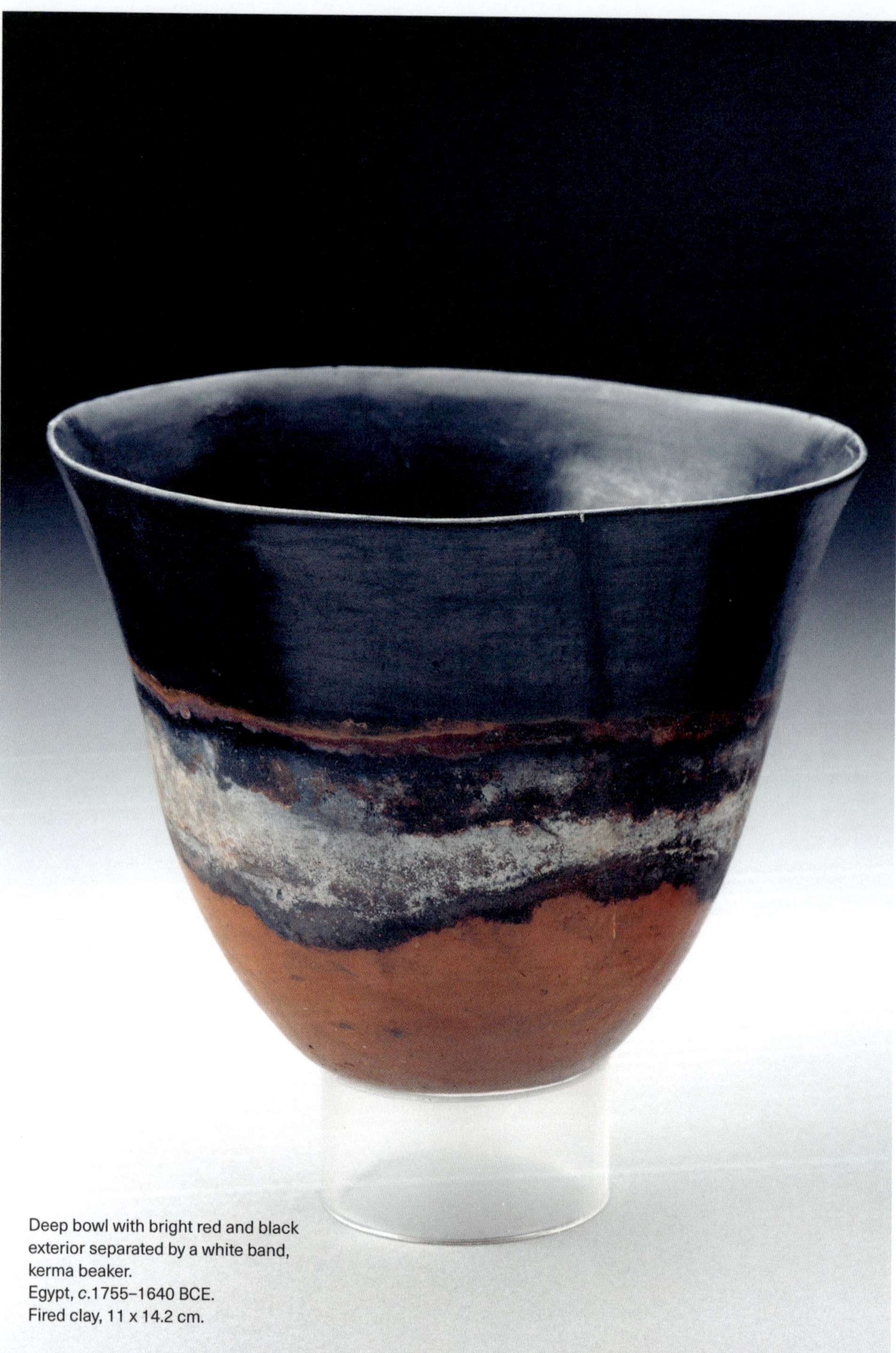

Deep bowl with bright red and black
exterior separated by a white band,
kerma beaker.
Egypt, *c.*1755–1640 BCE.
Fired clay, 11 x 14.2 cm.

Among the pieces of B-ware in the Ashmolean is a sherd, or broken piece, from quite a large jar, decorated with the outline of a form which is now known as the Red Crown of Lower Egypt. Ancient Egypt was divided into two political regions, Lower Egypt (in the north) and Upper Egypt (in the south), which were unified around 3100 BCE. The crown can be recognised by its distinctive flaring profile, rising high at the back, with a curling addition, like a springing tendril, that faces inwards. But it's unexpected to find this symbol in Upper Egypt – in fact, it's the only example from this time that has been found there. Did this shape already represent a ruler's headgear, much earlier than we like to think?

Black-topped ware, like other types of handmade pottery, was still produced in Nubia long after such technically accomplished vessels had disappeared from Egypt. It reached its peak in the "Kerma ware" of *c.*1800 to 1650 BCE, mostly thin-walled vessels where the black zone gradually passes to clear red through a band of intermediate colours. This decorative effect was the result of interaction between the clay and the mineral coating. It was clearly an intentional artistic effect, created by potters who were experts in the shaping and firing of clay.

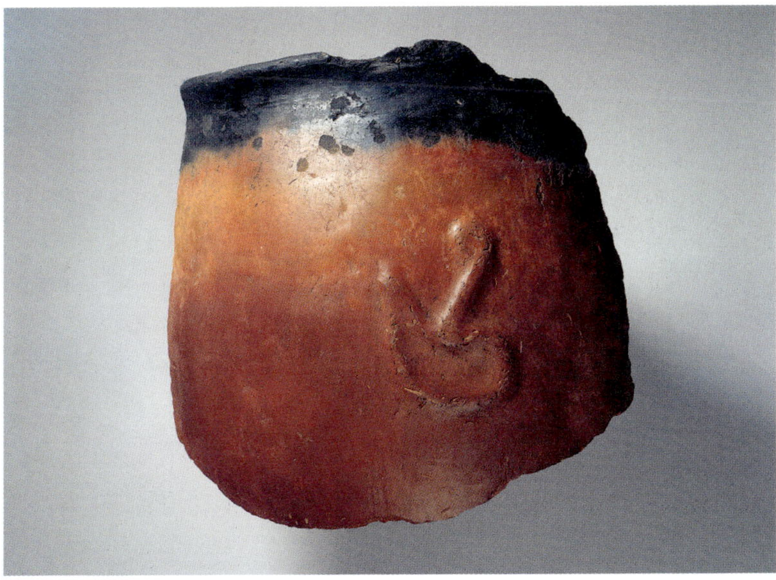

Rim sherd of necked jar.
Egypt, *c.*3600 BCE.
Clay, 25 x 19 cm.

3

"Skidding goat" pottery vessel

Possibly Iran
*c.*3500 BCE
Ceramic, pigment, 12.5 x 9.3 cm

Unfortunately, we don't know where this small jar – decorated in paint with lively animals – was found. However, it's similar to other vessels that were made in Iran midway through the fourth millennium BCE. Jars like this were made at settlements in the mountain valleys that connected the lowlands of Mesopotamia in the west and the high plateau of Iran to the east.

People moved between the regions, trading their farm animals and products in the small towns and villages. These communities flourished: for one thing, they controlled the supply of local timber and metals from Iran that weren't available in Mesopotamia. Their wealth might be indicated by the painted decoration on their pottery, which was extremely sophisticated. These fine vessels may have been used for special occasions. Some have been found in tombs and many of them have been linked to religious rituals or burial rites.

The quality of vessels like this one, with their thin walls and skilfully painted designs, demonstrates that from about 4000 BCE, potters were already specialist craftspeople. The pottery was turned slowly by hand on a wheel and then fired. In dark brown paint, artisans depicted birds, animals and geometric patterns, silhouetted on buff-coloured clay. Often, they chose to paint mountain goats. This goat, or ibex, is native to the Zagros Mountains of western Iran and may been an important symbol for the people of this region.

On this vessel, the painter has combined straight lines with columns of triangles, which may represent the mountain peaks, to frame the space on each side of the goat. This arrangement is repeated on the opposite side of the jar. The overall design is stylised, but it's far from static, thanks to the movement and dynamism that the artist has achieved when depicting the goat. The animal's long, wavy horns stretch up over its body to touch the vessel's rim. The most noticeable feature of the image is the way the ibex pushes its legs forward. Vessels decorated like this are sometimes described as showing a "skidding goat".

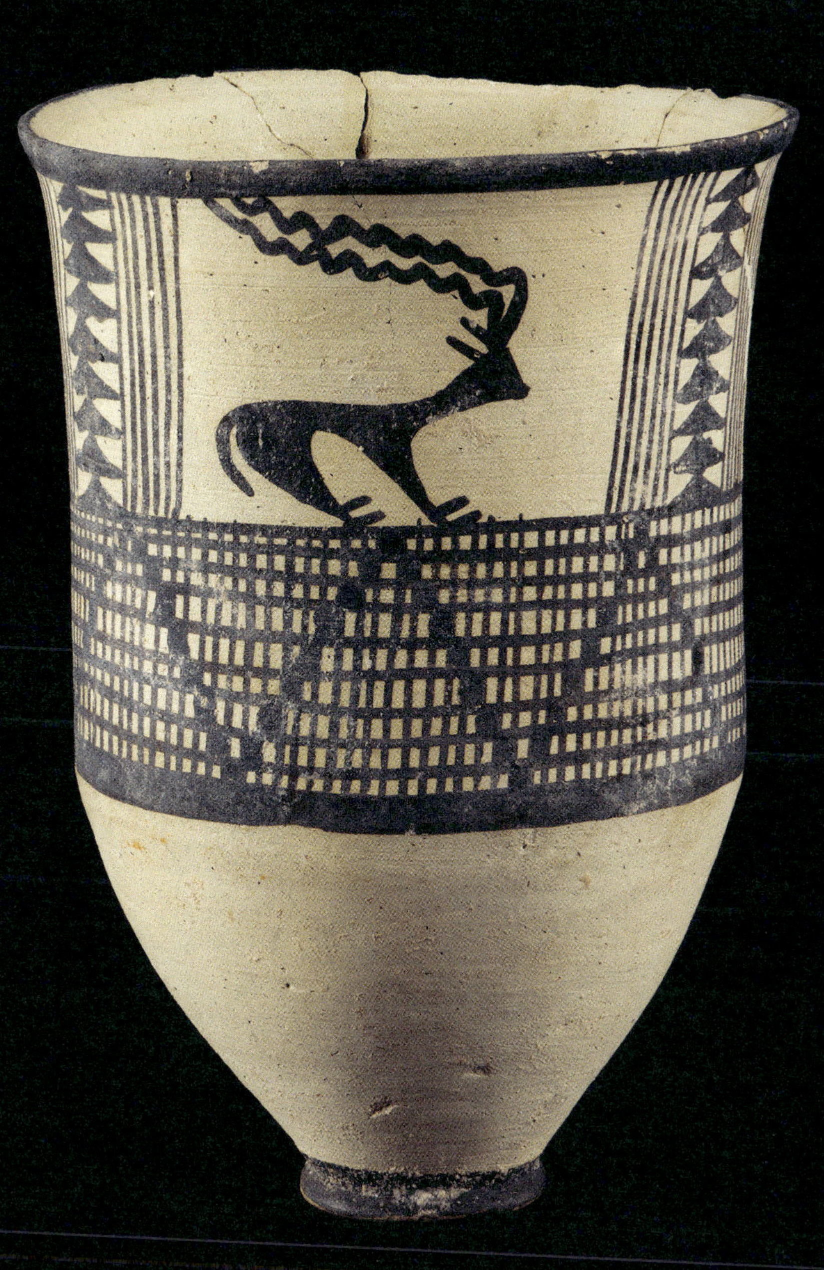

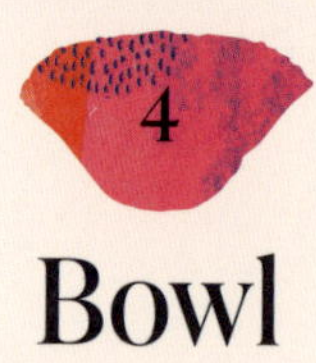

Bowl

Iraq
*c.*3100 BCE
Ceramic, 16.0 cm

While this may not be the most attractive bowl in the Ashmolean's collection, it's one of the objects that best helps us understand what life was like in the world's earliest cities, including how people were fed and organised. Known to archaeologists as a bevelled-rim bowl, it dates to the fourth millennium BCE.

Thousands of bowls like this one have been found at sites across Mesopotamia (modern Iraq and eastern Syria) and in the surrounding regions. At this time, pottery made on a wheel was widely available, used as storage and to serve food and drink, but bevelled-rim bowls served as cheap, convenient and disposable containers. They were made by pressing a lump of clay into a mould, possibly just a bowl-shaped hole in the ground. On the bottom of a bowl like this, we can often see the impression of the knuckles of the person who made it. Once moulded, bevelled-rim bowls were stacked in ovens to be baked hard.

The bowls were used for transporting and storing food and other substances, including offerings for the dead, stored in graves. However, we think that the main function of bevelled-rim bowls might have been baking bread. Many societies around the world bake loaves of bread to a standard dimension, and this might help explain why bevelled-rim bowls are all approximately the same size. In Mesopotamia, the loaves would have been distributed as rations to workers employed by large temples. These institutions received produce from the extensive agricultural estates that lay around the cities to which they belonged. Perhaps the produce was given as an offering to the gods, and once accepted, was distributed to the wider community. The bowls were a useful part of this process as they could serve as standard measures for the issuing of other types of rations to workers, such as barley, butter oil and even beer.

Remarkably, we know the role of these bowls because they are represented in some of the world's earliest writing. Using a sharpened stick or reed on small clay tablets, the temple administrators developed a method of recording the storage of agricultural products and the distribution of rations by drawing the details as pictograms and numerals.

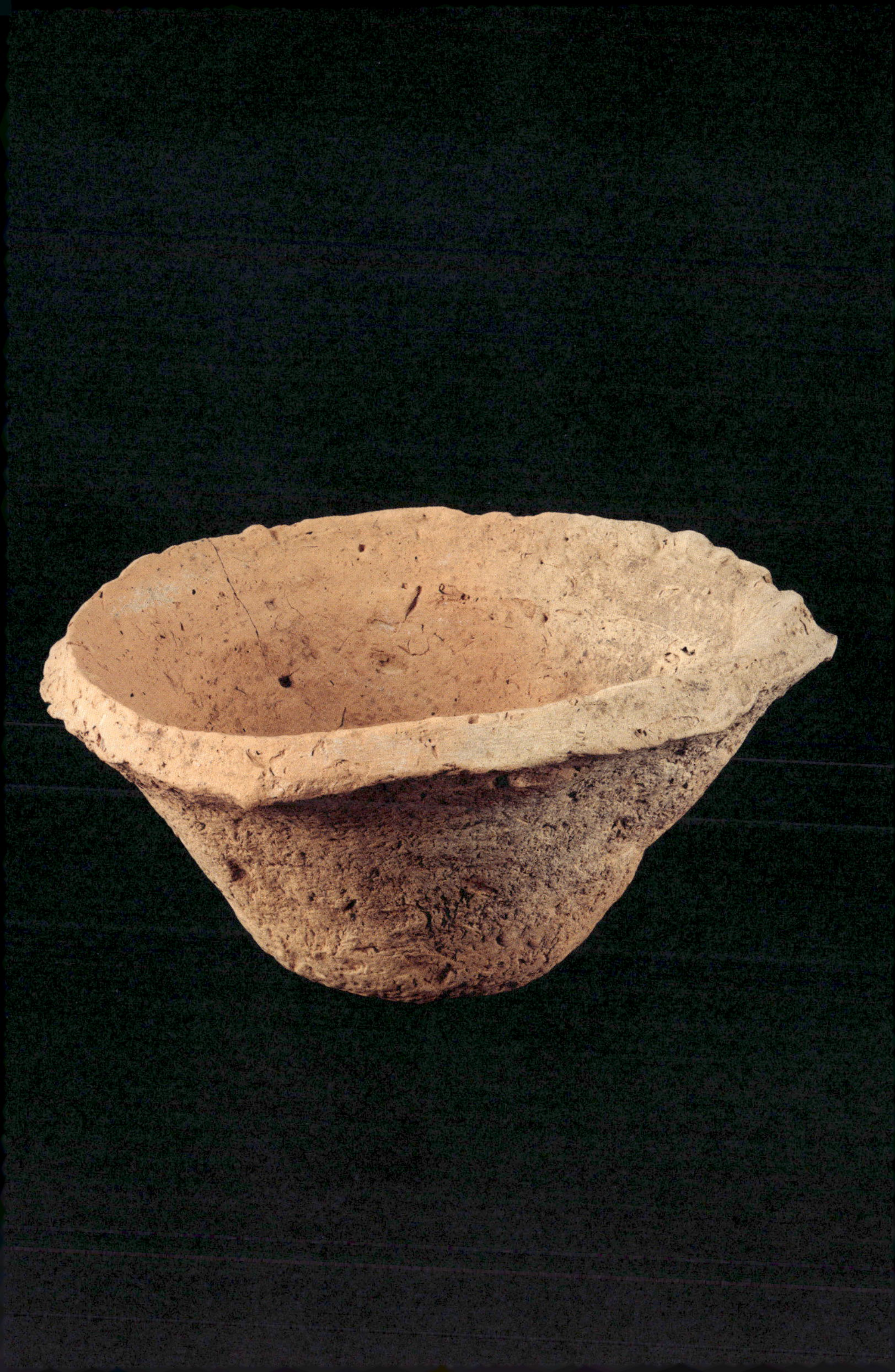

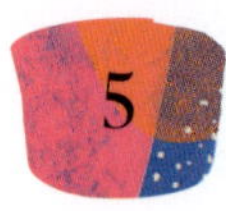

Polychrome Nal funerary pottery

Pakistan
3000–2001 BCE
Earthenware, with painted polychrome decoration,
8.9 x 15.3 cm

Polychrome Nal funerary pottery is one of the most lovely and unusual prehistoric ceramic wares that has been found across the subcontinent. It's yellowish beige in colour, sometimes slightly pink, with decoration in black or dark grey paint applied over the natural surface of the vessel. The pots are also decorated with red, blue or yellow pigment and repeated motifs. The use of these pigments makes Nal pottery stand out among prehistoric pottery. Its creators often decorated it with naturalistic depictions of fish and ibex.

These pots also have unusual shapes in comparison with other prehistoric findings. There's quite a variety, including disc-based open bowls, and narrow-mouthed, ovoid or ridged forms with a disc base. There are jars with a disc base that are almost straight-walled, but flanged slightly outwards. Some of the pots are carinated – they have a rounded base, joined to the sides of an inward sloping vessel. Other vessels are canisters with flat bottoms and round, straight-edged mouths. At the base of a Nal pot, the foot is turned (smoothed after the clay has dried) and usually meticulously finished.

The three pots illustrated here have been decorated in a similar way, with bands of motifs that look like the Greek letter *omega* running across the middle. This is one of the most commonly found patterns on Nal pots. We also frequently see bands divided with diagonal lines into triangles, or marked with semi-circles. Archaeologists often record an elaborate geometric pattern, like the one on page 32, with a central hexagonal medallion surrounded by borders and circular bosses radiating on each side of the hexagon. These bosses are raised and were obviously applied onto the wet pot before it was fired. Even though it's ornamental, this design probably imitates metal prototypes. Nal pots give us evidence, in their forms and motifs, for sophisticated metalworking, mining and smelting technology existing thousands of years ago.

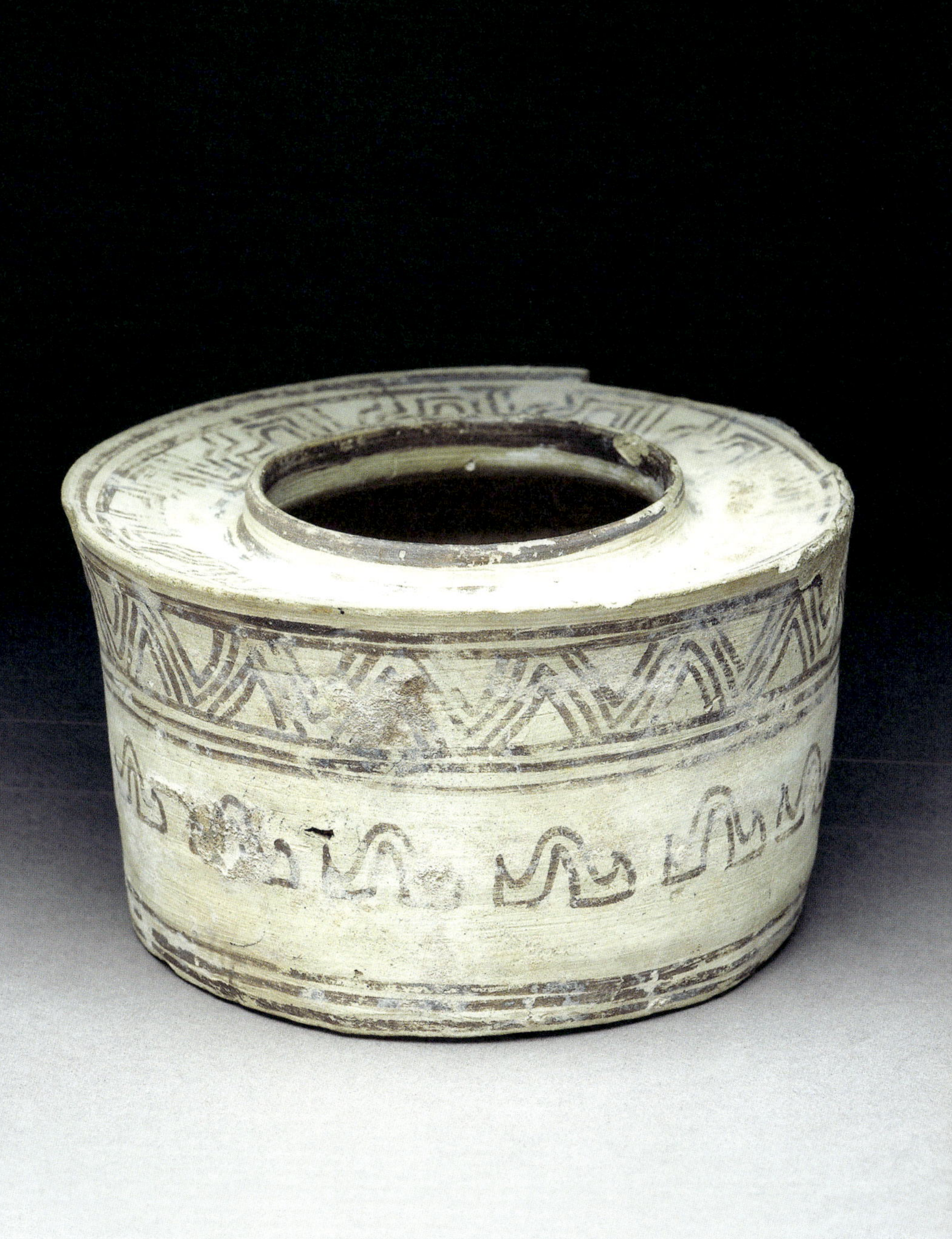

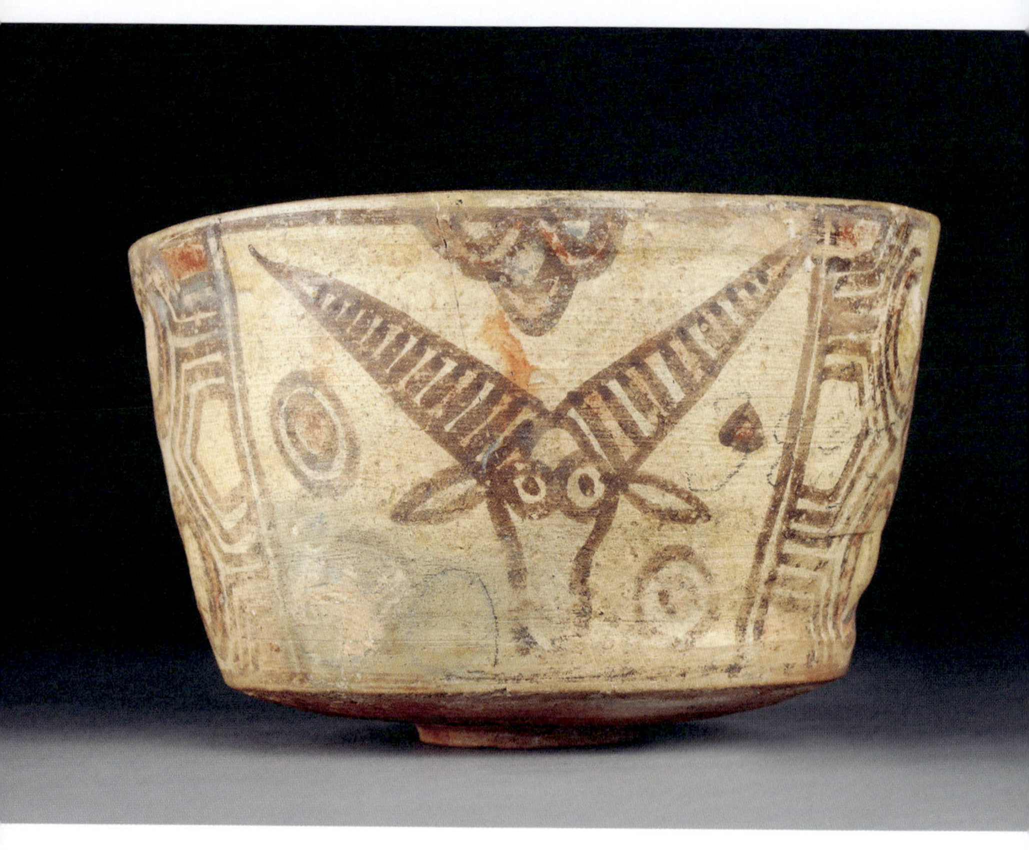

Bowl with geometric motif.
Pakistan, 3000–2001 BCE.
Earthenware, with painted polychrome
decoration, 8.7 x 14.1 cm.

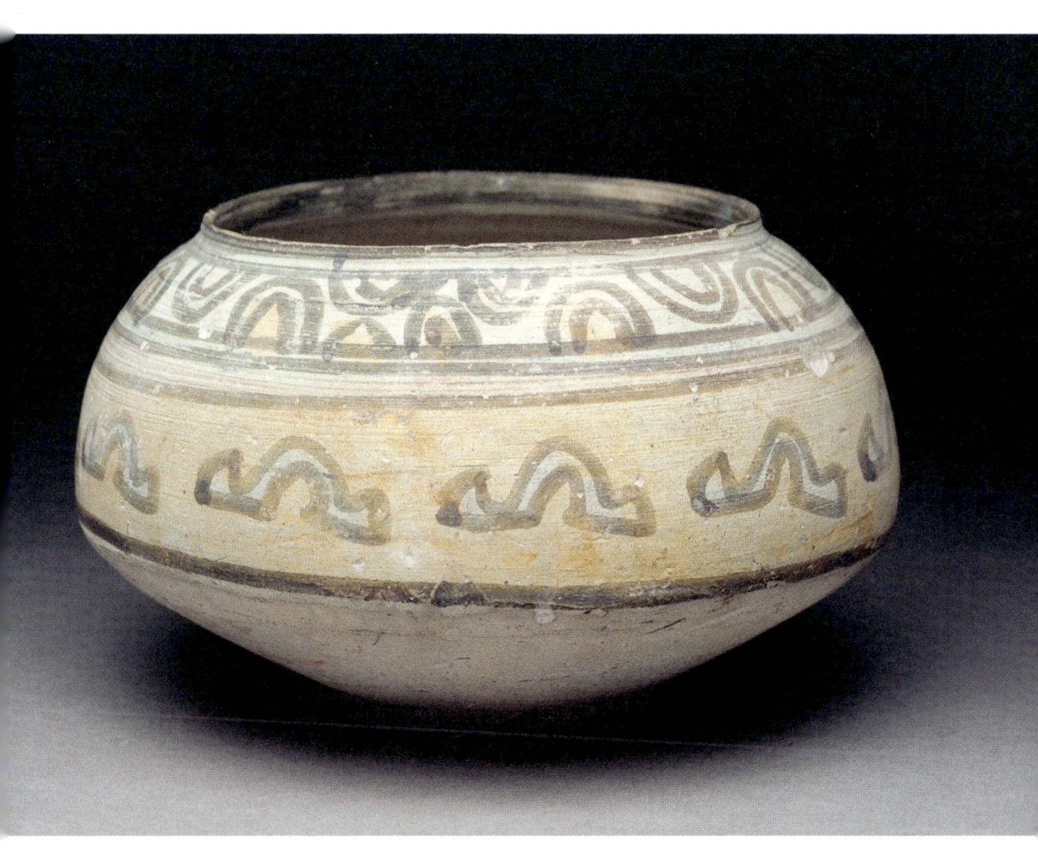

Bowl in the Nal polychrome style
Pakistan, 3000–2001 BCE.
Ceramic, part- or unglazed, 8.1 cm.

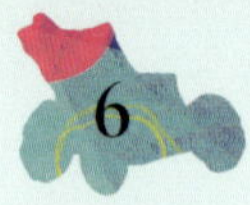

Model chariot with four-wheeler battle car

Iraq
*c.*2950–*c.*2575 BCE
Terracotta, 11.6 x 12.0 cm

This model of a four-wheeled chariot was made by hand from slip and brown fabric. Its surface is mottled in places and there are signs that it has been restored. Its maker smoothed its surfaces and edges to finish them, probably with a sharp blade.

Small ceramic models of chariots have been excavated at numerous sites in southern Iraq, dating to midway through the third millennium BCE. This was a time when the region was divided between city states ruled by kings who were competing for the control of land and trade. Warfare is a recurring theme in the imagery of the period. Carved reliefs depict armies of foot soldiers carrying spears, and spear throwers in chariots pulled by pairs of donkeys. We know of baked clay models of two-wheeled chariots, but four-wheeled vehicles like this one are usually shown in scenes of battle.

This hand-modelled chariot, decorated with red paint, was excavated within the remains of a palace at Kish, a site that is famous in the literature of ancient Mesopotamia as the place where kingship itself was thought to have been established by the gods (see page 58). The chariot is constructed as a box, open at the top where the driver sat. He would have held reins that were fed through a hole, shown at the top of the front panel. Another hole, further down the front of the model, is for a wooden pole, to which the donkeys would have been tethered. The driver would have been accompanied by a spear thrower whose weapons were stored in the two-holed quiver at the left front of the vehicle. Four solid wheels, which would have been made of wood if this were an actual chariot, have been fitted to the model with restored axles. Archaeologists have also uncovered miniature clay images of both humans and animals, some of which would presumably have been associated with these chariots. Figures like these have sometimes been found with stone cylinder seals, clay seal impressions and written tablets, suggesting that they played a part in managing state resources.

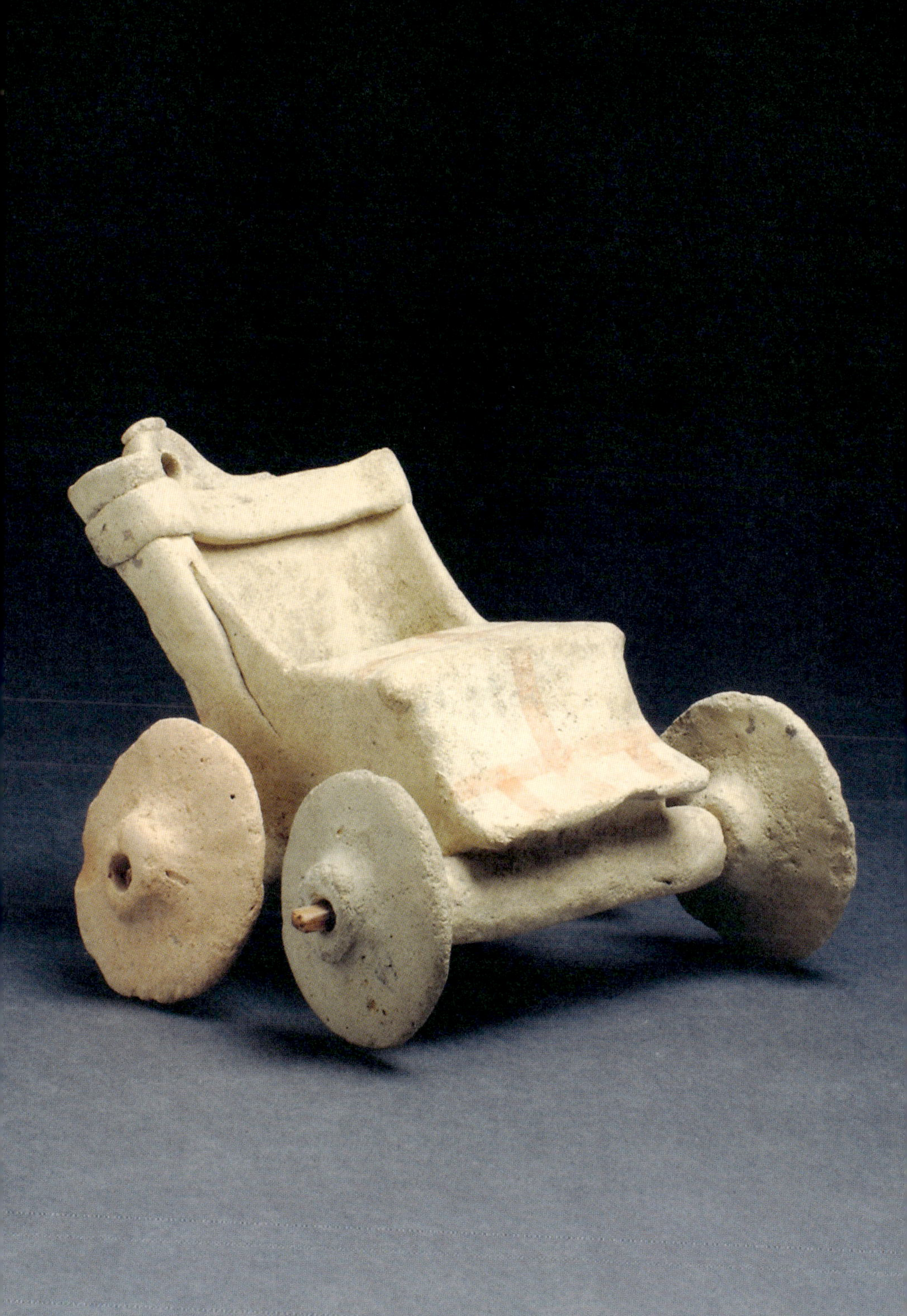

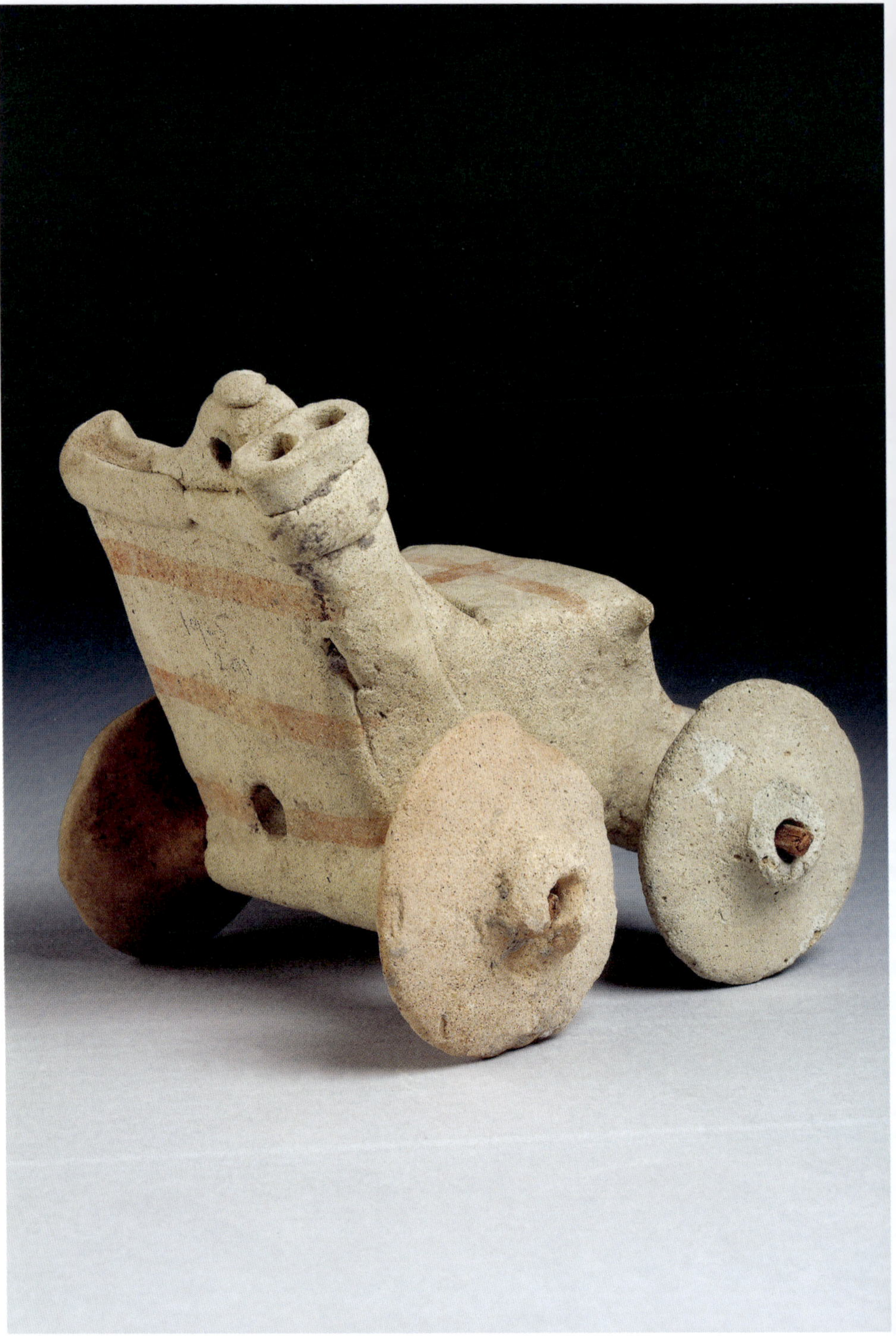

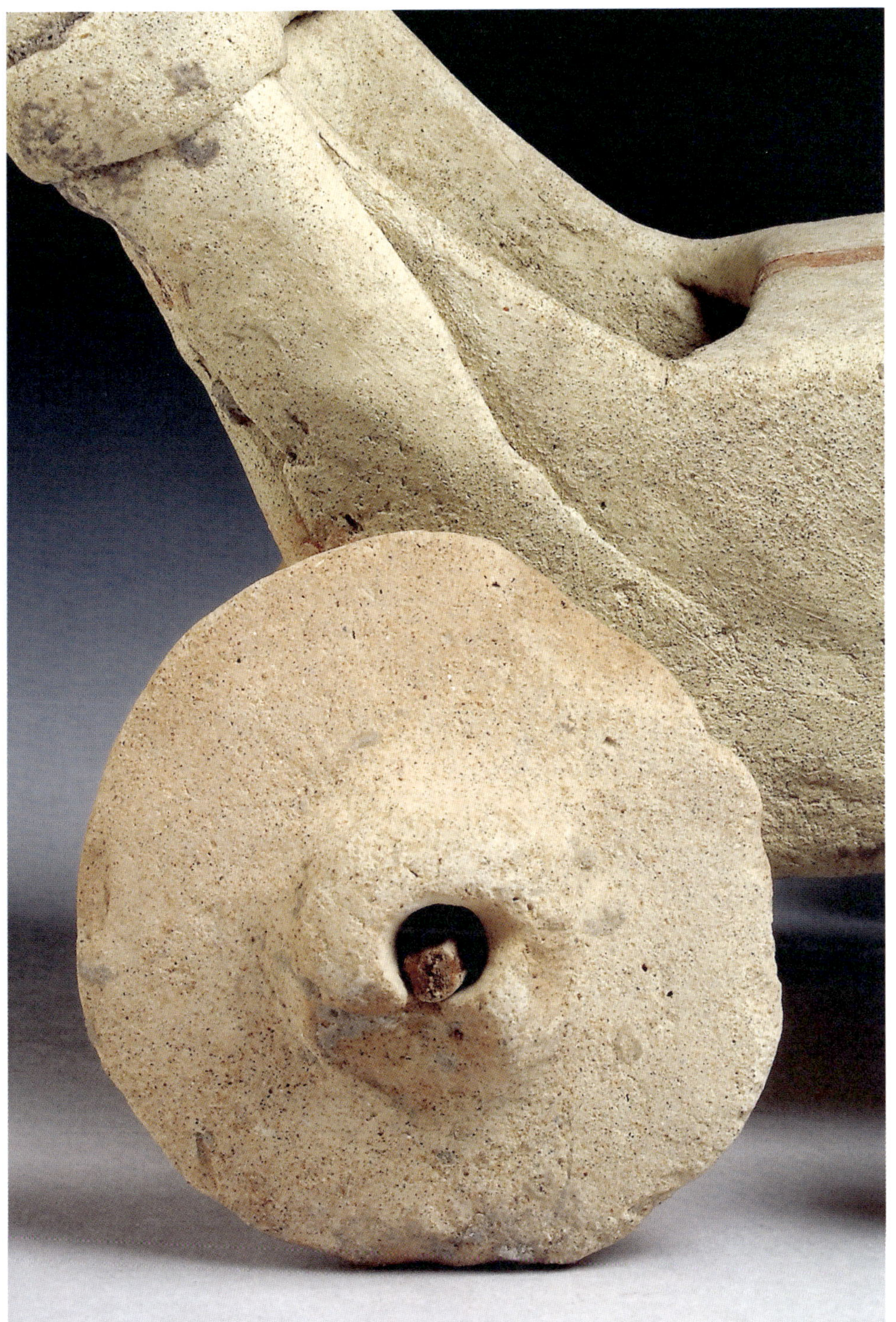

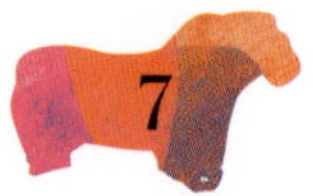

Terracotta figure
of a bull or ox

Pakistan
2500–2300 BCE
Terracotta, 5.4 x 9.2 x 3.5 cm & 3.5 x 6.5 x 2.0 cm

Archaeologists working at Indus Valley sites have recovered few sculptures of humans, but numerous figures of terracotta animals. Usually, these are bovine animals such as bulls, water buffaloes and zebus. Some of the creatures have movable heads, fixed to their bodies by a bolt. Often, they form part of a group of animals pulling a toy cart: a depression in the neck of the animal pictured shows where the yoke may have rested.

Since there are sometimes holes in their hooves or base plate, we can guess that the creatures could be used as wheeled toys. Other animals found at Indus Valley sites include rams, monkeys, elephants, turtles, felines, rhinoceroses, markhors (wild goats), dogs and hares. We don't know what their function might have been.

We think that the earliest evidence for carts in Harappa is at about 3700 BCE. The model carts that the animals pull were made true to life, and the large number of these items, presumed to be toys, found is a reflection of the importance of carts at the time. Archaeologists have found several accurately portrayed carts at Harappan sites that reflect the different types of vehicle in use. Some were designed for speed, but a four-poster canopied cart on a plinth seems to have been ceremonial or, like a palanquin, designed to shield its inhabitants. Carts not only conveyed people, but also carried goods. They must have been viewed as an important technology since they could transport commodities, such as bricks to construct cities and grain to support the inhabitants therein.

We believe that bulls held an important ritual position in Harappan culture. Their horns are frequently used as a motif on pottery and as a headdress for men in "yogic" postures. A common motif on Indus seals is a decorated bovine animal, often dubbed an Indus "unicorn". It's hardly surprising to see toys made in the likeness of bulls in an agrarian economy, in which these creatures would have been highly valued, and were the most common domestic animal.

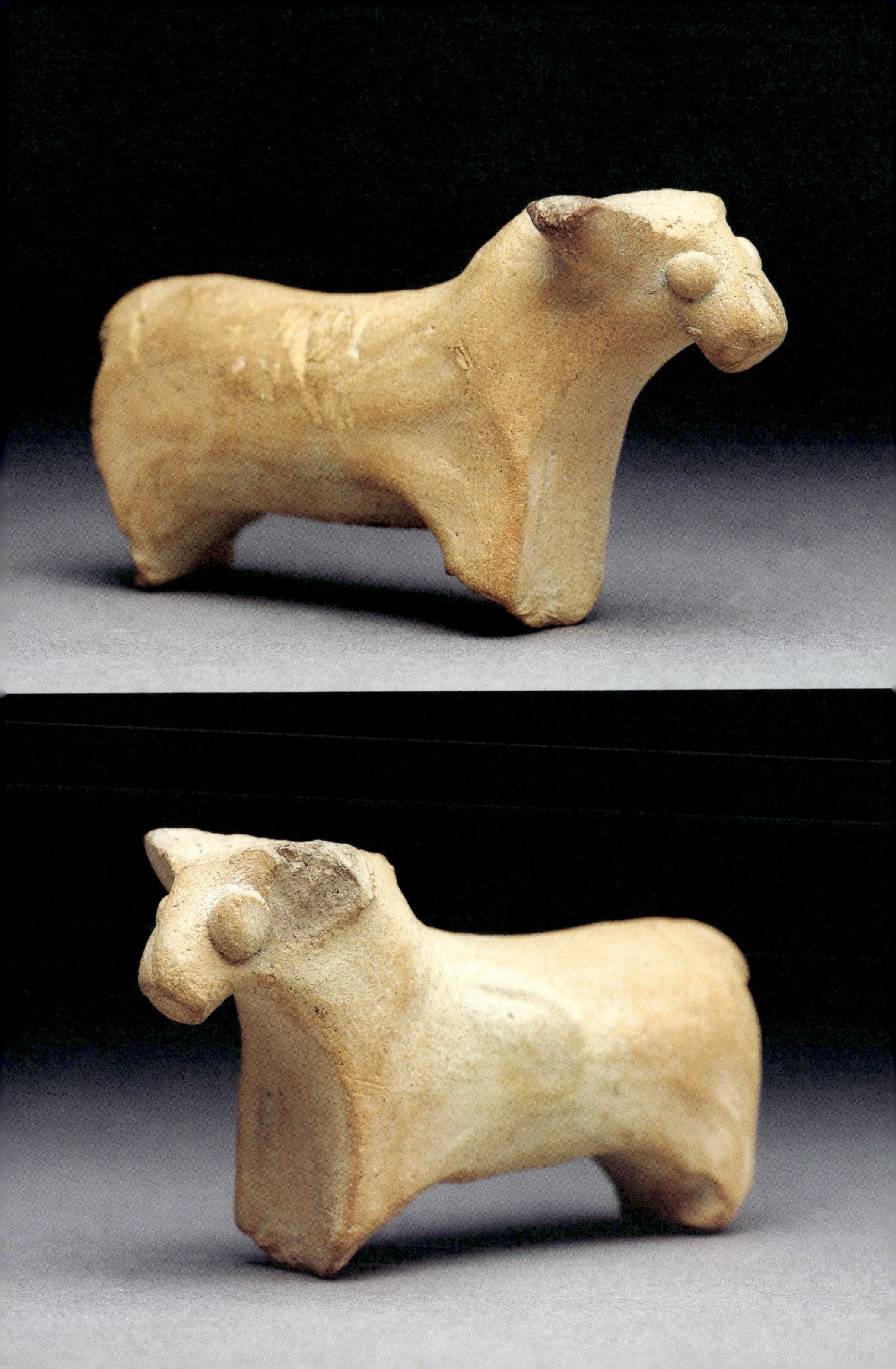

In Indian mythology, the bull is a symbol of virility. Shiva, one of the principal Hindu deities, rides a bull named Vrishabha ("the white bull"), who is sometimes also called Nandi ("the joyful"). In an enduring legend, the warrior goddess Durga battles and defeats a buffalo demon named Mahisha. We know that bovine animals have been ritually sacrificed since antiquity.

These objects, then, might not just be gaming counters or toys – they could just as easily have a ritual purpose. We could speculate that the carts were intended as a land dweller's means of transporting spirits to the afterlife, comparing them to the tomb offerings of boats found in Egypt or the figures of horses in Han Chinese burials. The only thing we know for sure is that these animals and their carts were made across the subcontinent for thousands of years, from before 3300 BCE, when the Indus Valley civilisation began, and later than the post-Mauryan period, in the second century BCE. What we can observe is that the animals are modelled with tremendous feeling, in contrast to the stylised forms used to depict humans in the same period.

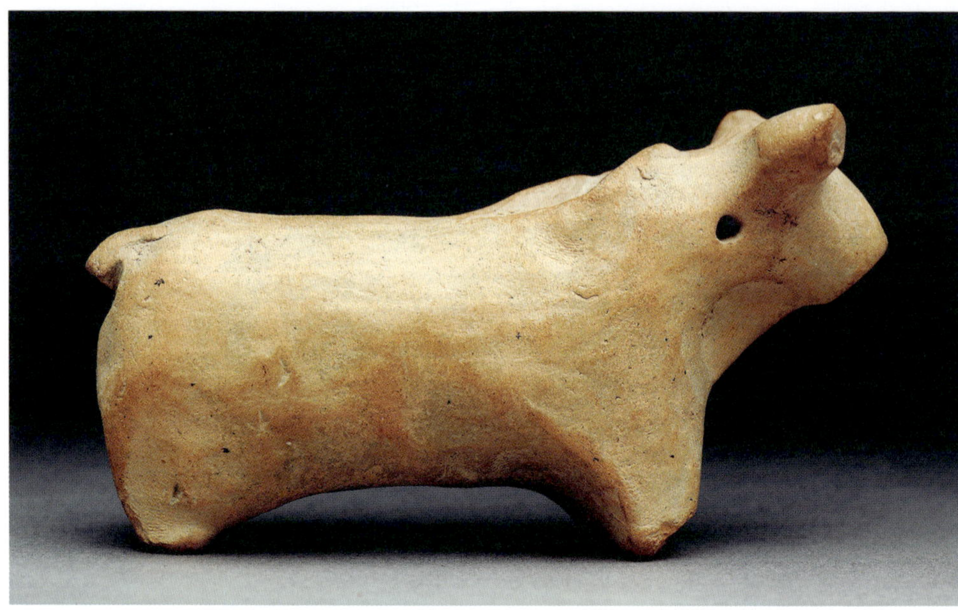

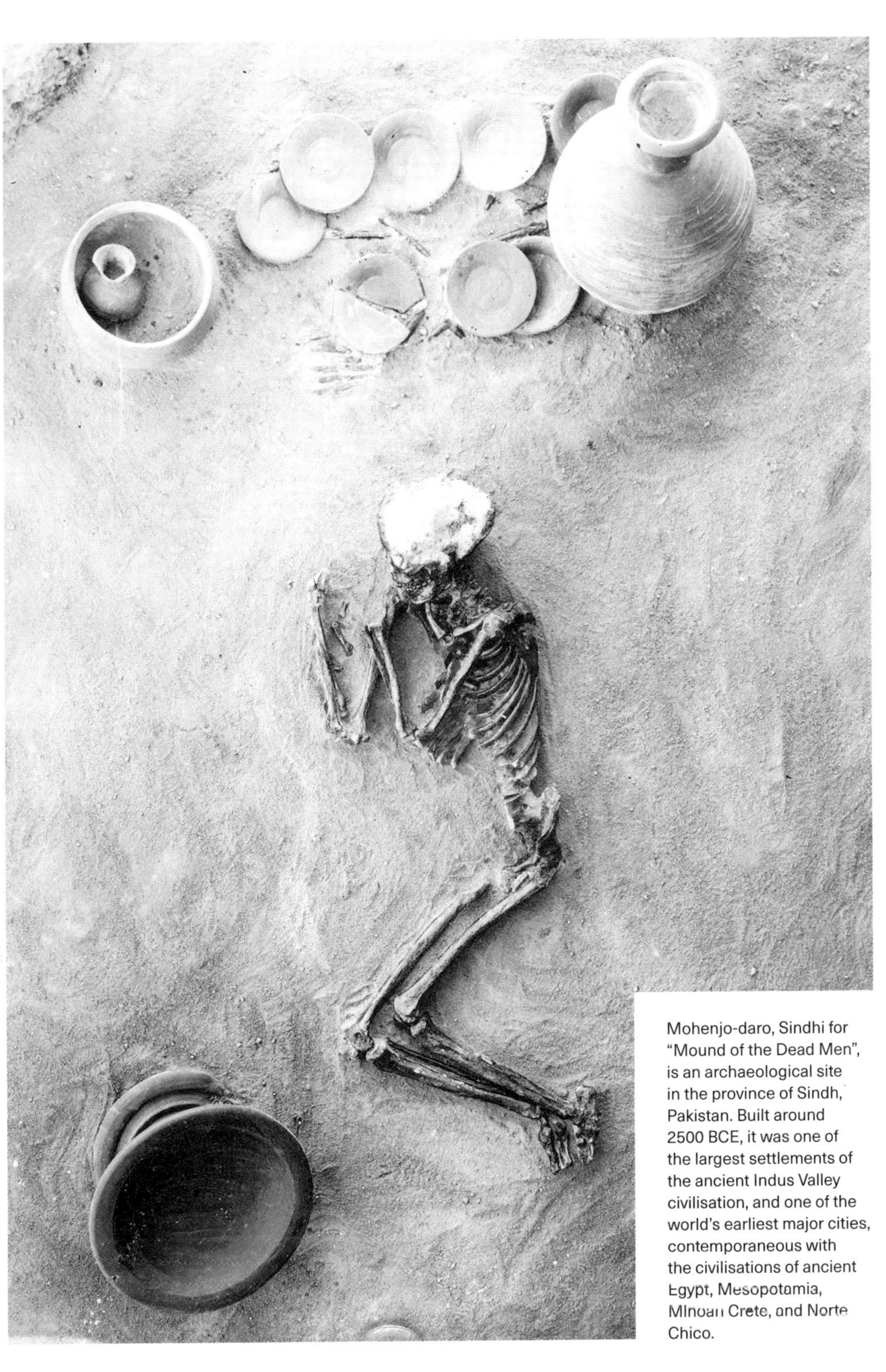

Mohenjo-daro, Sindhi for "Mound of the Dead Men", is an archaeological site in the province of Sindh, Pakistan. Built around 2500 BCE, it was one of the largest settlements of the ancient Indus Valley civilisation, and one of the world's earliest major cities, contemporaneous with the civilisations of ancient Egypt, Mesopotamia, Minoan Crete, and Norte Chico.

Indus Valley pots

India
2500–1900 BCE
Terracotta, 10.2 cm & 31.7 cm

Around 4,000 years ago, these Indus Valley pots were fashioned, in most cases, on a fast wheel. More complex pieces (such as pedestalled dishes with their supporting bases and very large jars) were made of separate, wheel-thrown parts that were luted together with a wet clay slip before firing.

At a few sites, archaeologists have found vertical kilns with a subterranean firing chamber, and on top, perforated bricks to draw up the heat. These kilns can only hold smaller wares, so they must have been used alongside Indian pit or open-fired kilns, which were more common. We think that the expanded rounded bases of large jars and pots may have been made by beating them out with a paddle and anvil. The bases of round-bottomed Indian water pots are still made this way today.

The long jar pictured is an example of a type that archaeologists have uncovered in many sizes. Given its shape, it was probably used for storage. Some experts have speculated that, if lined with muslin, the jars could have been used to thicken yogurt or prepare other dairy products. Others have suggested that they may have been refuse jars, draining water and collecting compost. The jars have even been associated with rituals that might have involved filtering substances.

Our understanding of Indus pottery is limited because of the very different procedures for data collection and analysis today compared to when the sites were excavated. These days, we even attach different values to the data. The early excavations at Mohenjodaro, Harappa and Chanhudaro in the 1920s and 1930s produced masses of sherds and pots, but not enough quantitative analyses to identify patterns, associations and relationships. While today we know that different types of pots are found in different parts of a site – since each pot had a specific function and was used by a specific community, or perhaps even by a different ethnic group – these details were not always systematically recorded. To understand more about these pots, we have to rely on the work of scholars who have tried to reopen and reinvestigate the original excavations.

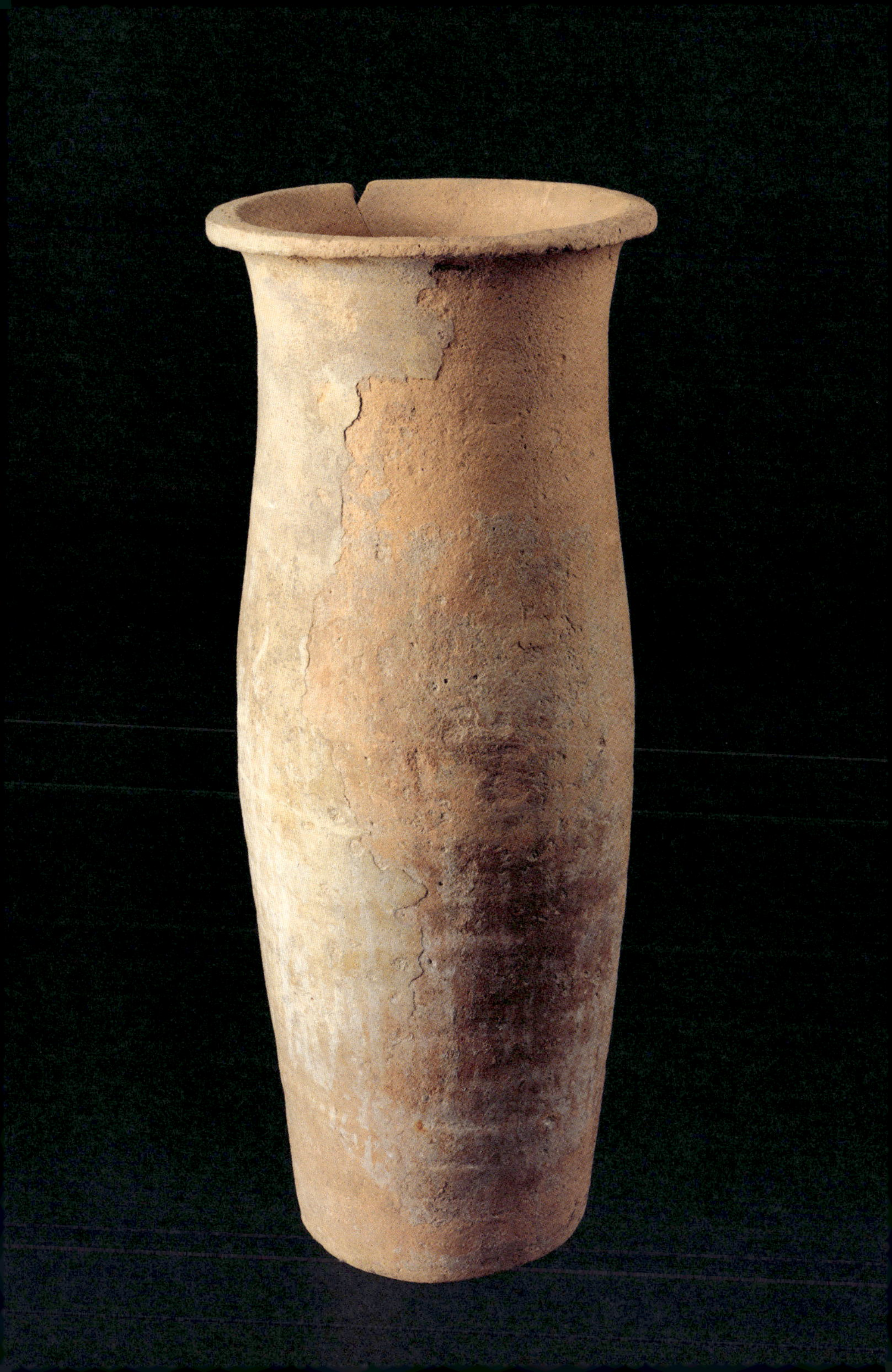

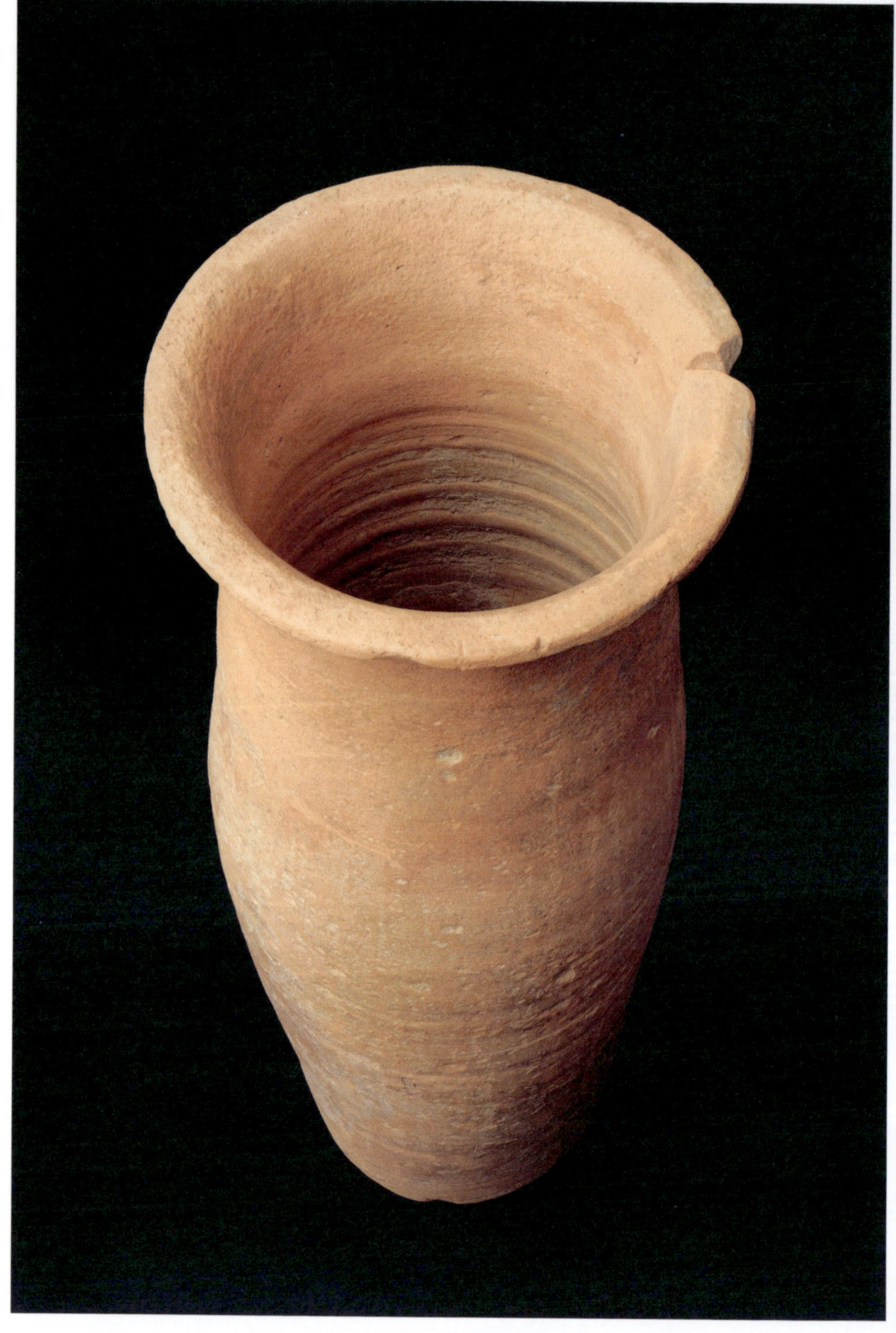

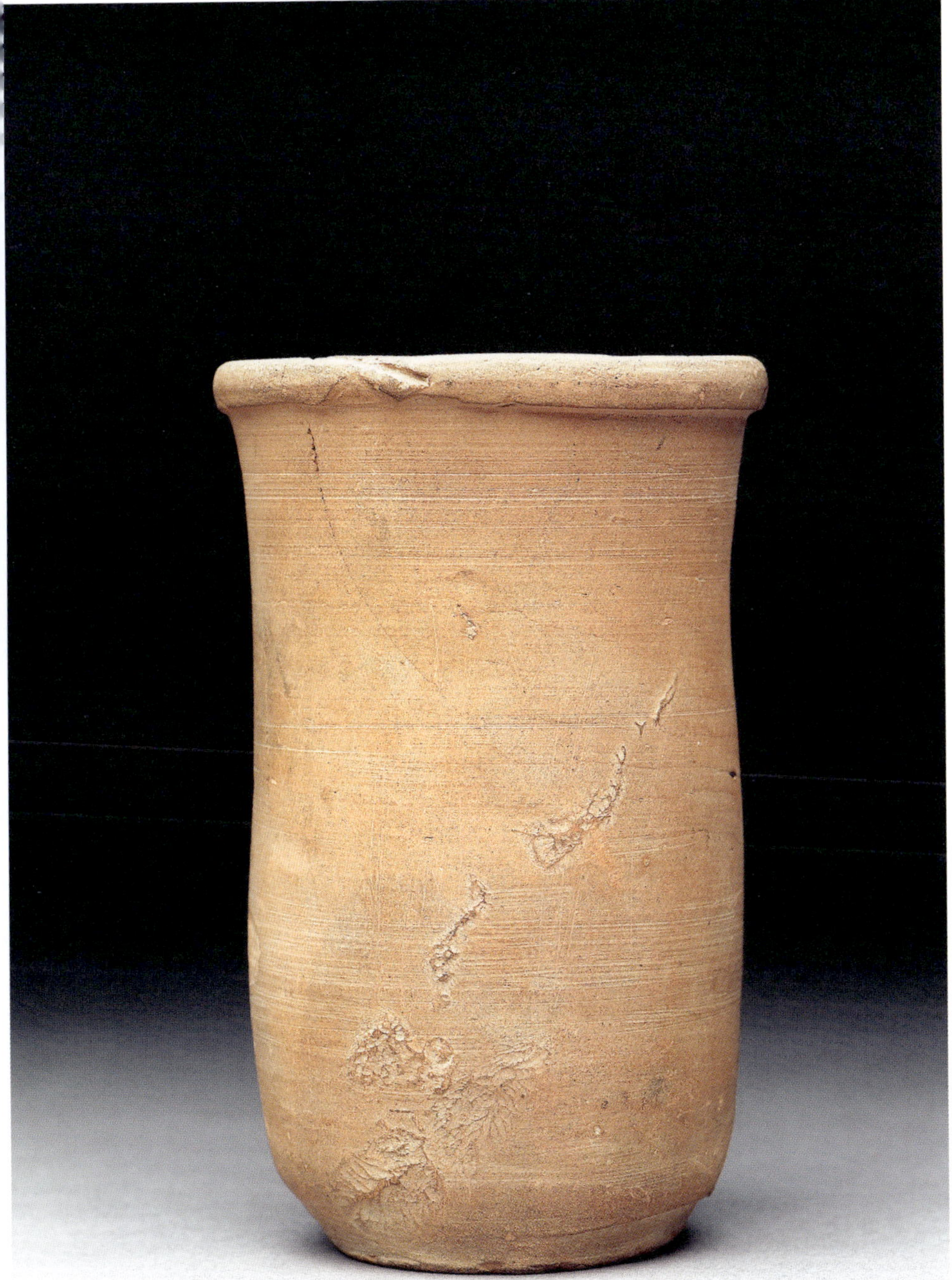

9

Pottery lion

Egypt
*c.*2325–2175 BCE
Fired clay, 42.4 cm

This pottery statue depicting a lion on an oval base was found with a cache of royal sculpture in the temple enclosure at Hierakonpolis, a religious and historic centre of early Egypt, and is a rare surviving example of Egyptian sculpture in clay.

The modelling and firing of large clay sculptures posed serious artistic and technical challenges to their creators. The artist who depicted this lion has approached the task with a blend of realism and stylisation. The beast's face and muscular body are vividly natural. The same goes for its paws, the toes curled in repose yet promising powerful action with their claws. The lion's mane, however, is rendered somewhat geometrically. The ruff around the face forms a circle continuous with the ears and the fall of hair on the chest is shown as a bib-like square. The lion's body is proportionately small for its head, probably because of the technical limitations of the medium, but the overall impression achieved by the artist is one of dignity, strength and majesty. The oval base with a rim may have been added for both technical and functional reasons.

Lions were significant creatures in ancient Egypt, symbolising royalty, protection and ferocity. The big cats were even kept as pets by the pharaohs – images exist of both Ramses II and Tutankhamun accompanied by pet lions. Paired lions often served as the guardians of entrances and, at temples, sculpted lions performed the same function. The British Museum's Prudhoe Lions, a pair of eighteenth-dynasty red granite sculptures, are a magnificent example, and also share this pottery statue's combination of abstraction and realism. This pottery figure may have had a companion – the excavators found fragments of another pottery lion elsewhere at the site. The two probably served as guardians within the temple precinct.

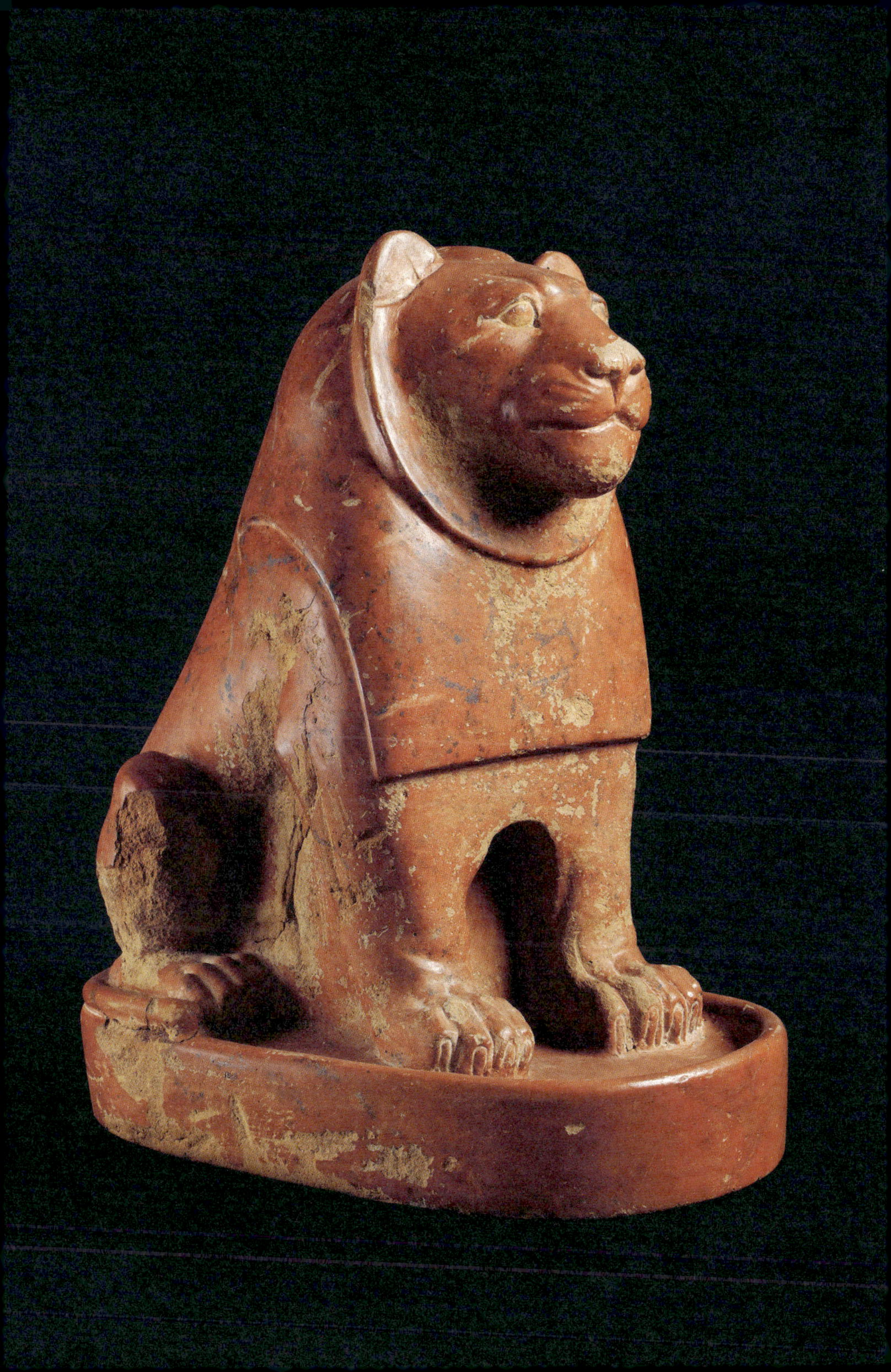

10

Terracotta votive figurines

Crete
c.2100–c.1090 BCE
Terracotta

In prehistoric Crete, there was a long tradition of using clay to make figurines depicting humans and animals. The earliest ones that have been found were in the Early Neolithic levels of Knossos in Central Crete.

They were made by some of the first inhabitants of the island, in the seventh millennium BCE. The figurines pictured here were made much later. They belong to the Bronze Age of Crete, known as the Minoan period, which was named by the British archaeologist Sir Arthur Evans after the mythical king Minos. During the Minoan period, objects and architecture are similar across the island, and these are typical examples of Minoan figurines.

The two male figurines were both found at the site of Petsophas in East Crete, a sanctuary on a hilltop overlooking the Minoan harbour town of Palaikastro. Petsophas was first excavated in 1903 by John Myres, who discovered a stone building surrounded by deposits of clay figurines of both humans and domestic animals. An experienced archaeologist, he was visiting Palaikastro as part of a cruise party but, as the director of the excavation, Robert Carr Bosanquet, wrote in the report of the year's discoveries, Myres "was induced to exchange the luxuries of ocean travel for the privations of the excavator's lot". Since Myres's discovery, more "peak sanctuaries" have been found across Crete: deposits of figurines found in high places. These sanctuaries probably served as a ritual focus for the surrounding agricultural land, since archaeologists have found in them large numbers of animal figurines, particularly cattle. The human figurines are often making particular gestures, frequently with their elbows out or arms raised. We could interpret them as gestures of prayer. The male figurines pictured are shown wearing codpieces, a typical Minoan garment, and one has a dagger in his belt.

It looks similar to the others, but the female figurine was made around 500 years later. It was found at Knossos in a shrine which was established after the Bronze Age palace was destroyed. The Petsophas figurines are handmade, but the skirt of the female figurine was made on a wheel, and the horizontal striped decoration is similar to that on contemporary pots. It's likely that all these figurines were made by potters who also made vessels and had the kilns to fire

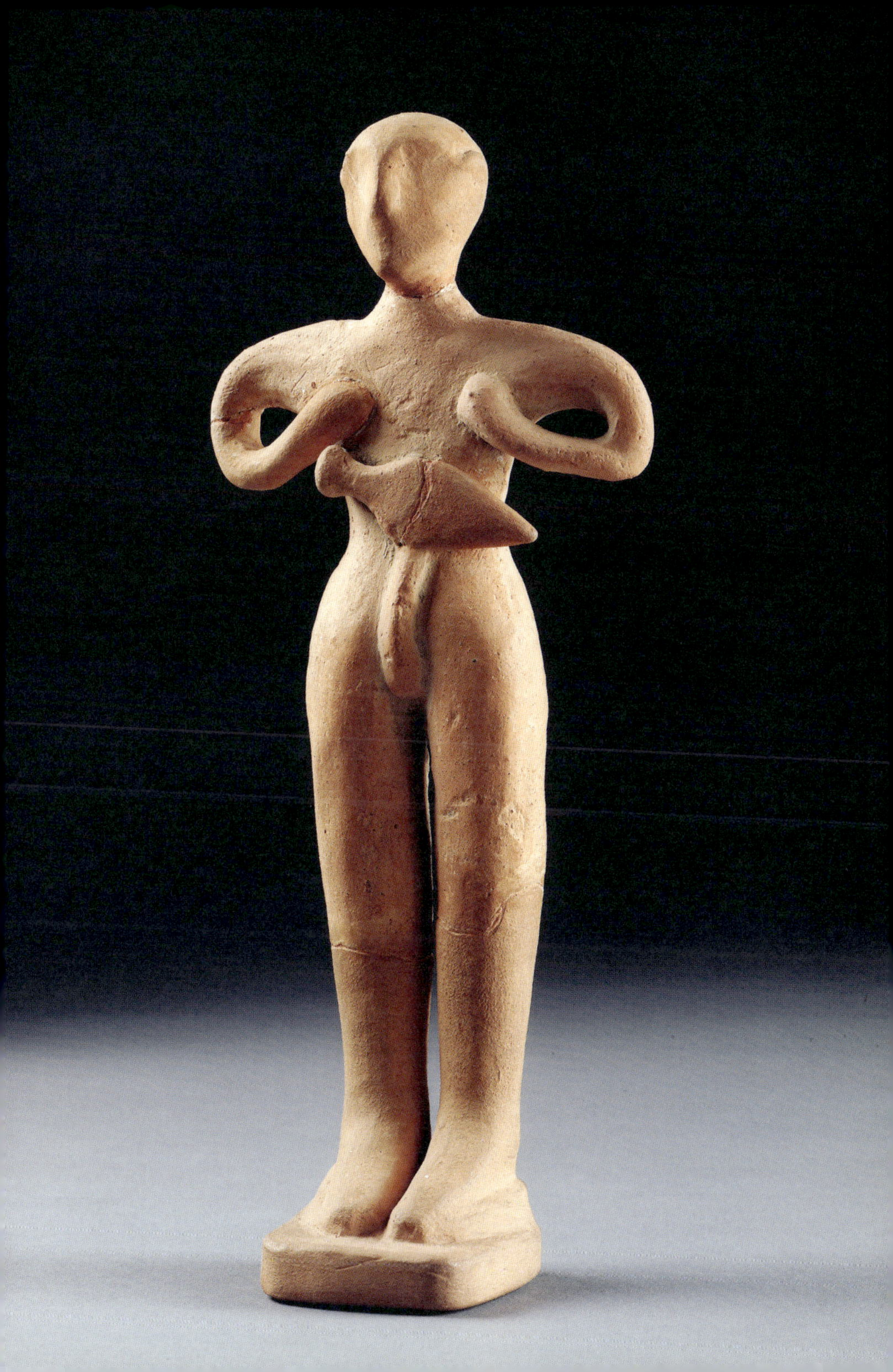

them. While peak sanctuaries were connected with agriculture, the sanctuaries that emerged on former palace sites on Crete seem to have commemorated the important centres of the Minoan civilisation that had now fallen into ruin. Despite the change in focus, the form of worship – the offering of figurines making gestures of prayer – remained remarkably similar over time.

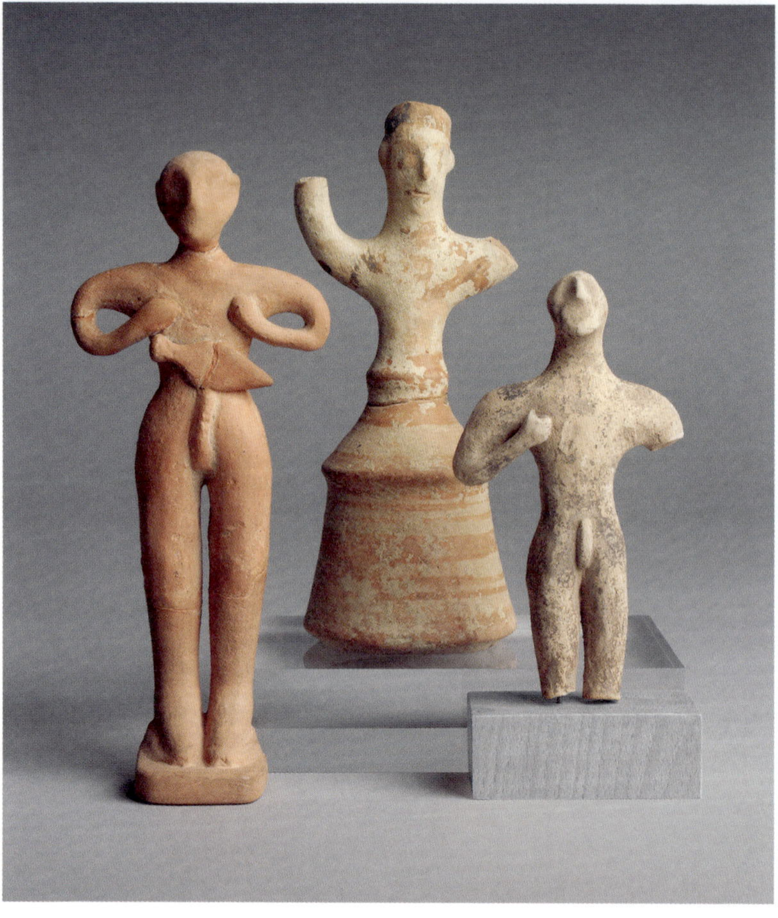

(above left and previous)
Human figurine, male
with dagger. Crete,
*c.*2100–*c.*1700 BCE.
Terracotta, 19 cm.

(above right) Terracotta
votive figurine of a man.
Crete, *c.*2100–*c.*1700 BCE.
Terracotta, 13 cm.

(above centre, and opposite)
Terracotta figurine of female
in bell skirt with raised arms.
Crete, *c.*1400–*c.*1090 BCE.
Clay, pigment, 7 cm.

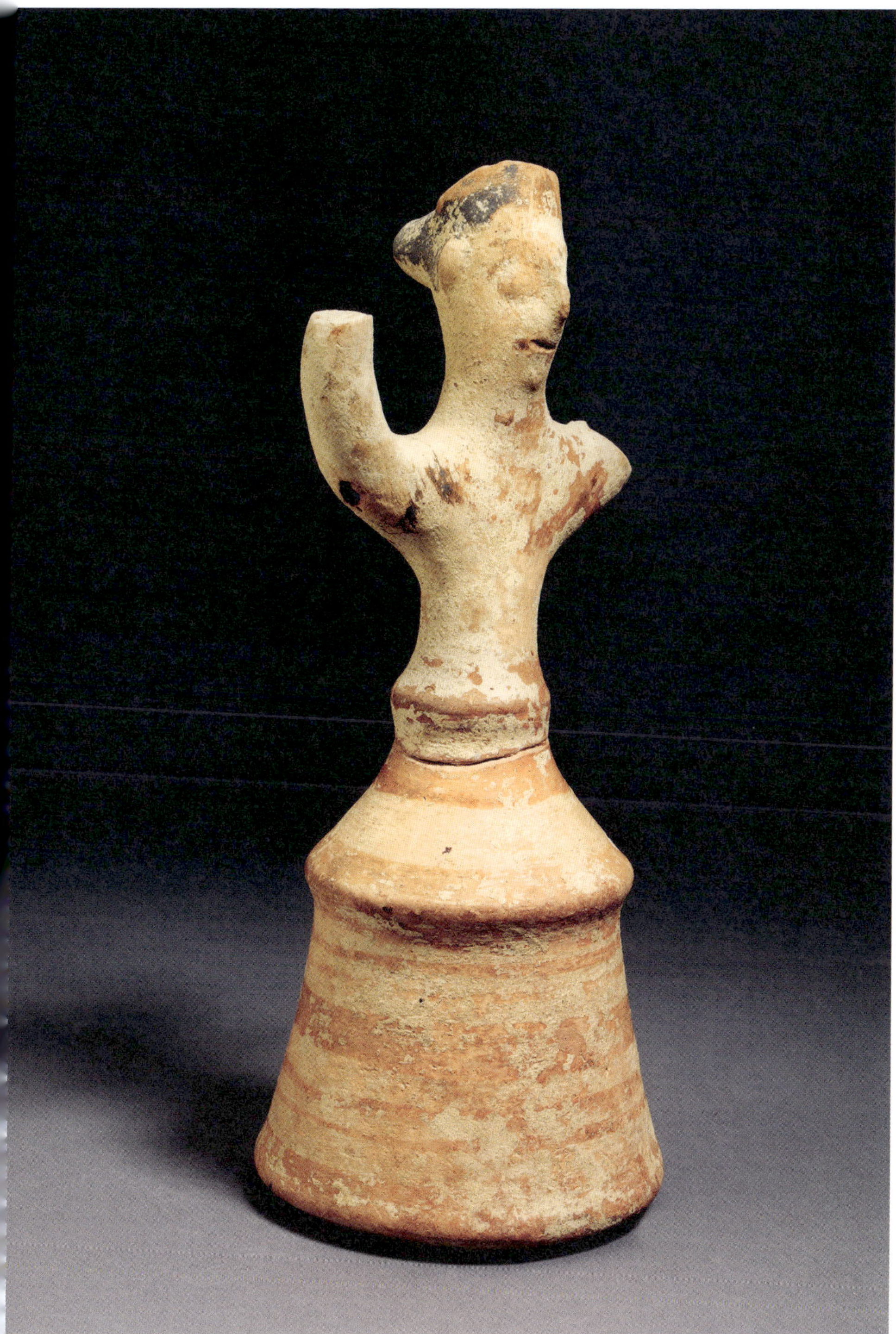

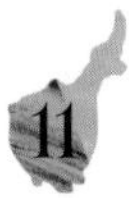

Red polished stag-shaped figurine with three legs

Cyprus
c.2100–*c*.1850 BCE
Ceramic, 16.8 x 12.5 x 7.5 cm

Alongside the usual bowls, jars and jugs, potters in ancient Cyprus created vessels and figurines shaped like animals and people. In the early periods (or from the Neolithic Period onwards, *c*.5500 BCE) they were always handmade, while from about 1200 BCE, they were also produced on the potter's wheel.

Often, both pots and figurines were decorated in the same colours and patterns and fired in the same kiln. In the late third millennium and second millennium, such figurines retained the shapes and features of everyday ceramic vessels, even though they couldn't really be used as, for example, a cup or a jar.

This stag is a perfect example of a vessel–figurine hybrid. It was produced in the Early Bronze Age in Cyprus (2300–1900 BCE) in Red Polished Ware, the predominant and most elaborate type of decorated pottery of the time. This ware is characterised by a shiny red polished slip and incised and pierced decorative patterns. These are often highlighted by a white lime-paste filling. Small lug handles are attached at various different points of the vessel. The only feature which identifies this figurine as a stag or deer is its crudely formed antlers. Its body has the shape of a slightly tilted bulbous bottle, giving the stag a bloated and sagging appearance. It stands on just three stumps, which don't look anything like the long, slim legs of a real deer. Atop the stag's hollow neck sits the head: a vertical, pierced lug handle forms the muzzle, and eyes that are emphasised by circles are incised around them. The stag's ears are indicated by bits of clay pulled out on either side of its head. Its antlers were formed separately and attached on top, where the bottle spout would have been. All that's left of the spout is a barely visible almond-shaped hole in between the base of the antlers. It couldn't have functioned as a bottle mouth – only as a vent hole preventing the hollow figure from exploding during the firing process.

Like most Red Polished pottery that has been uncovered, this well-preserved stag figurine must have been found in a tomb, where it would have served as a burial gift.

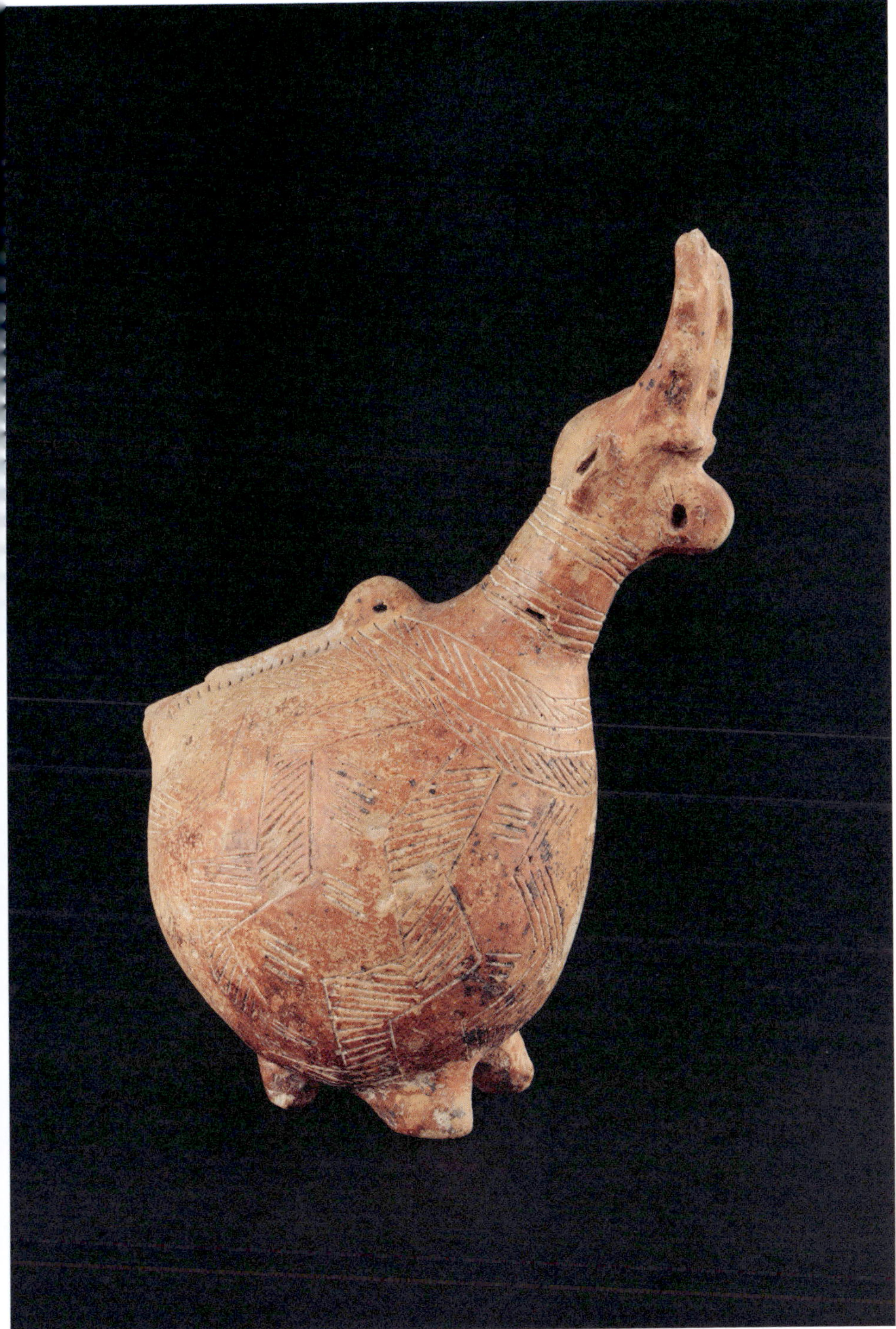

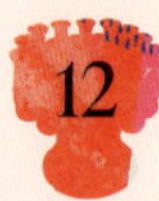

Kernos with two rows of containers and central bowl

Greece
*c.*2300–*c.*2100 BCE
Ceramic, pigment, 32 cm

A *kernos* (plural *kernoi*) is a ceramic vessel onto which other smaller vessels have been fixed, usually in a circular arrangement. This *kernos* is an especially complex one. Inside its outer ring of 15 small, tubular vessels there is another ring of 10 more. A slightly larger globular vessel is nestled at the centre.

Every component is handmade: the pottery wheel only arrived in Greece after the Early Bronze Age (*c.*3000–2000 BCE). Whoever decorated this *kernos* put in plenty of effort. They used a white slip to cover the reddish local clay, and matt black paint to add a variety of geometric patterns. The diagonal stripes create the impression of movement. The slip and the paint were applied in different stages before the piece was fired. It would have needed somebody very skilled to fire such a large and complicated vessel.

This *kernos* belongs to the Cycladic culture of the third millennium BCE. At this time, trading links between the Cycladic islands in the Aegean Sea enabled a complex society to develop, which produced an array of decorative objects. The most famous are marble Cycladic figurines, with their unique, simplified depiction of the human body. Widely admired, they have inspired many artists from the twentieth century to the present day, including Amedeo Modigliani, Barbara Hepworth and Henry Moore.

Back in the third millennium, clay was another material used by artisans to experiment with form and decoration. Phylakopi, on the island of Melos, was an important centre of pottery in this period. Archaeologists have found vessels decorated in a similar way to this one at Phylakopi, including *kernoi*. We don't know where this vessel was discovered because it was part of a nineteenth-century collection before it came to the Ashmolean, but it is likely to have come from Melos. Since it isn't broken, we can guess that it was probably uncovered in a tomb, but down the centuries, enthusiastic collectors of Cycladic antiquities have left very few tombs intact for present-day excavation.

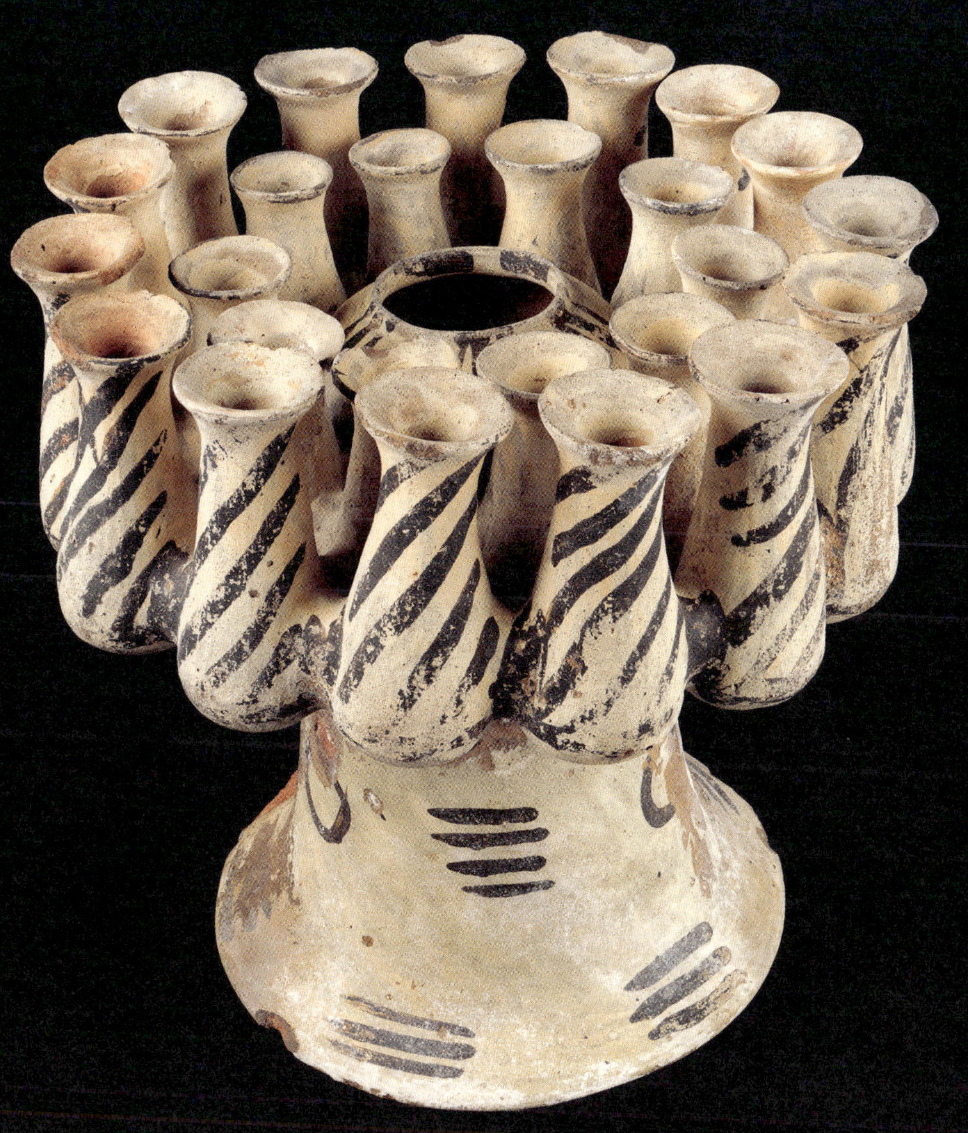

We don't know what people on the Cycladic islands used these vessels for. In the later years of classical Greece, vessels like this one were used to hold and carry religious offerings, particularly the first fruits of the harvest. When this *kernos* arrived at the Ashmolean, plant matter and the seeds of flowers were found inside some of the smaller vessels. It could therefore be deduced that it had been used as a flower vase in a Victorian country house, but ironically, the seeds were also a clue as to how such an elaborate vessel might have been used in the long distant past.

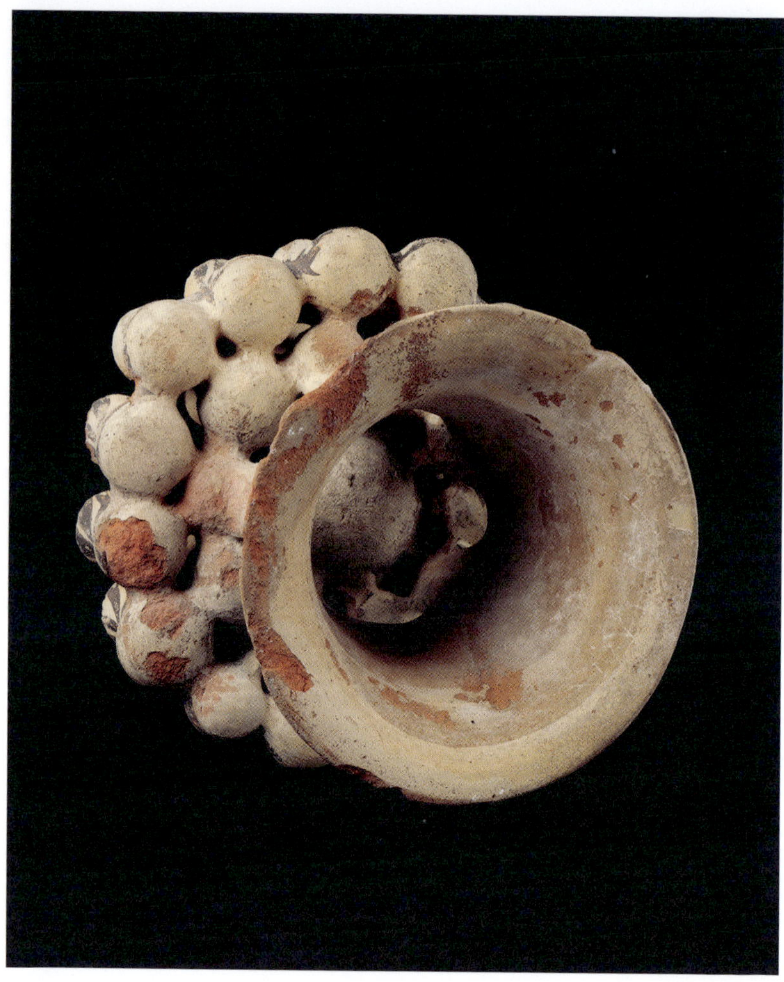

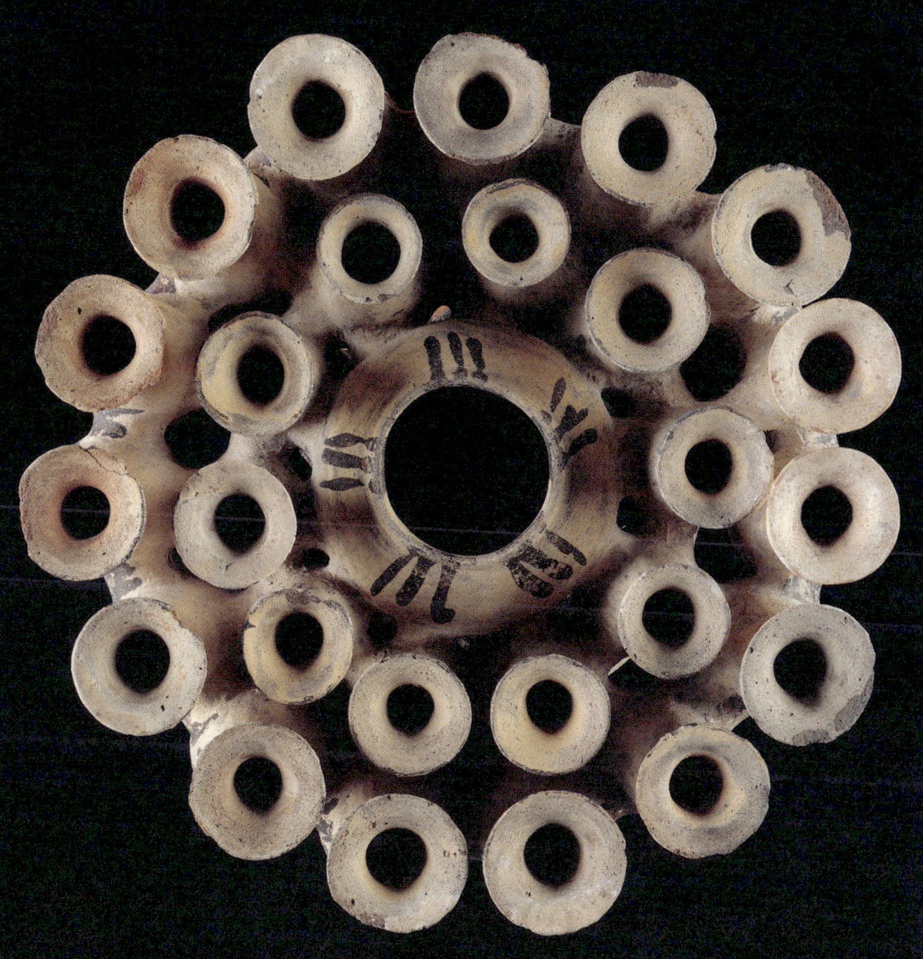

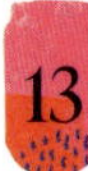

Sumerian king list

Iraq
*c.*1800 BCE
Clay (baked for conservation), 20 x 9.1 cm

This baked clay block with eight columns of dense wedge-shaped writing (cuneiform) is a list of cities and their rulers that claims to stretch back to the beginning of time. King lists were created by learned scholars to suggest that the whole of Mesopotamia (modern Iraq) had always been ruled from a single city – though not always the same one.

This wasn't actually true: Mesopotamia was often divided between independent city states. Scholars took lists of the kings from several cities and placed them end to end. We know that in fact, some of the kings' reigns had overlapped.

This list sets out to present a sequence of these cities and their rulers from the very beginning of time to around 1800 BCE, when this version was compiled. The scribes who copied and evolved these lists were consciously creating something that fell between a work of literature and a list. The person who wrote out this version signed their creation at the very end: "by the hand of Nur-Ninsubur". His list is the most complete and best-preserved king list ever discovered.

The list begins in a mythological time, when kingship was "lowered from heaven" by the gods, and the world is said to have been ruled, sometimes for many thousands of years, by individual kings. For example: "In the city of Zimbir, En-men-dur-ana became King; he ruled for 21,000 years." This mythological time was separated from a more recent historical time by a great flood: "Then the flood swept over." The scribe writes that after the flood had swept over, and kingship had descended from heaven, kingship was in Kish. We can understand him to mean that kingship had been lowered from heaven for a second time and restarted in a new city.

Legendary kings of the distant past are followed in the list by kings we recognise from historical documents. Some of the earliest names include remarkable or unusual rulers: a shepherd, a fisherman, a boatman, a leatherworker. A female tavern-keeper is said to have been king – and for quite a long time. Among these kings is listed the great hero Gilgamesh, who appears in several stories preserved in the cuneiform documents of Mesopotamia.

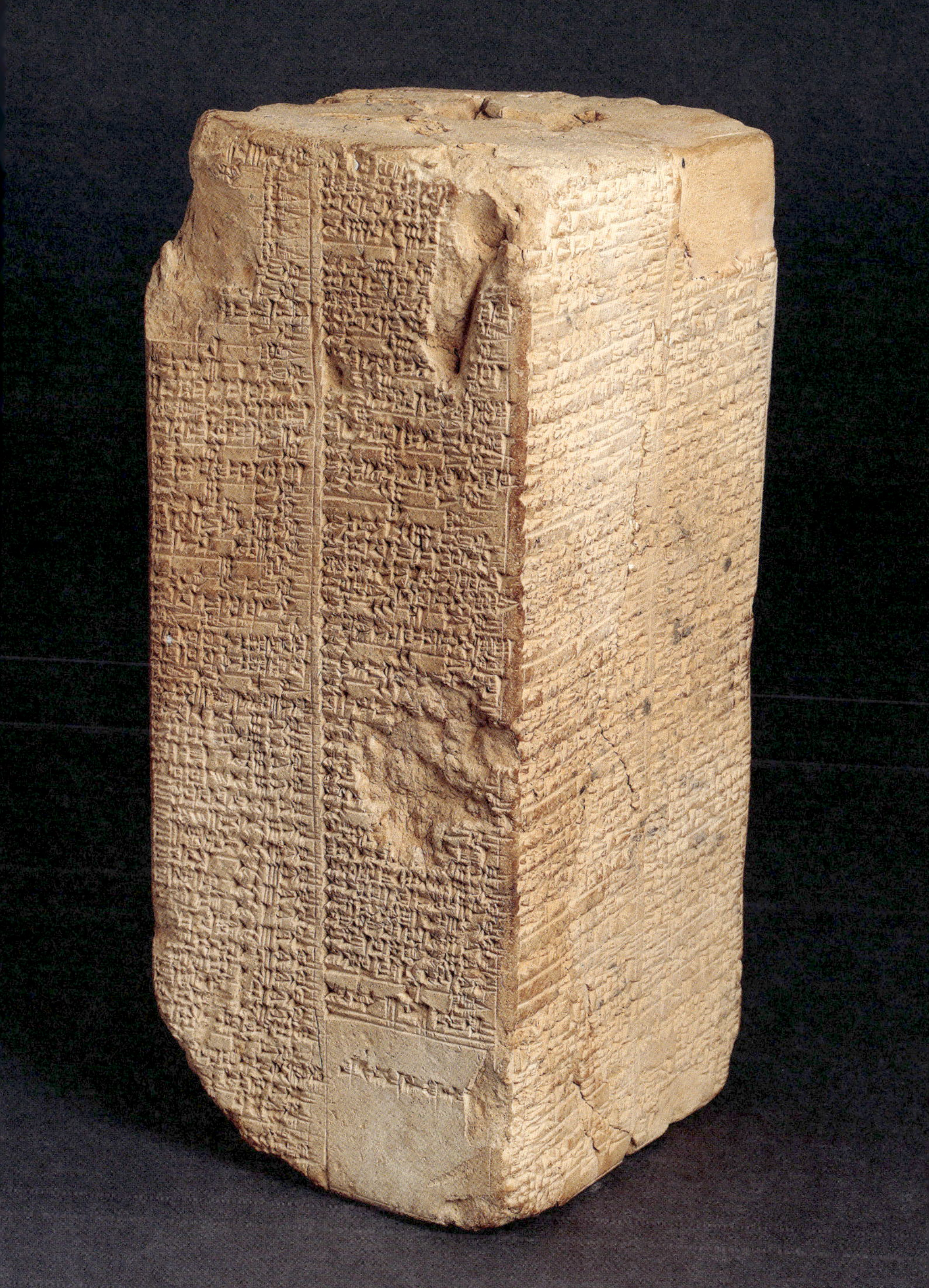

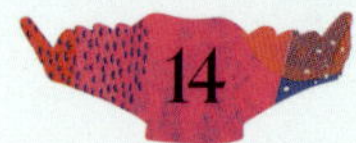

14

Two-handled spouted cup

Crete
*c.*1900–*c.*1850 BCE
Clay, pigment, 6 x 18 cm

Kamares Ware is a black-slipped and intricately decorated style of pottery from Bronze Age (Minoan) Crete. Its name comes from the place where it was first discovered in large amounts: Kamares Cave, in southern Crete.

Its archaeological significance was uncovered by coincidence. An English archaeologist, John Myres, happened to be visiting the museum at Heraklion in 1893 when a Cretan villager brought in pottery from Kamares Cave. Myres realised that he had helped to excavate the same type of pottery at a site called Lahun in Egypt, a few years before. This realisation was important for the development of Aegean archaeology: the pottery could be dated using the well-established Egyptian chronology. It also proved that there were trade connections between Crete and Egypt in the early second millennium BCE, the period known as Middle Kingdom in Egypt and Middle Minoan in Crete.

Large-scale excavations only began in Crete in 1900. British and Italian teams began to uncover two complex buildings they came to call "palaces" – at Phaistos in the south of the island and Knossos in the north, where this cup was discovered. A large quantity of Kamares Ware was excavated at both palaces. Because of the connections with Egypt, archaeologists were able to date the layers of earth within which this pottery was found. This meant that it could serve as a chronological marker for other finds. This type of black-slipped pottery was made at the same time that the palaces were emerging, at the start of the second millennium BCE. It was frequently used to make drinking and pouring vessels to be employed in the ceremonies that took place there.

The form of Kamares Ware vessels, particularly the very thin walls and black shiny slip, might have been designed to imitate silver vessels (which rapidly blacken when they tarnish). This theory has especially clear evidence in this vase, with its crinkly rim, which would have been incredibly difficult to make and fire using clay. The shiny black surface was created by firing the vessels at a very high temperature (around 900 °C) in a reducing environment so that the slip blackened and developed a glossy surface. The same technique was refined and used later on in Greece for black- and red-figure pottery.

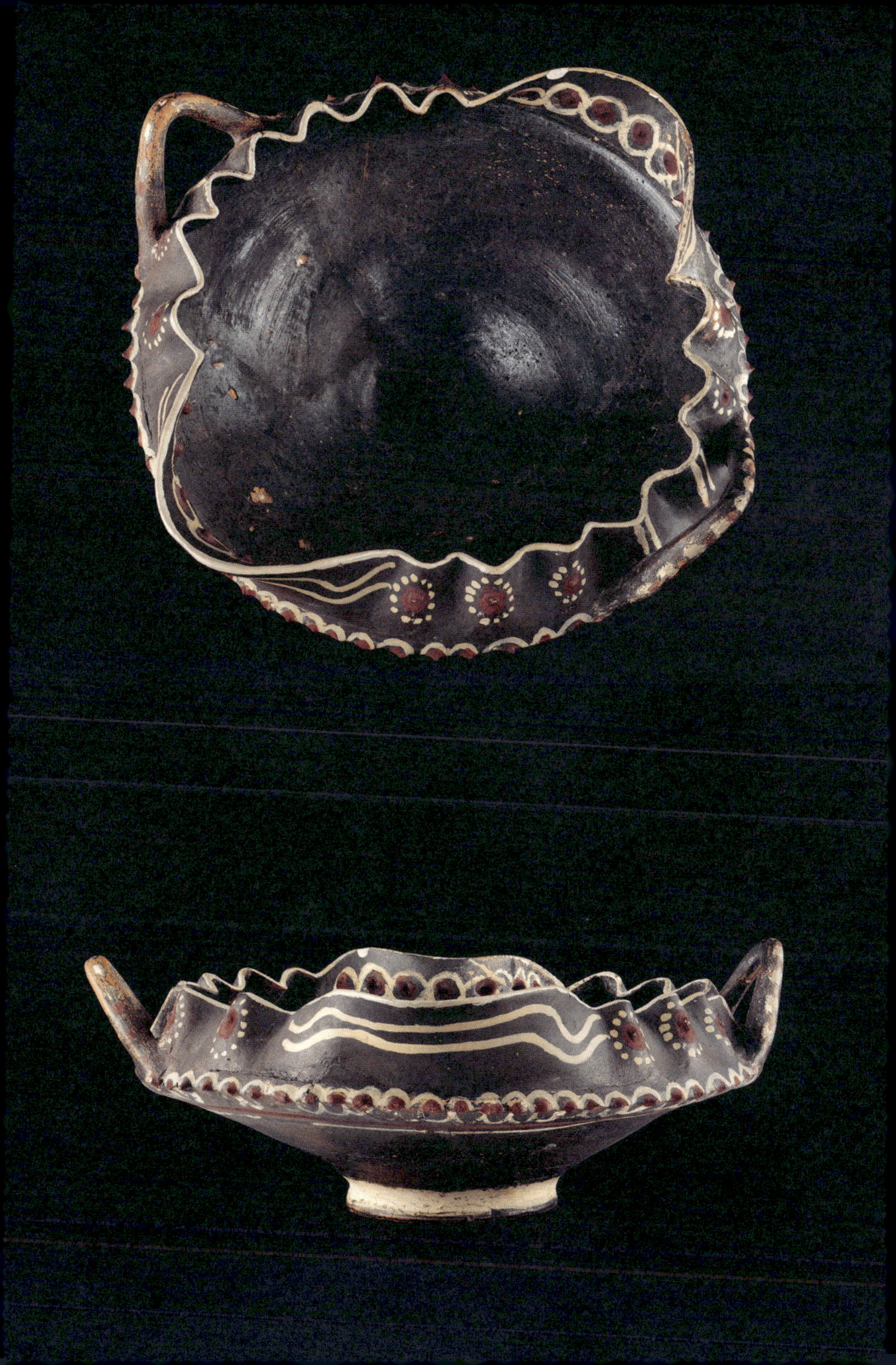

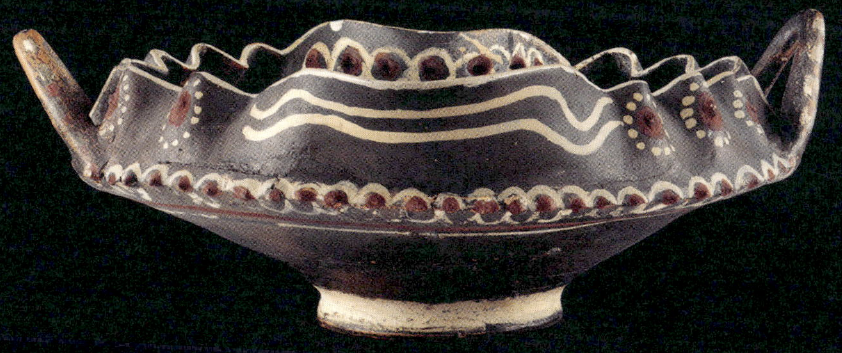

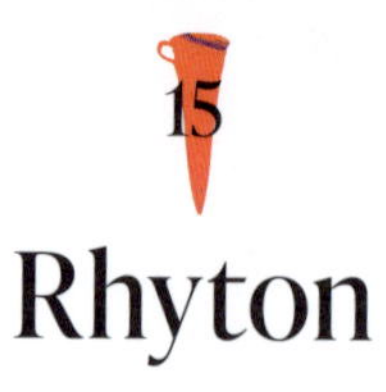

15

Rhyton

Crete
1600–1450 BCE
Ceramic, 35.4 cm

This painted ceramic rhyton was made at the height of pottery production at the Minoan Palaces in the New Palace Period (1750–1500 BCE). This period was a golden era of Minoan civilisation in which the great palace at Knossos was hugely expanded.

Rhytons like this one were frequently exported to Cyprus, the Cyclades, Egypt and the Mycenaean mainland of Greece. Its shape is associated with procession and cult practices.

Rhyta are pierced vessels that were used for pouring liquid offerings (libations) to the gods, such as wine, milk, blood or perfumed oil. They come in a variety of shapes and sizes, but this one, conically shaped with a handle at the top to ensure a steady hold, is an example of what might be the most common type. You can see very similar vessels depicted in frescos at Knossos and in the Tomb of Rekhmire at Thebes. These paintings help us understand how rhyta might have been carried in antiquity, and how important they were. Recently, archaeologists discovered a painted jug at the site of Akrotiri, on the island of Thera, which contains the only known illustration of a rhyton being used for pouring in the Bronze Age Aegean. It shows two youths pouring liquid into a conical rhyton, which then directs the liquid into a sacred plant between them.

The bands of decoration that run up and down the length of this rhyton heighten the impression of length. The hanging, three-petalled flowers might be crocuses. They alternate with festoons of dotted semi-circles, all suspended from horizontal bands of dots. This band alternates with a looping meander line which holds dots within each loop. Looking closely at the garland bands, we can see a resemblance to the garlands of flowers on the decorated ships in a fresco called the Miniature Fresco, painted in the same house at Akrotiri. Experts have also recognised these garlands on the *ikria*, the temporary, lightweight walls erected on ships in the Aegean Bronze Age to offer elite passengers privacy. These connections may suggest that this rhyton has something to do with maritime life.

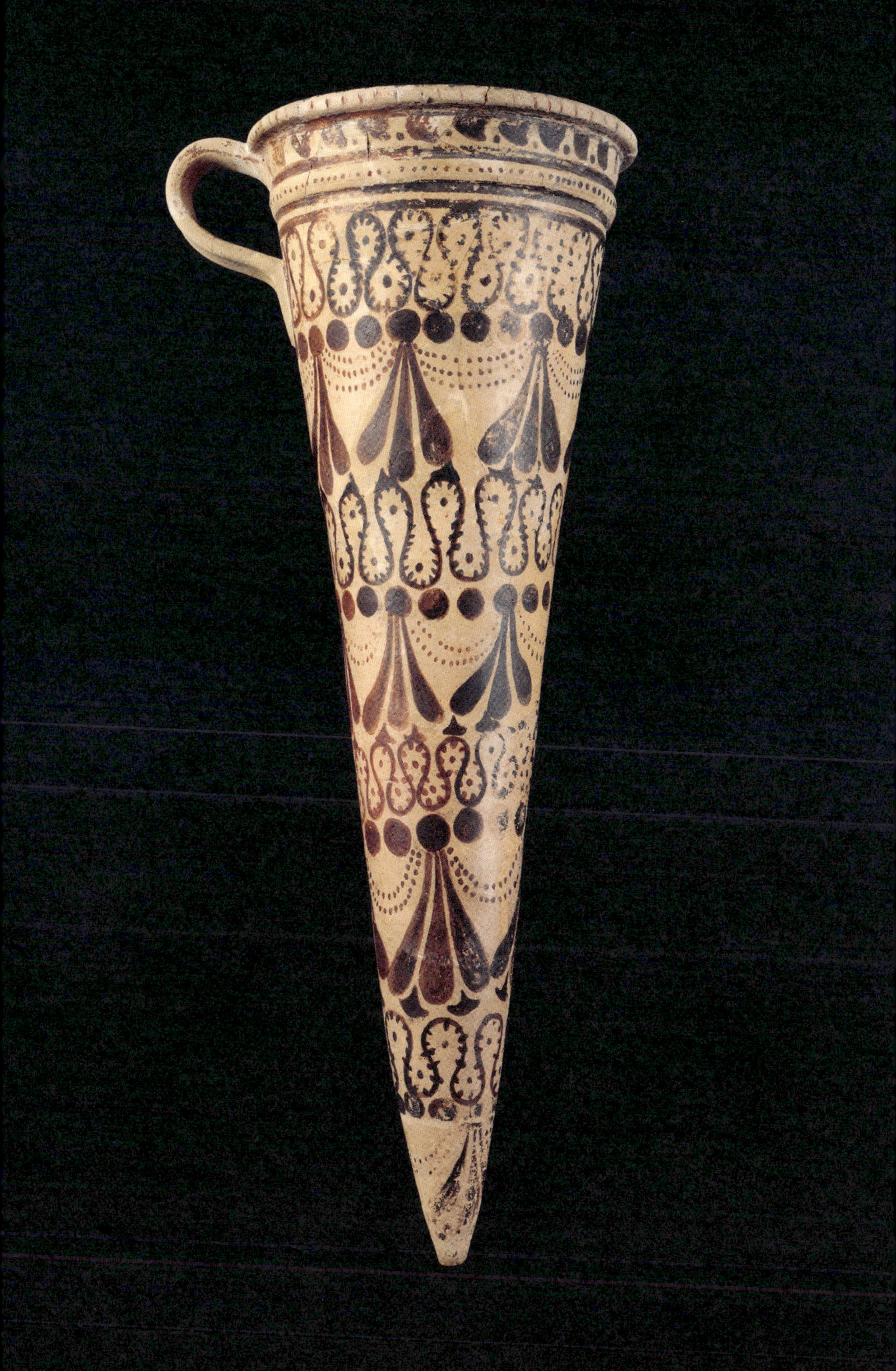

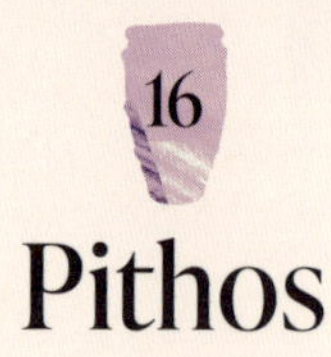

16

Pithos

Crete
*c.*1700–*c.*1400 BCE
Clay, 139 x 88 cm

This large storage jar is known as a pithos (the plural is pithoi). It was one of hundreds found in the storerooms of the Bronze Age palace at Knossos. It's decorated in relief with bands, which have sometimes been compared to ropes, and circular "medallions".

The handles would have probably been threaded with ropes to help move the *pithos* into place, which might explain this choice of decoration. The potters who made these large vessels would have built them up with coils of clay and surrounded them with fuel to be individually fired. Both forming and firing would have been tricky jobs that required some skill.

A number of palaces were built across Crete from 1900 BCE, but following widespread destruction in around 1450 BCE, only the palace at Knossos remained. From this point there are clear connections between Knossos and the Mycenaean culture of mainland Greece.

The first *pithoi* at Knossos came to light during an excavation by the Cretan amateur archaeologist Minos Kalokairinos in the 1870s. Due to this discovery, the site was known as *"Ta Pitaria"* (the place of the *pithoi*) until Sir Arthur Evans's excavations in 1900. Evans (Keeper of the Ashmolean from 1884 to 1908) was able to secure the rights to excavate the palace by purchasing a share in the land in 1894 and then buying the remainder in 1900. The government of the newly independent Crete was in favour of archaeological exploration. Evans continued the excavations started by Kalokairinos in the same area of the palace and soon found more storerooms and more *pithoi*, including this one.

This huge *pithos* has a capacity of around 550 litres. It shows us that the West Magazines of Knossos, the 19 storerooms along the palace's western wall, would have provided storage on a grand scale. The great jar would have been filled with agricultural produce, perhaps grain or olive oil, and this tells us more about what went on in the palace. Grain would have been used to feed the palace workforce, many of whom, according to the Linear B tablets also found at Knossos, were textile workers. The palace dwellers would have used olive oil for religious offerings or in the manufacture of perfumed oil.

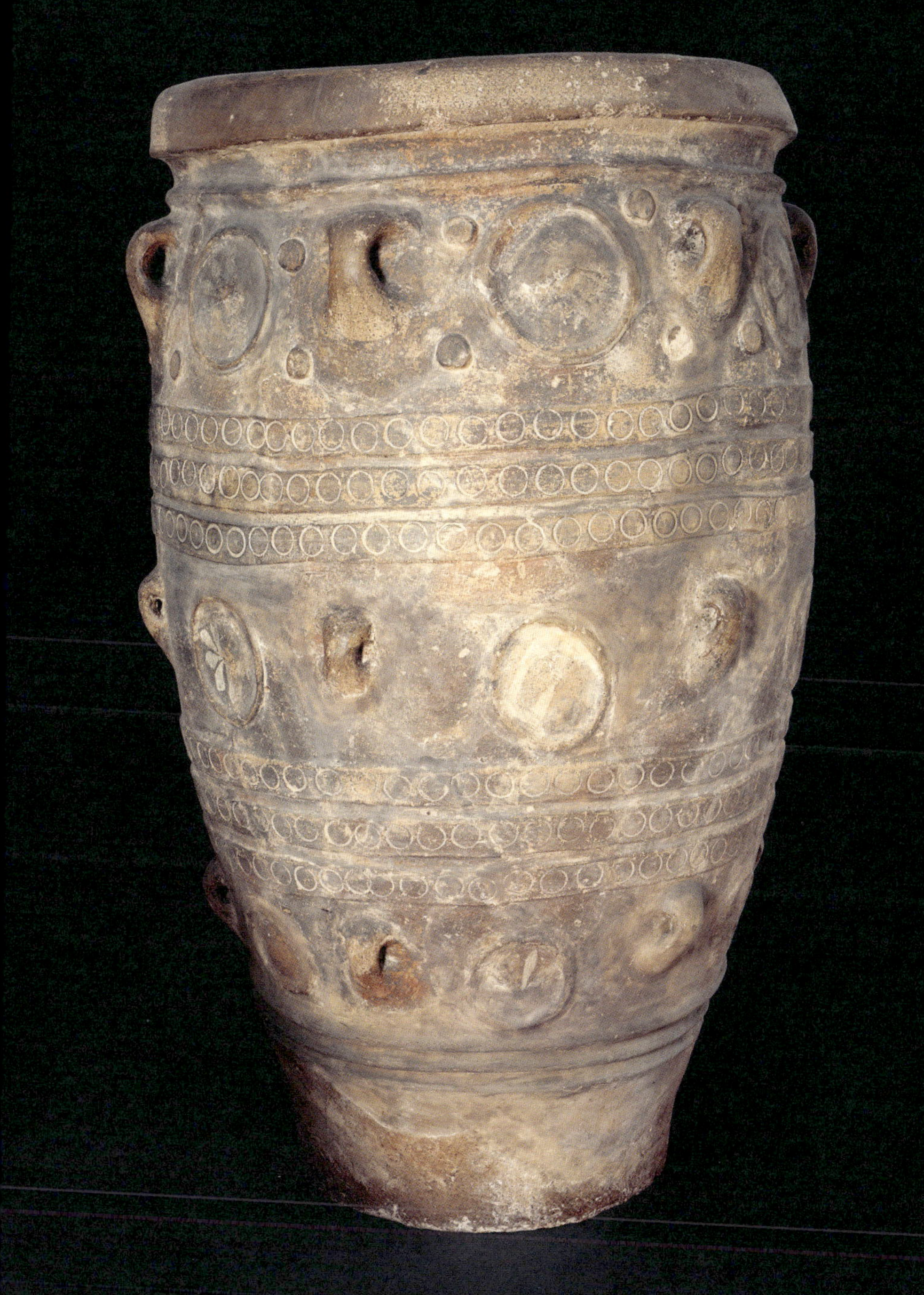

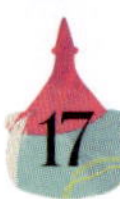

Megalithic small pots with finial lids

India
1500 BCE
Black and red ware

Black and red ware is one of the types of pottery most frequently recovered from ancient Indian sites. The vessels have a dark interior and lip and their red outer surface is occasionally marked with black patches. Alongside medium and small pots like the ones pictured, archaeologists often find larger, handmade pots of a plain red ware, which were used for grain or water, or as sarcophagi.

The smaller pots usually consist of dishes, lidded bowls, pedestalled bowls, tulip-shaped vases, conical vessels and jars of various sizes. They often have rounded, tapering or conical bases that would sit securely in the large numbers of ring stands that accompany them.

Their lids fit well: some are tall and conical, featuring finials (ornaments used as terminating motifs) with a knob on top, while others are shallow bowls that could also serve as supping cups. On the whole, these pots are wheel thrown, made from finely textured, well-levigated red clay and fired to produce a distinctive burnished surface, often with patches of red and black appearing at random. Seldom are the pots painted, but they are decorated: their creators have incised simple lines into their surfaces, applied slip (liquified clay) and polished them before firing. Some rare pots show signs of salt glazing: texture created by the addition of salt during firing.

Many pots like this also show traces of graffiti. These marks have not been deciphered but some of them turn up over and over again, such as the letter *ma* in the Brahmi script, and various oblique lines. The marks are a bit of a mystery. They were scratched on after firing, so it's unlikely that they are potters' marks. And we know they aren't owners' marks, since pots from a variety of graves and sites bear the same graffiti. The same marks also occur on pots of different shapes and sizes, which must have different uses, so it seems unlikely that they are meant to indicate their contents, either.

Pots like these have been found in excavations at some of the most ancient sites in India: Harappa, Lothal, Rangpur, Rojdi and Surkotada. However,

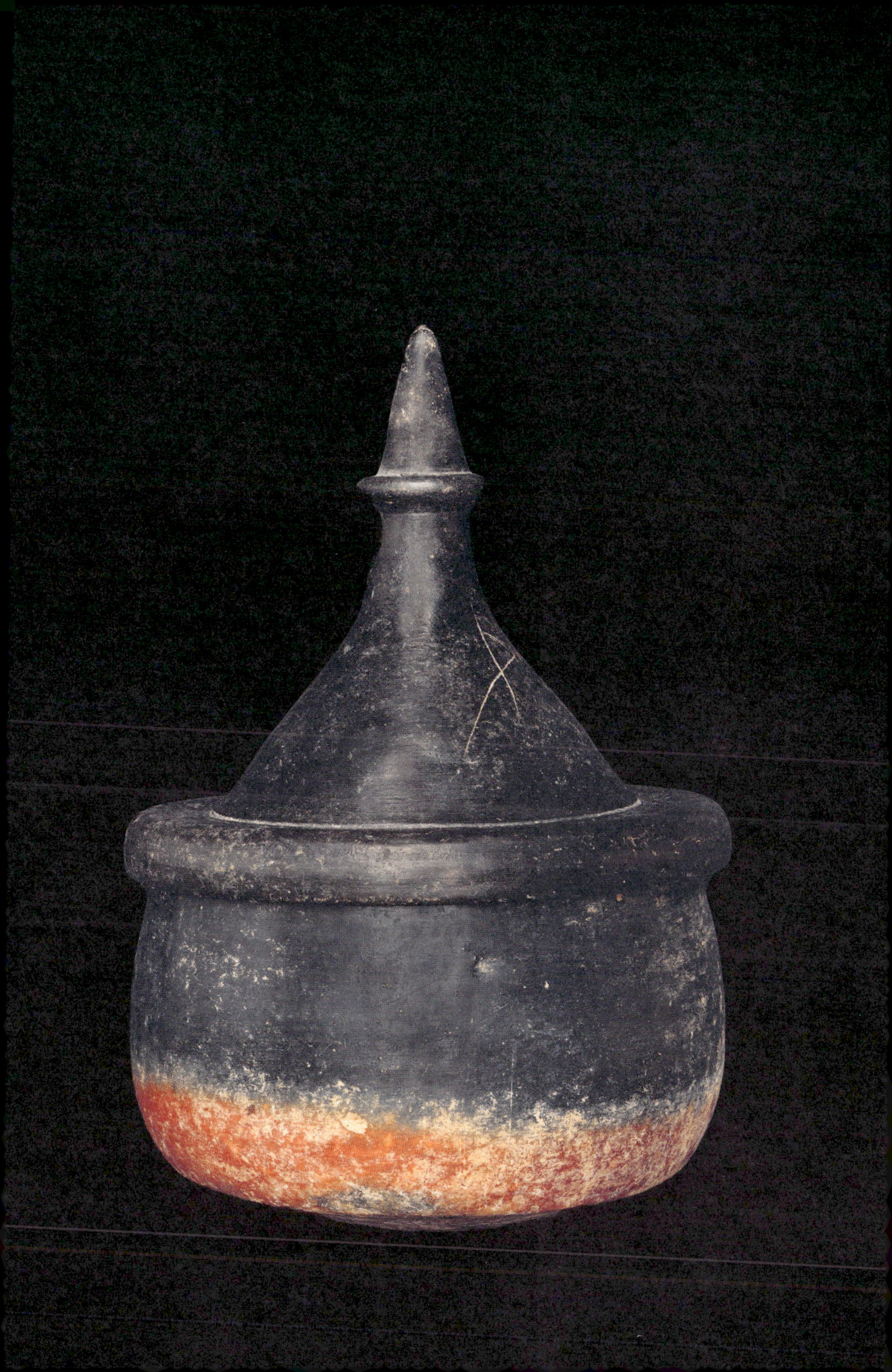

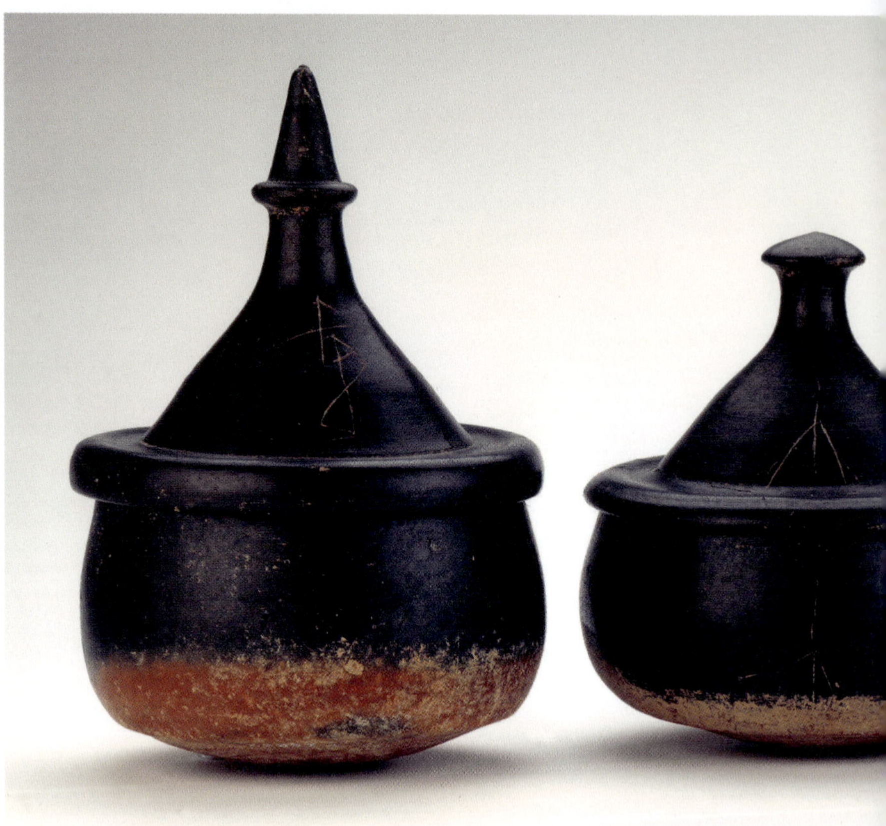

these pots are most closely associated with the Megalithic cultures of Peninsular India which flourished over 3,500 years ago, and this is the origin of the pots pictured here. Most of these pots come from excavations conducted at sites in the south-east by Dr E H Hunt (1874–1952), a Hyderabad state medical officer. Like so many other people who are interested in archaeology, he was fascinated by the Iron Age Megalithic burials in the area.

Iron Age burials are found across India, and took place over a broad time frame. The burials have been found as far north as Burzahom in Kashmir, Khandesh/Maharashtra in the northern Deccan Plateau and in the hills (such as Kotia in the Allahabad district) south of the Ganga Valley. However, the greatest number of burials, and the most varied, come from South India. While some of the most remarkable Megalithic complexes have been found in Kerala, most are spread over Tamil Nadu and Karnataka, and they're especially dense in Andhra Pradesh. We think they indicate settled Iron Age

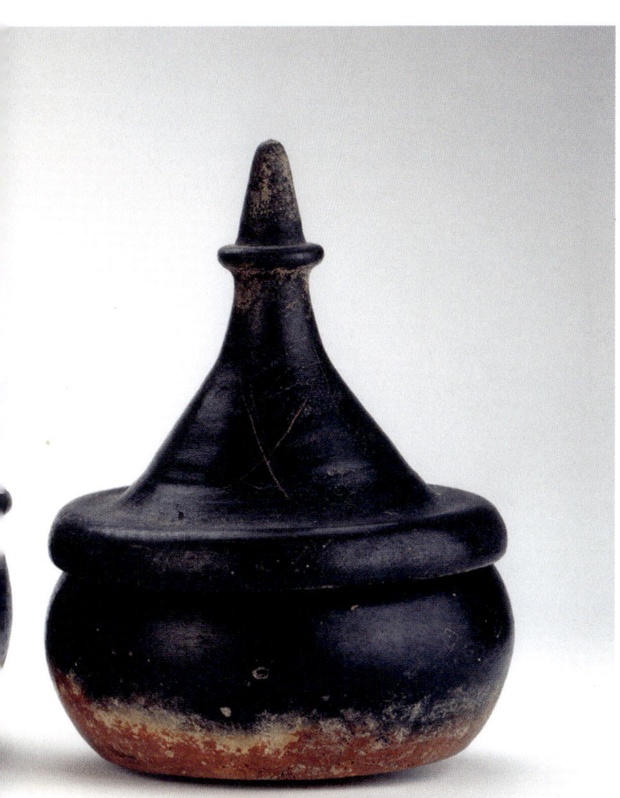

Pot. India, 1500 BCE.
Black and red ware, 14.5 cm.

Pot. India, 1500 BCE.
Black and red ware, 11cm.

Pot. India, 1500 BCE.
Black and red ware, 13.5cm.

cultures dated between 1500 and 200 BCE. In fact, the dates established by the thermoluminescence testing of one of the pieces in the Ashmolean's collection has corroborated these dates. It was found to have been fired 2,200 to 3,400 years ago: that is, between 1600 and 200 BCE.

It's difficult to work out an internal chronology within the Megalithic clusters, as many different types seem to occur simultaneously and in the same place. Archaeologists have not found ossuaries at all of the Megalithic complexes, but those that have been discovered consist of chambers that contain single or multiple skeletons. Sometimes remains are interred in urns, or large terracotta troughs. Often the grave contains iron implements such as tools, weapons, pots, stands and tripods, and etched carnelian beads. Sometimes, there are metal objects: bronze, copper and gold. Almost always, there is black and red ware. Presumably, these grave goods were intended for use in the afterlife.

18

Jar with octopus motif

Crete
c.1450–*c*.1400 BCE
Clay, pigment, 74.5 x 54 cm (restored)

An impressive 75 centimetres tall, this three-handled jar depicts an octopus with six arms swimming in an abstract seascape. It was presented to the Ashmolean in 1911 by the then independent state of Crete (it became part of Greece in 1913). The jar comes from Sir Arthur Evans's excavations at the palace of Knossos.

Evans used the term Minoan for the culture he uncovered there, after the legendary King Minos. Alongside the *pithoi* found in its storage rooms (see page 64), he discovered this octopus jar. It had fallen into the area where the pithoi were stored when an upper floor collapsed in the catastrophic fire that destroyed the palace.

There was a long tradition of depicting octopuses and other cephalopods in Minoan art, such as on seals, frescoes and jewellery. Most notably, an octopus appears on a stone vessel from the palace at Knossos – carved in relief and known as the "Ambushed Octopus", the animal stares out from a coralline underwater scene. Following the catastrophic volcanic eruption on the island of Thera around 1600 BCE, which would have affected the trade and production taking place at the palace at Knossos, designs similar to the "Ambushed Octopus" were transferred to painted pottery, which was more readily available and an easier material to work with than stone. This type of decoration is known as the "Marine Style", arguably the most famous pottery produced in Crete in the Bronze Age. The pictured vessel, despite its cephalopod subject, belongs not to the "Marine Style" but to what archaeologists call the "Palace Style", which dates to the period of Mycenaean influence on Crete. Compared to the earlier, more naturalistic depictions of octopuses in Minoan art, this octopus is rather stylised, notably in the number of tentacles. Despite the less realistic approach, the animal's open eyes and curling limbs bring it to life. The sponge-print background and faint depiction of a murex shell – which was used for making purple dye – on the vessel's shoulder also evoke the sea. The jar indicates the importance of the sea to the palace, both because of the marine products used by the people who lived there and the significant impact of overseas trade on their daily lives.

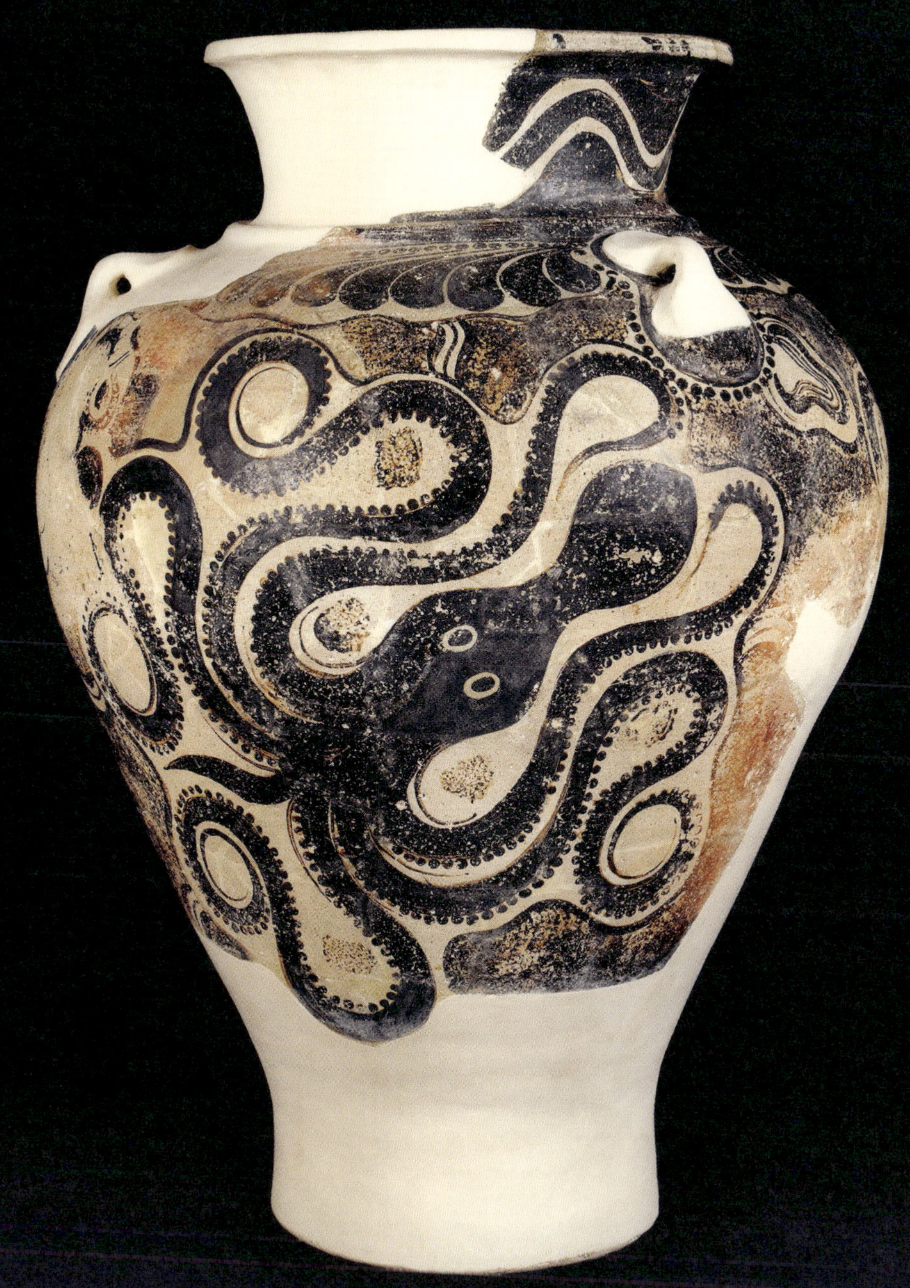

Vessel in the form of a zebu

Iran
*c.*1200–900 BCE
Fired clay, 26 x 39 cm

Very little is known about the culture of the people who made this vessel. In the late second and early first millennium BCE, Iran was in the midst of a period of transformation. It is possible that these centuries saw the arrival onto the plateau of populations speaking Iranian languages, which would ultimately come to dominate the region.

Local identity seems to have shaped these people's lives, and in north-western Iran there existed a distinctive ceramic tradition in which craftspeople created hollow containers fashioned in the shapes of animals. Most of them are humped bulls (zebus), like this one, and stags, some mounted on wheels. There are also pottery figures of men and women, who are often shown carrying spouted vessels.

The culture that produced these objects is known by the term "Amlash", which is simply the name of a market town at the south-western corner of the Caspian Sea. It's where vessels like this one were brought in the decades after 1930 to be sold to antiquities dealers. These objects had almost certainly been plundered from graves. We can reach this conclusion because comparable examples, along with weapons and jewellery, have been excavated by archaeologists in burials at the site of Marlik Tepe, in the same region as Amlash. This might also suggest that the vessels were intended for ceremonial use, and perhaps associated with burial rituals. We don't know where this particular curvaceous bull, made of reddish-brown clay, was found. Like other creatures of its type, it has a muzzle that projects as a beak-like spout. This feature finds parallels in contemporary metal vessels from the same region.

If we compare the zebu to a real animal, it's clear that its hump is very exaggerated. Its maker might have intended to convey the impression of strength and power. The hump is decorated with a pattern of five impressed circles, with two more circles representing the zebu's eyes. The people who made it did not produce written records, and there are no references to the area at this particular period in any accounts from neighbouring regions. Only future excavation will help to illuminate the culture from which this personable animal derives.

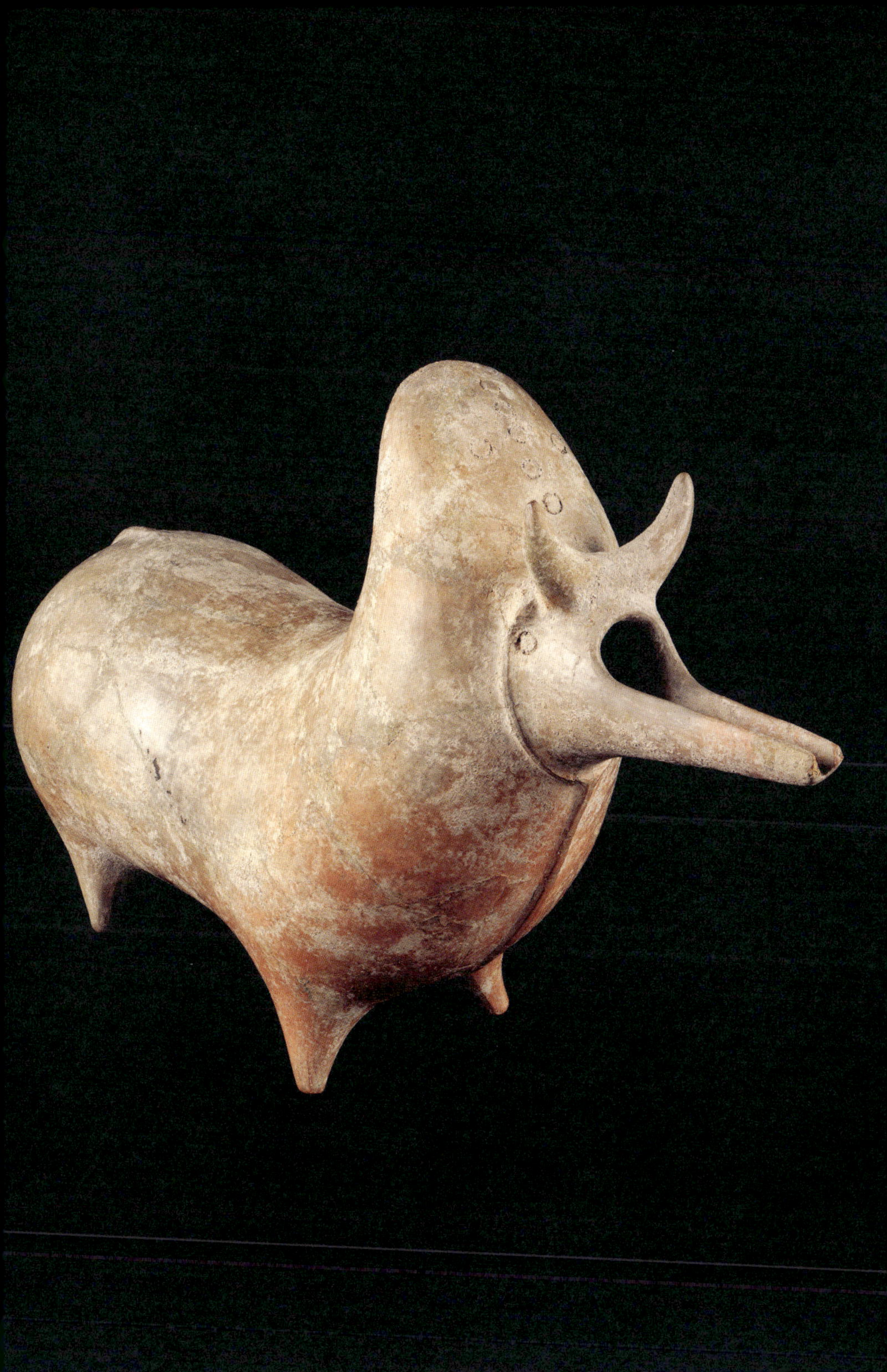

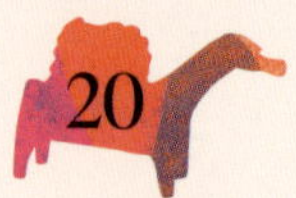

Terracotta horse rhyton

Cyprus
*c.*1100–1050 BCE
Ceramic, 22.5 x 37.2 x 8.5 cm

In ancient Cyprus and its neighbouring cultures, potters produced vessels designed for very specific purposes. This unique rhyton was created for the offering or pouring of liquid in religious or burial rituals. It's shaped like a stumpy-legged horse, with a nude male rider sitting sideways on its back.

It has two disguised openings or spouts. The one for filling up the vessel leads vertically from the top of the rider's head, through his body and into the body of the horse. The one for pouring out goes through the horse's neck, head and mouth. The pourer would have held and tilted the vessel using the handles attached to either side of the rider.

This wheel-made rhyton and its hand-formed rider were created in the last phase of the Late Bronze Age, or Late Cypriot IIIB period (1100–1050 BCE), in Proto White Painted ware. This ware was the prototype of White Painted pottery, one of the most popular decorated pottery wares in Cyprus in the first half of the first millennium BCE. Proto-White Painted ware was characterised by motifs and patterns painted in black-brownish paint on a white or cream-coloured surface. This rhyton is decorated with black-brown painted hatched lozenges and triangles on the horse's body and legs, a large star in a circle at the front of its chest, and zigzag lines along its mane.

The naked rider has eyes indicated by two piercings. His beard, and the entirety of his upper body and thighs, are painted black. The elliptical concentric lines on his shins might indicate shin protectors, part of a warrior's armour. He holds an almond-shaped object, which might be a bottle, in his raised right hand, as if he is pouring from it. Along the backs of each of the handles, towards the upper spout, crawls a hand-formed snake.

We can imagine what its purpose might have been from three clues. The horse-and-rider motif and the presence of snakes — which were an ancient symbol of death, burial and renewal — and the fact that the vessel is very well preserved all lead us to believe that it served as a libation vessel during a burial ceremony and then was left as a burial gift for the deceased, who may have been a horse rider and warrior.

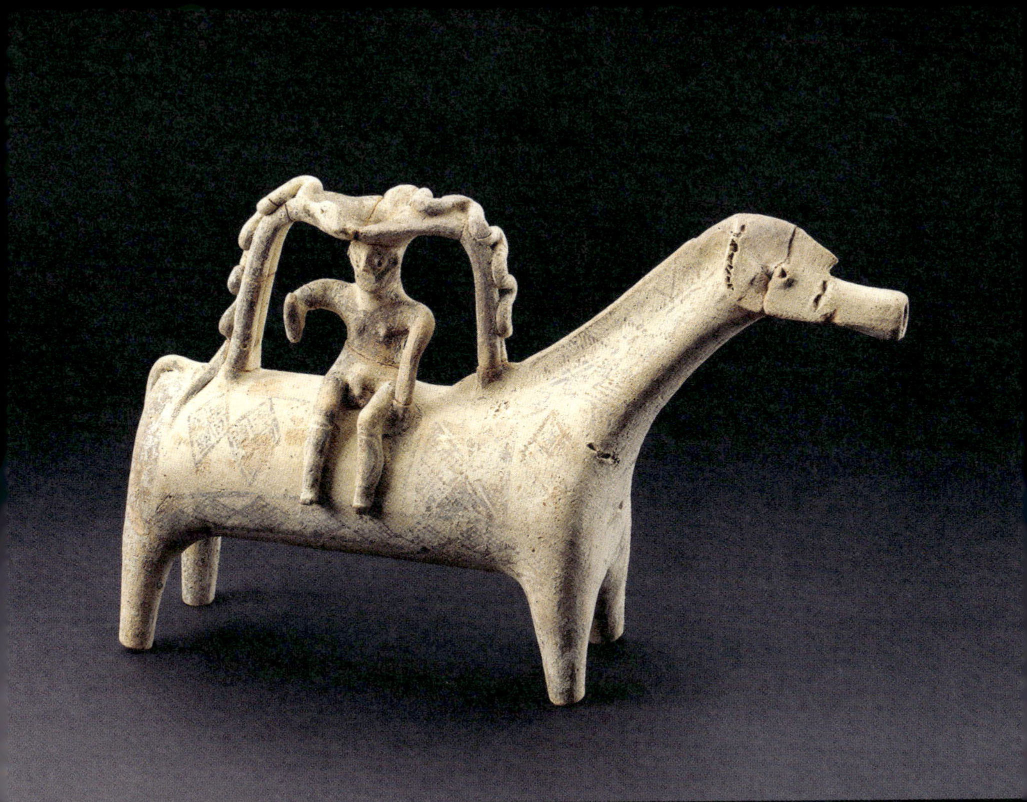

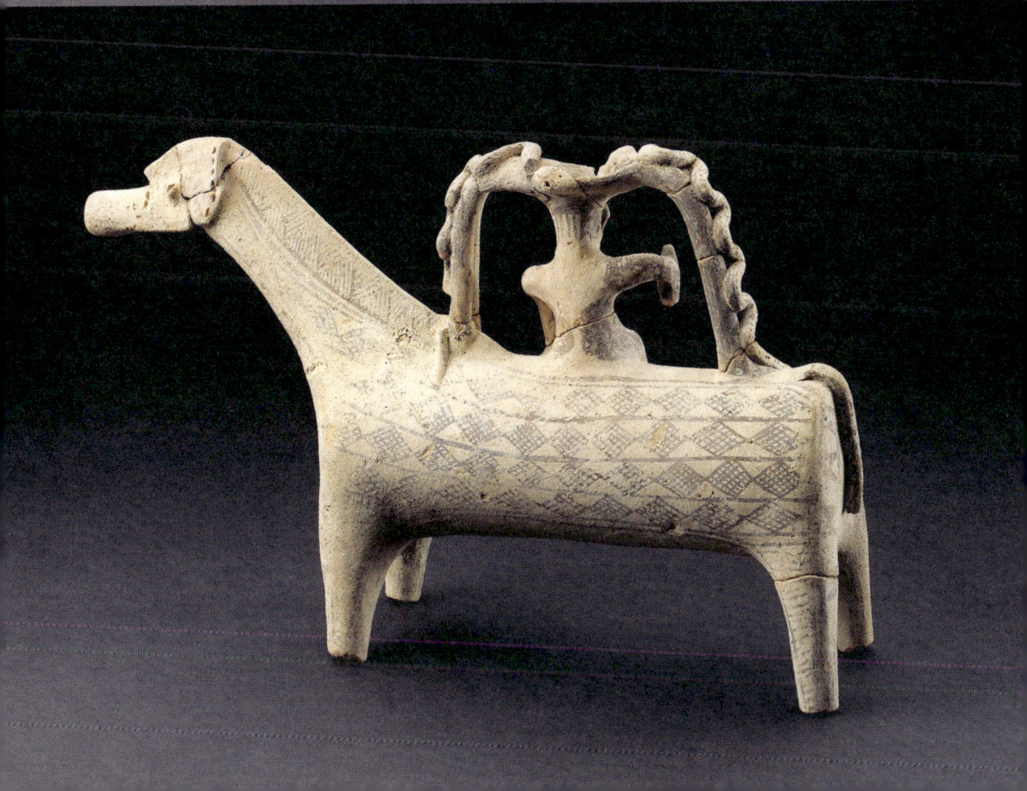

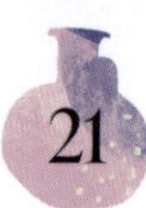

21

Barrel-shaped jug

Cyprus
*c.*700–600 BCE
Ceramic, 28.5 x 27.9 x 22.3 cm

This barrel-shaped water jug is an example of Bichrome Ware, one of the most popular, most elaborately decorated wares produced by Cypriot potters during the first half of the first millennium BCE. This ware typically has figurative and geometric motifs and patterns painted in brown-black and red (bichrome means "two colours"). The light white or cream-coloured background is achieved by a slip covering the vessel's entire outer surface.

Originally produced in the seventh century BCE, the jug was acquired by the Ashmolean in 1885 from the German excavator Max Ohnefalsch-Richter. According to his own article published in the *Journal of Hellenic Studies* in 1884, he found it while excavating one of the many tombs in the necropolis near Larnaca, the site of the ancient city-kingdom of Kition. To this day, the jug is like nothing else in any collection of Cypriot antiquities anywhere in the world. It's one of the highlights of the Ashmolean's Ancient Cyprus gallery.

The jug is a one-off, first of all, because of its shape. But what is most noticeable about it is another unique quality: its prolific painted decoration. It depicts almost the entire repertoire of figurative motifs that were ever painted by Cypriot potters and vase painters on their vessels during this time. This looks very different from the more numerous and conventionally decorated bichrome jugs of this period. These usually feature an egg-shaped or globular body and a trefoil mouth (a shape like three overlapping rings). But, most obviously, these other jugs have painted on their bodies just one or two motifs – for example, a bird or a flower. The images float freely on the otherwise empty or undecorated vessel surface, which is why this decoration technique is called Free Field Style.

This jug is different. It's covered with a variety of repeated and symmetrically arranged motifs. The wide body front displays a highly stylised but elaborate tree with a rosette-decorated trunk and curly palm leaves. Remarkably lifelike stags nibble on either side of the tree, two crested waterbirds (which might be ibis) arranged heraldically above them. A simpler variation of this

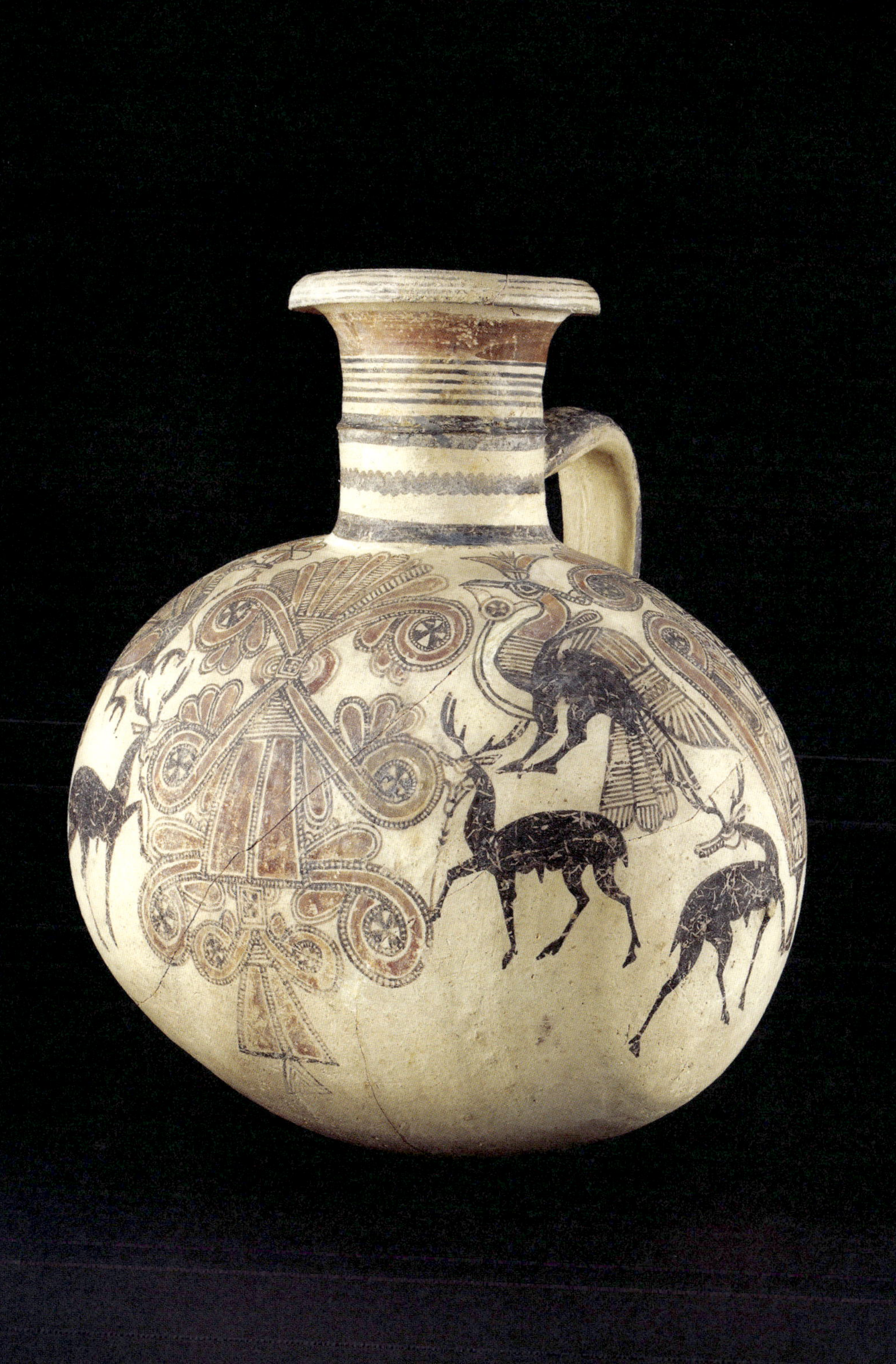

composition, with climbing stags but without birds, is repeated on both sides of the body. A smaller tree fills the gap underneath the handle.

All these motifs, including the "tree of life" (with or without animals), are widely attested to being (or can often be seen) on pottery and other objects found in the ancient Near East and in Cyprus in the first millennium BCE. They refer to the life, prosperity, and fertility of vegetation, animals and, ultimately, people. In such an arid region, without rainfall for many months over the summer, all living things heavily depend on rain and the availability of water. It's no accident that Cypriot vessels decorated with these motifs are invariably the water jugs which would have held this precious liquid.

This jug must have been made by an especially artistic potter and painter in the ancient city-kingdom of Kition. Kition was largely inhabited by Phoenicians, who had immigrated to the island from their homeland in nearby Lebanon since the ninth century BCE. The unmistakably Phoenician shape of the vessel, with its array of motifs originally from the Near East, suggests that its creator was a resident Phoenician who, with this water jug, created a true masterpiece of Cypriot art.

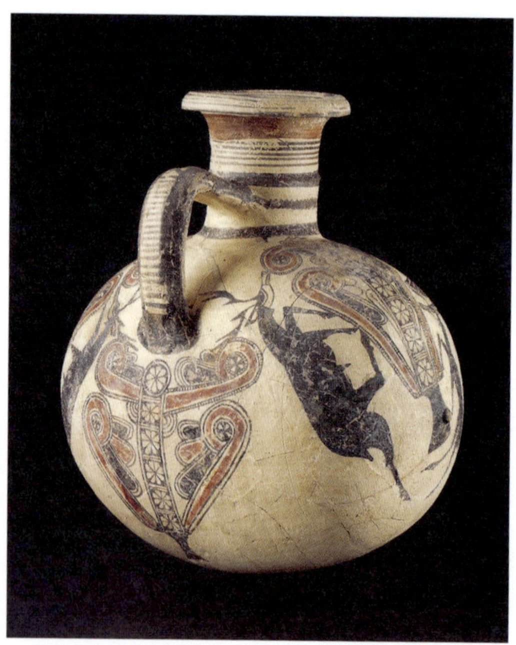

(left) This image of the handle side of the jug shows the painted motifs of climbing stags and a smaller tree of life.

(opposite) Example of a Free Field Bichrome Ware jug with image of an ibis bird and reeds. Sinda, Cyprus, Ceramic, 24.7 x 19.6 cm.

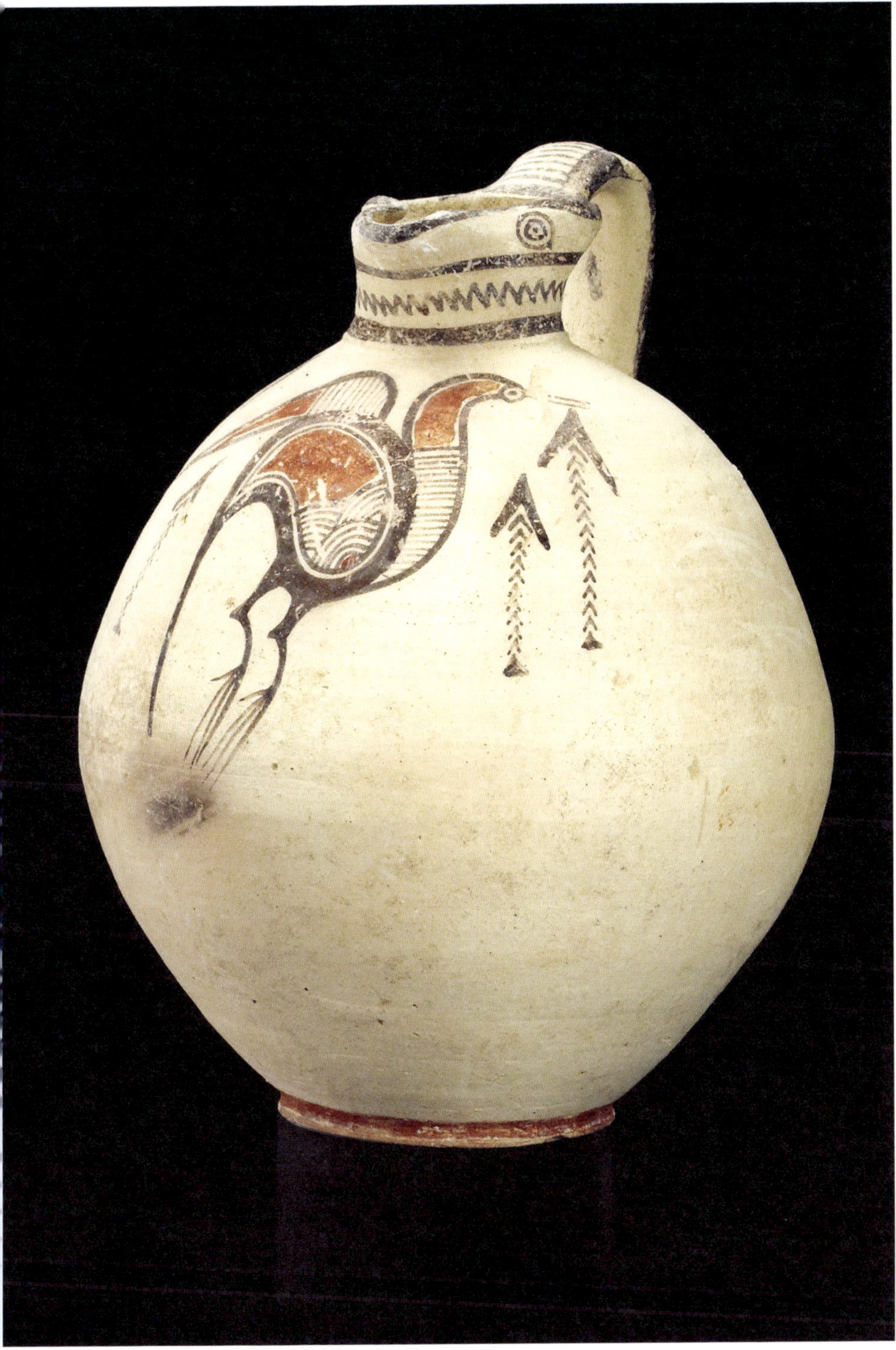

Athenian amphora

Greece
*c.*720–700 BCE
Clay, 67.7 x 39.5 cm

The style of Greek vase-painting in the ninth and eighth centuries BCE gave its name to the era: the "Geometric Period". The Geometric style consists of dense bands of intricate decoration, in particular the meander motif, occasionally interspersed with angular figures of humans and animals.

Humans are shown in a contorted perspective, in which each part of the body is displayed with its most characteristic viewpoint: heads in profile, chests from the front and legs from the side. Only later did Greek vase painters begin to systematically depict the body naturalistically, from a single viewpoint.

This large amphora (storage vessel) is decorated with lozenges on the neck and with several bands of geometric decoration on the body, including zigzags, triangles and stylised waterbirds. More unusually, this vase also has moulded decoration on the rim, shoulder and handles in the form of snakes. The crowded designs suggest that the maker has been influenced by woven textile patterns. Archaeologists believe this vase was created at "the Tapestry Workshop". We don't know the names of the painters and potters who belonged to this workshop, but the attribution allows art historians to group together stylistically similar vases and date them.

Some of the best examples of Geometric pottery survive from Athens. There, vases of this shape often had a funerary use, as a grave marker or as a container for a cremation burial. This is probably what this amphora was used for, because it is largely intact and the focal scene shows the laying out of the body at a funeral (known as *prothesis*). Male and female mourners, arms raised in grief, surround the deceased, who lies on a bier covered by a shroud. Warriors and chariots evoke heroic deeds and glorify the status of the person who died. The scene suggests that the burial with which this vase was associated was that of a male. The snakes coiling around the vessel symbolise the underworld.

It's possible this amphora came from the area of Athens known as Kerameikos, where, beyond the Geometric Period, there were pottery workshops. The word "Kerameikos" means "potter's quarter", and it's the origin of the term "ceramic".

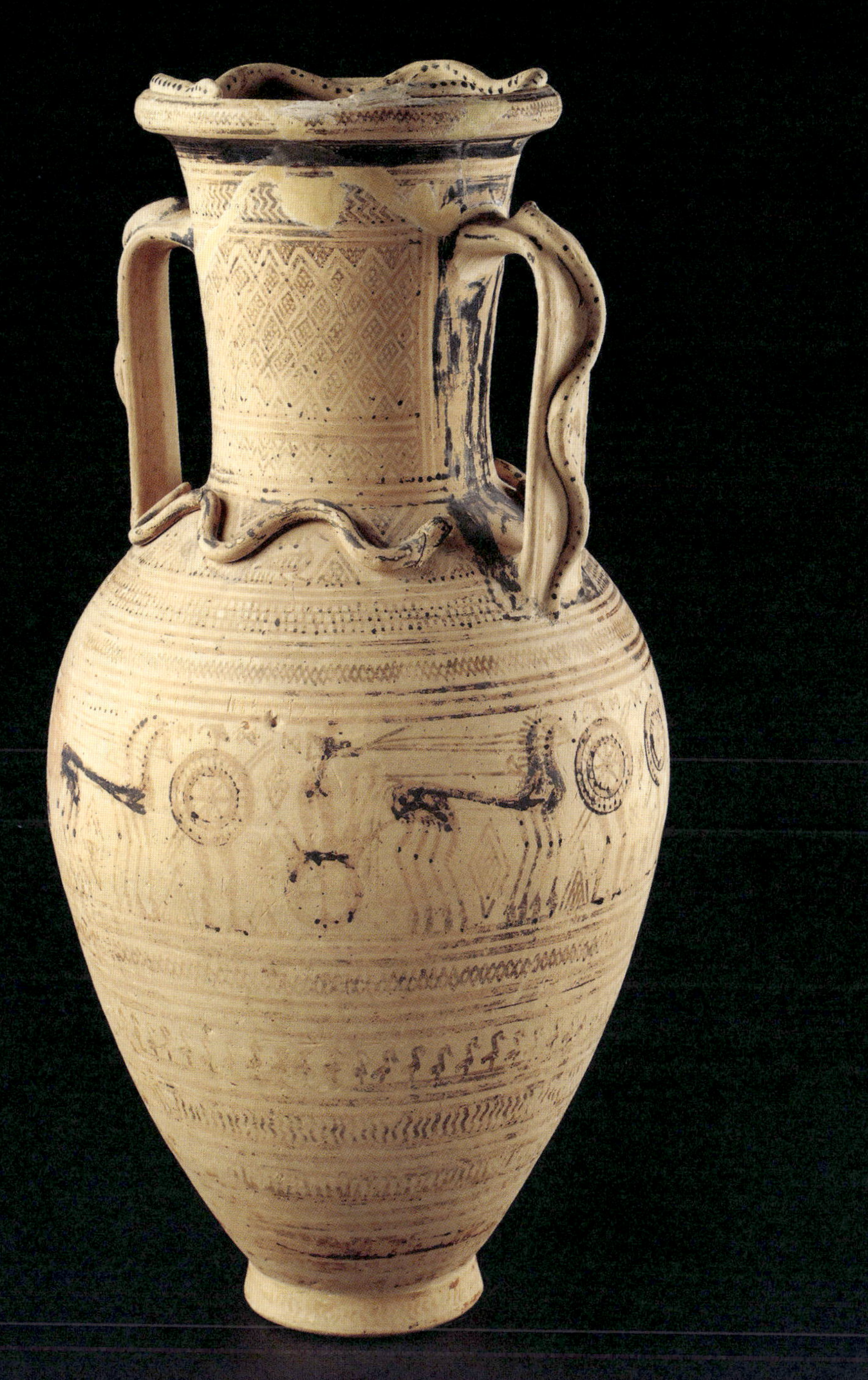

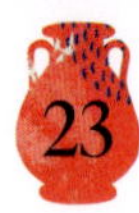

The Shoemaker Vase

Greece
500–470 BCE
Terracotta, black-figure, 40 cm

This scene of a shoemaker cutting out a leather sole around the foot of a customer offers an intriguing glimpse into everyday life in fifth-century BCE Athens. Scenes of craftsmen are relatively rare on Athenian pottery, which usually depicts heroic, mythical and elite subject matter, especially in the black-figure technique, but the Ashmolean holds a notable collection of vases showing men at work.

In the image on this vase, the craftsman bends intently over his task, while the customer balances on the table, resting his hand on the shoemaker's head. Another man looks on, leaning on a staff and holding branches – he may be the owner of the workshop. Draped in *himatia* (cloaks), all three men are bearded, indicating that the customer is not a child, but has simply been shown on a smaller scale to fit within the scene. Below the table is a bowl of water that was used to soften the leather before cutting. A shelf above displays the cobbler's tools – a knife like the one being used, and a curved so-called "clickers knife" for cutting heavy leather. The remains of an ancient shoemaker's establishment has actually been found near the agora in Athens.

Originating in Corinth in the seventh century BCE, the black-figure technique of vase painting, in which the figures are silhouetted in black paint with incised details against an unpainted clay background, was adopted and refined by Athenian craftsmen in the subsequent century. The results are remarkable. This *pelike*, a jar used for storing wine or other liquids, is attributed to a craftsperson known only as the Eucharides Painter. White paint has been added to pick out details, such as the block supporting the leather, which was presumably made of wood, and the joins of the metal folding stool. The other side of the vase depicts an enigmatic scene of satyrs, the followers of the wine god Dionysus, with a goat and the figure of an individual who is possibly Hermes, the messenger god (he was also associated with commerce). This Dionysiac imagery may have a link to the branches held by the workshop owner and the wreaths worn by two of the men in the shoemaker scene on the other side of the vase.

The status of craftsmen in antiquity is generally thought to have been fairly low, although some achieved great fame, and may have acquired wealth.

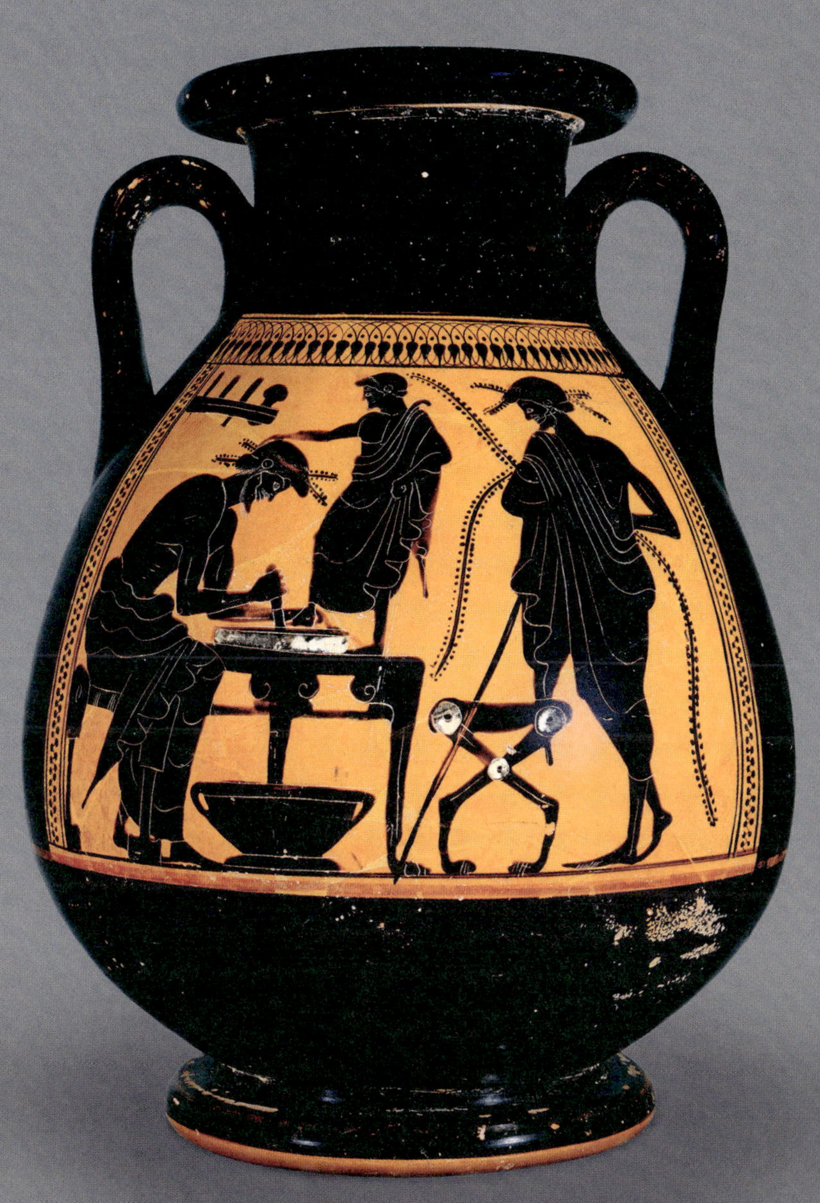

Celebrated sculptors were known by name, and vase painters and potters also occasionally identified themselves as having made or decorated their vases. Scenes of craftsmen at work may not have been as popular with the local Athenian market; examples have been recovered from Etruria, and this *pelike* was found in Rhodes.

Alongside this example, the Museum has on display a red-figure Athenian cup showing a metalworker fashioning a helmet. There is also a large up-turned bell-shaped vase or bell krater depicting the interior of a potter's workshop, with a seated painter balancing the very same vessel, a bell krater, on his knees. He is shown decorating it with the clay-based slip that produced the shiny black gloss of the vessels. The Museum's collection also holds vases that illustrate women at the loom, weaving textiles, in scenes that could represent commercial as well as domestic production.

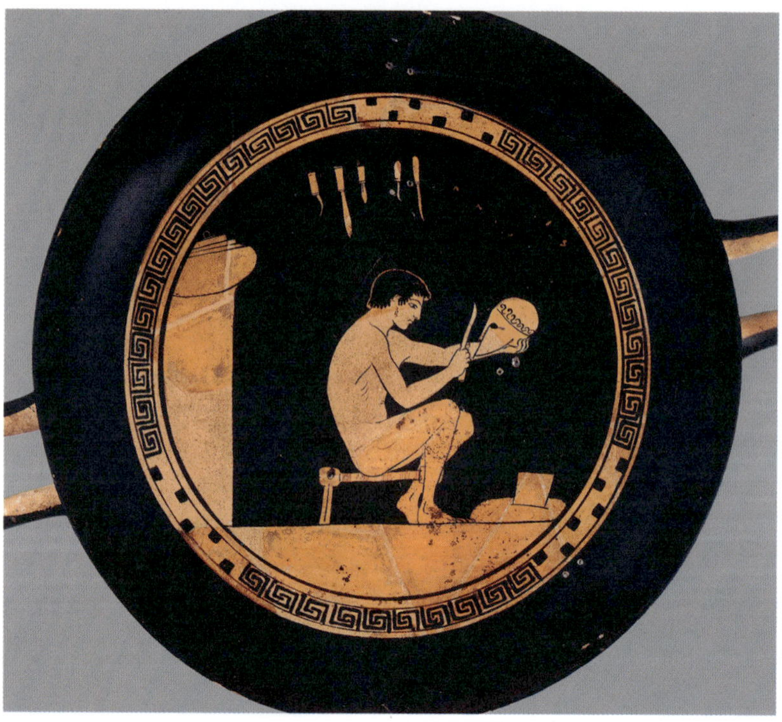

(left) Attic red-figure pottery stemmed cup depicting a helmet maker, attributed to the Antiphon Painter (490–480 BCE). Italy, 10.4 cm.

(opposite) Attic red-figure pottery bell krater depicting a potter's workshop, attributed to the Komaris Painter, *c.*430–42 BCE, 35.7 cm

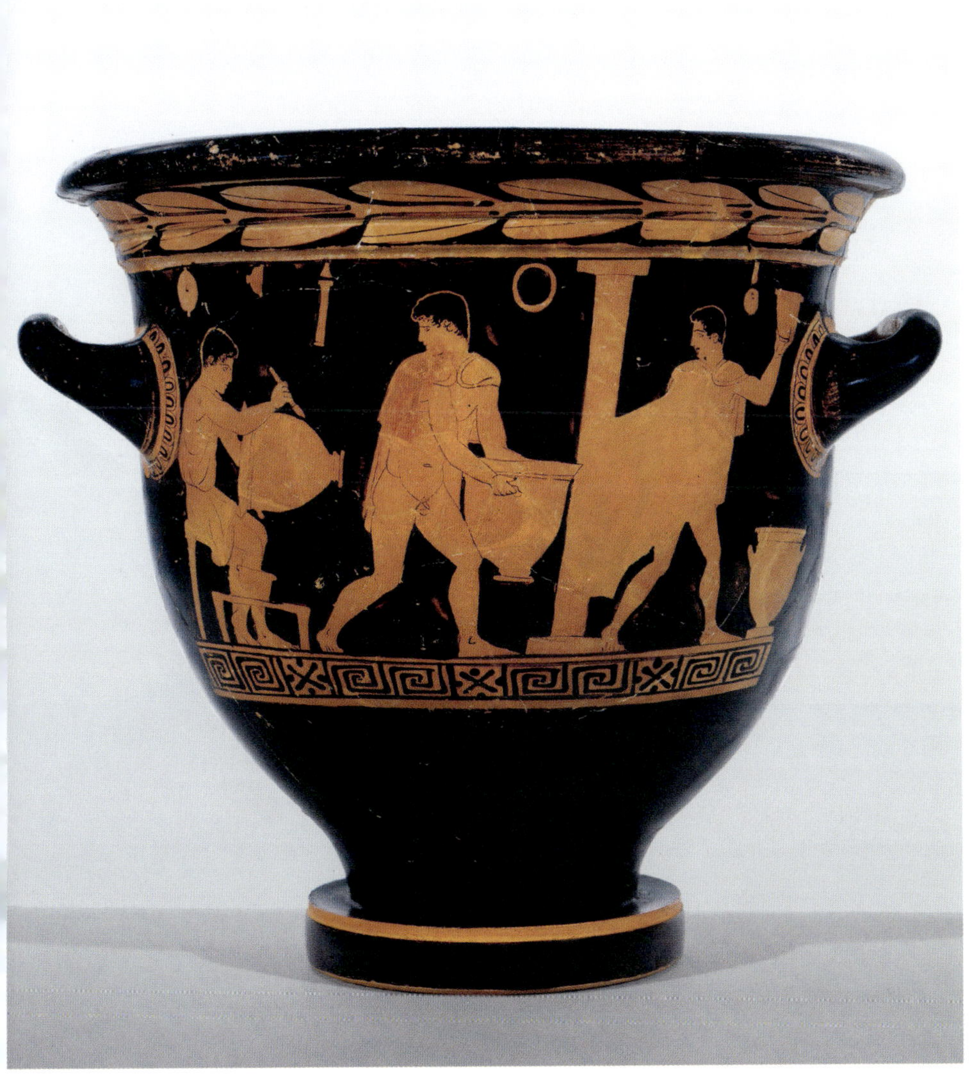

Attic black-figure pottery stemmed cup

Greece
530–515 BCE
The Andokides Potter (530–500 BCE)
Terracotta, black-figure pottery, with painted
and incised decoration
12.3 x 42.6 cm

An approach now known as the black-figure technique was often used to make Greek pottery in the Archaic Period (750–480 BCE). The figures and decorative motifs were painted in an iron-rich slip which was then fired in a three-stage process. In the kiln, the potter created an alternatively oxidising and reducing atmosphere, which left the areas covered in slip a glossy black and the underlying clay red.

The pot's creator incised figural details through the slip into the clay below. In this period, we can identify individual potters and painters by their craftsmanship and occasionally through their signatures. This cup is not signed but has been associated with a sixth-century Athenian potter called Andokides and an anonymous painter whom art historians call the Lysippides Painter. The rich orange colour of the clay indicates that the cup is an Athenian product.

A two-handled cup like this is called a *kylix*. It captures the typical elements of the Greek symposium, a boisterous drinking party for elite men, which featured wine, music, humour and sex. In the image inside the cup, six men recline outdoors, some wearing ivy wreaths, others eastern-style turbans. One plays the double pipes (*aulos*) – a lyre hangs nearby – while another prepares to spank a boy serving wine with his slipper. The boy is naked, but the reclining symposiasts wear a typical Athenian garment known as the *himation*. The vines, heavily laden with grapes, allude to Dionysus, the god of wine, and the copious consumption of the symposiasts. On the underside of the cup are the heads of satyrs, the followers of Dionysus, who were known for sexual excess and causing mischief.

Drinking vessels were used in party games such as *kottabos*, where wine dregs were flicked from a cup at a target. Many of them also employ visual

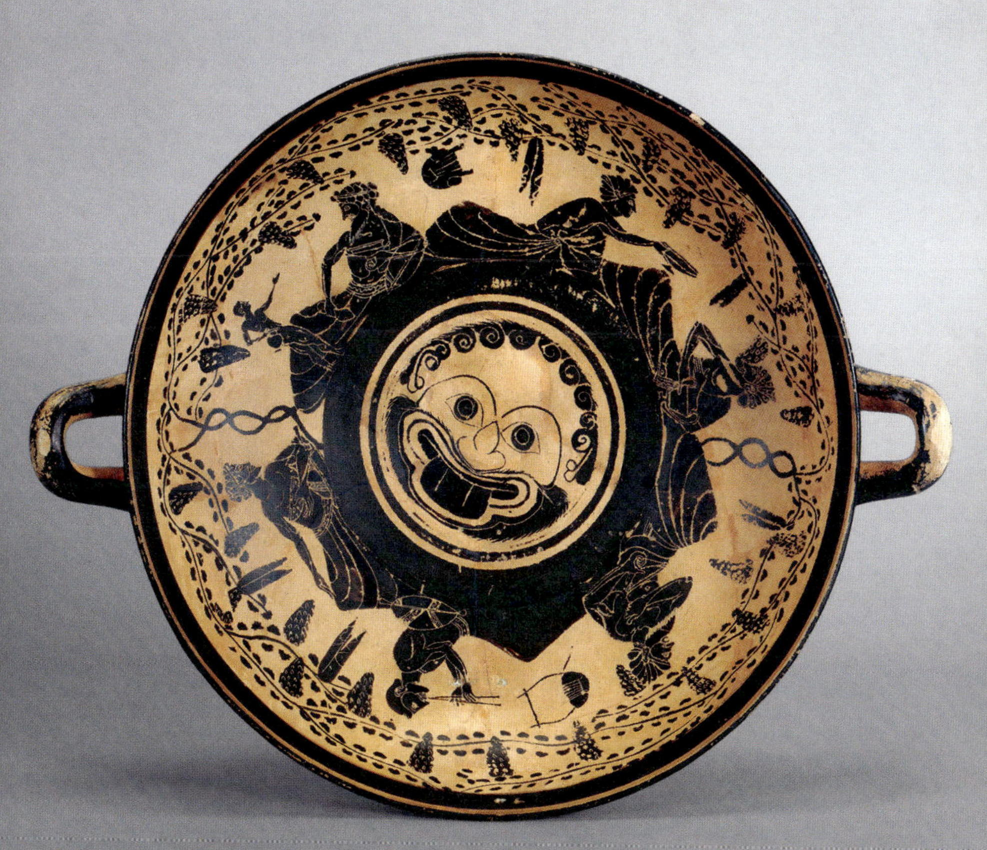

jokes or play tricks. Here, the drinker, draining the last drops of wine, comes face to face with the grotesque head of a Gorgon sticking out its tongue. Lifting the cup reveals to the drinkers that its foot is fashioned in the shape of a large penis and testicles. Commentators have even suggested that the big eyes painted on the underside turned the *kylix* into a grotesque mask when raised in the act of drinking. Some experts, however, have noted that staring eyes and penises were both used in the ancient world to ward off bad luck, suggesting that these details that seem light-hearted might have had another more serious purpose.

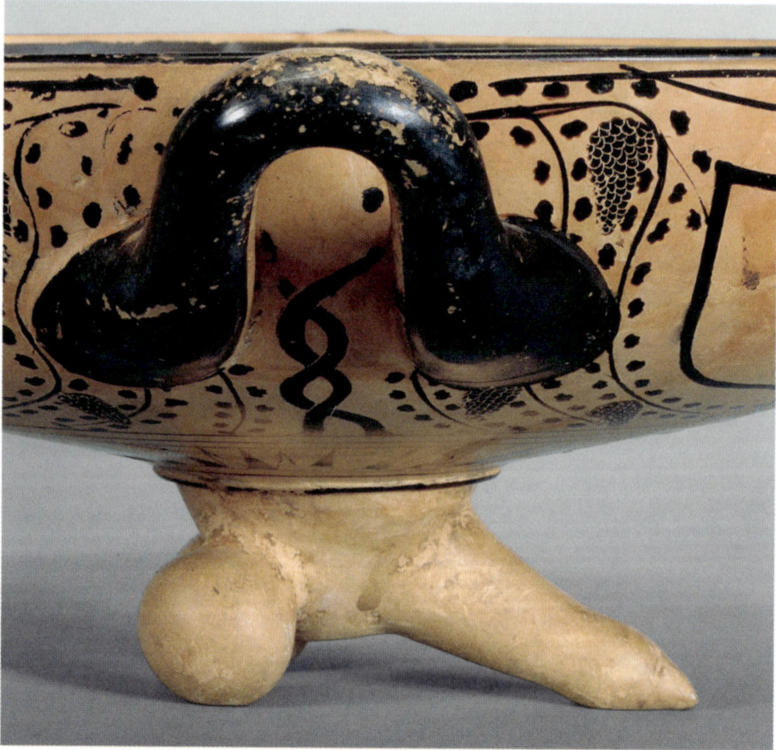

The underside of the drinking cup serves as both a crude drinking joke but also to ward off bad luck in the ancient world.

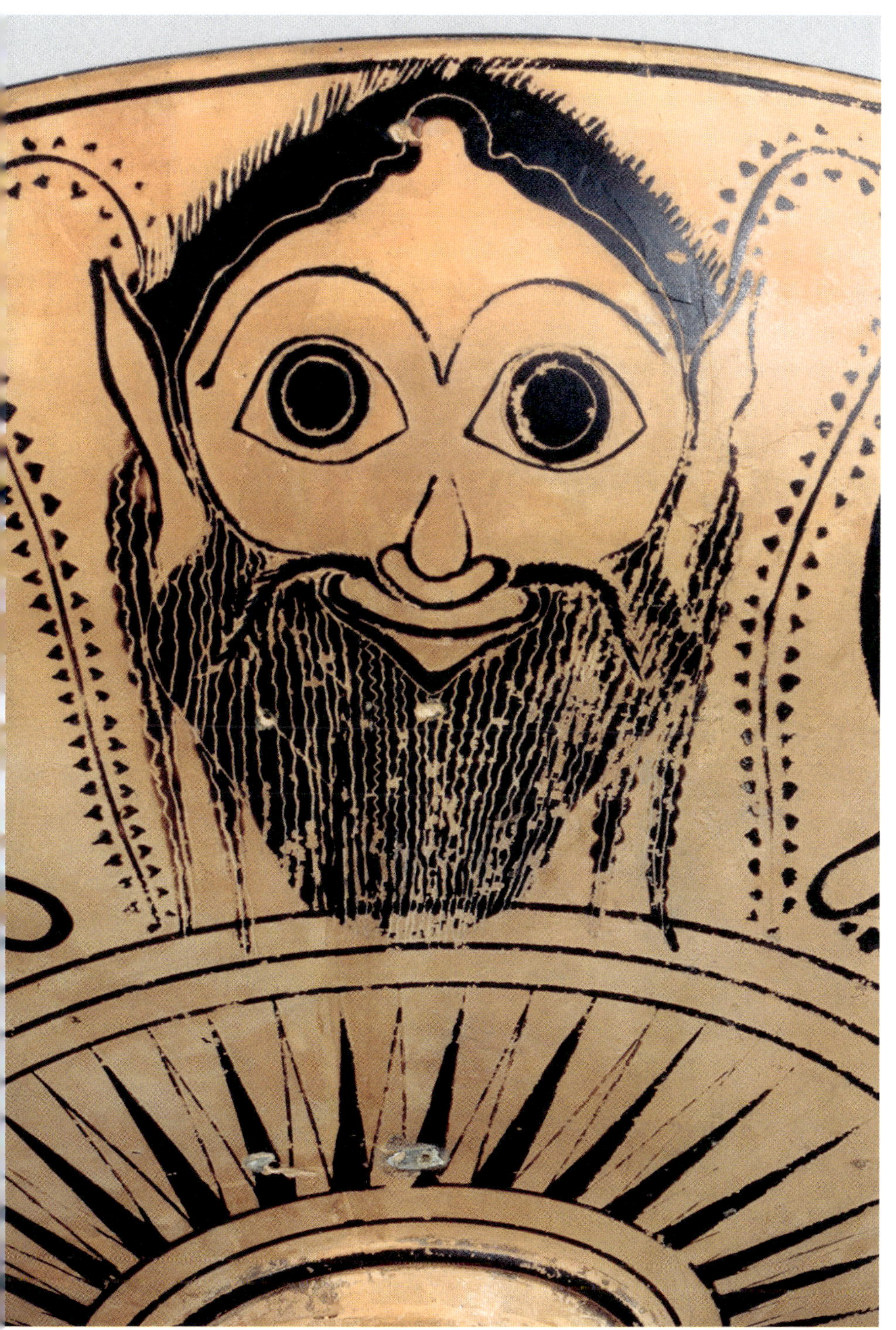

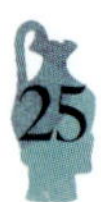

Attic red-figure pottery head vases

Greece
*c.*500–460 BCE
Pottery, with painted, incised and moulded decoration,
approx. 14.7–14.9 cm

These unusual head vases were made somewhere around Athens in the fifth century BCE. At this time, pottery vessels were generally made on the wheel, but these vases were made using moulds, probably ceramic and in two pieces, which were joined while the clay was still damp. The spouts and handles would have been made separately.

Next, their painter would have decorated them in the same way as other red-figure pottery, with details such as the eyes and eyebrows painted on in a slip that turned glossy black when fired. The lips of the female head were painted red using other pigments, and the teeth of the grinning bearded figure were painted white. The female head is decorated with a palmette design which is typical of vases at this time. The head vase was to be used as a pottery vessel rather than admired as a piece of terracotta sculpture.

Head-vases were made in several forms in this period, but satyrs and maenads (female followers of Dionysus) are some of the most common. Vessels like these were used for serving and drinking wine. A satyr's head makes sense in this context because satyrs were the attendants of Dionysus, the god of wine. Maenads were the constant companions of satyrs, and there are even some head vases with satyrs and maenads back to back. The maenad here is especially attractive, with her stylised curls, elegant eyebrows and carefully outlined lips. The satyr, with his snub nose and animal ears, is not quite as refined. His most striking feature is his spreading beard with its row of incised curls.

Wine drinkers congregated at the symposium in Athens, a formal gathering of male citizens that often descended into drunkenness and debauchery. Women generally took part as entertainers or courtesans rather than being allowed to recline on couches like the male symposiasts. The drinking vessels used in the symposium often referenced Dionysus in some way and their imagery was not intended to be taken entirely seriously.

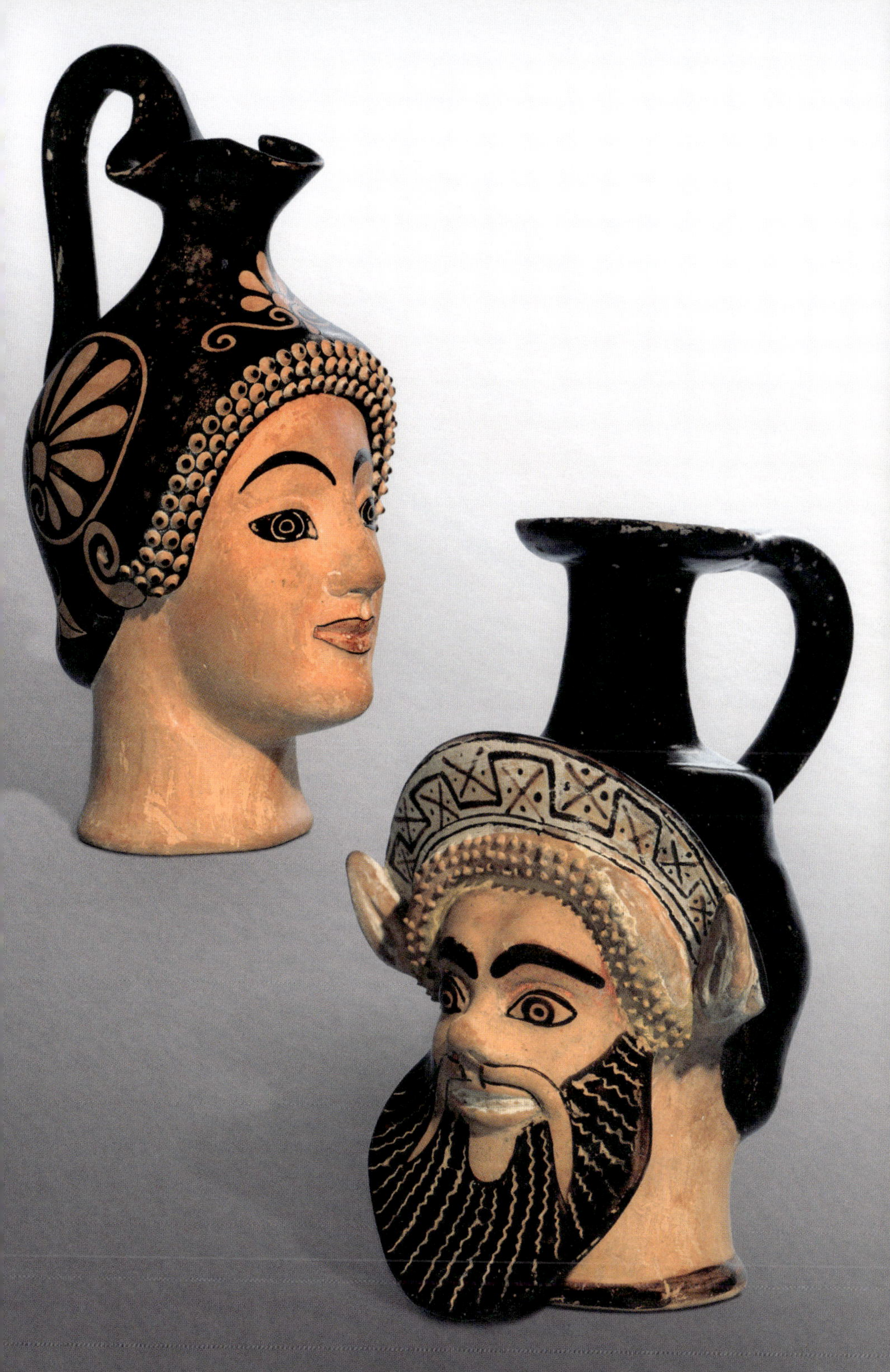

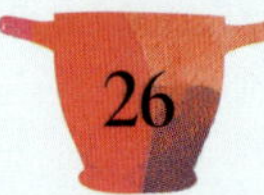

Boeotian black-figure *skyphos* depicting Odysseus at sea and with Circe

Greece
400–301 BCE
Black-figure pottery, 15.4 cm

This *skyphos*, a type of drinking cup, is one of the finest surviving examples of ceramics decorated in a comic style of the black-figure technique, which is known as "Kabiric".

In Athens, the black-figure technique was barely used beyond the middle of the fifth century BCE, apart from for the decoration of vessels made for ritual purposes, and prize vases. The vessels made in Boeotia, the region to the north of Athens, seem to have served a similar purpose. The black-figure scenes appear on vases made in connection with the Kabirion sanctuary near Thebes, which was dedicated to a mystery cult. Most vessels which were found there are drinking cups of various shapes. They are often decorated in a particular comic style known as "Kabiric", after the sanctuary, which flourished in Boeotia in the late fifth and fourth centuries BCE.

On one side of the cup, we see the ancient Greek hero Odysseus (also known by the Latin variant Ulysses), the cunning and resourceful king of Ithaca, known to us as the star of Homer's epic poem, *The Odyssey*. Depicted partway through his perilous decade-long journey from the war in Troy to his home in Ithaca, he wears a traveller's hat and holds a sword in one hand and its scabbard in the other. The other figure is the seductive enchantress Circe, who is about to dose him with a drugged potion. Behind the sorceress is her loom bearing a section of finished cloth, its yarns held taunt by loom weights.

On the other side of the cup, Odysseus balances confidently astride two amphorae (storage vessels), which he is using as a raft. With a foot on the belly of each clay pot, his cloak streaming out behind him, the master mariner surfs the choppy sea, holding a trident. He is blown along by the god of the north wind, Boreas, whose puffing face appears in the top corner of the scene. Both Odysseus and Boreas are identified by name. The scene could be a depiction of the hero's escape from the clutches of another character in the

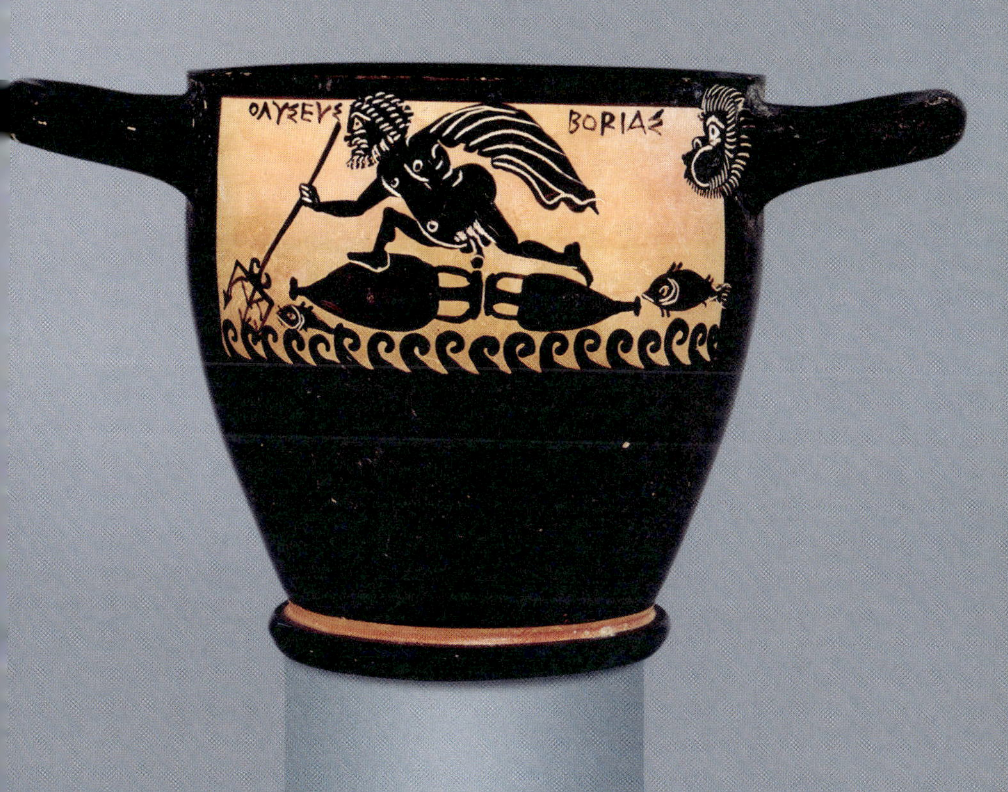

ΟΛΥΣΕΥΣ
ΒΟΡΙΑΣ

Odyssey, the nymph Calypso, who eventually helped to build him a raft to sail home.

The figures on the cup are very different from Athenian black-figure depictions of mythological scenes. Odysseus has a pot belly and a look of surprise – he doesn't look much like the intrepid, wily figure of the Homeric poems. Circe, as in other Kabiric depictions, appears to be a Black African woman. This could be a perceptive depiction of the mythical sorceress said to live on a far-off island. However, in this context the painter might be setting up a contrast between Circe's appearance and the female bodies more usually depicted on Greek vases.

Perhaps these drinking cups were made for a festival where a lot of wine was drunk and social norms were temporarily turned upside down. That might explain this dramatic, surprising treatment of a well-known Greek myth.

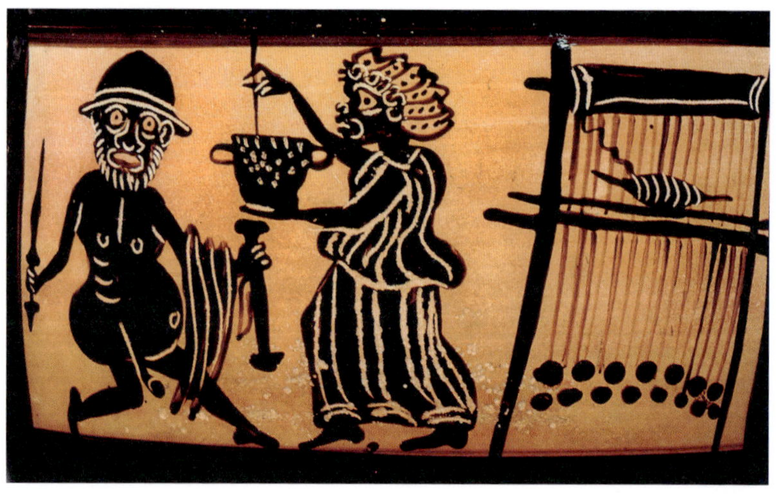

(above)
Odysseus and Circe.

(opposite)
Circe giving the potion to turn Ulysses into swine, anonymous Italian printmaker after Parmigianino (1503–40). Engraving, *c.*1520–1600.

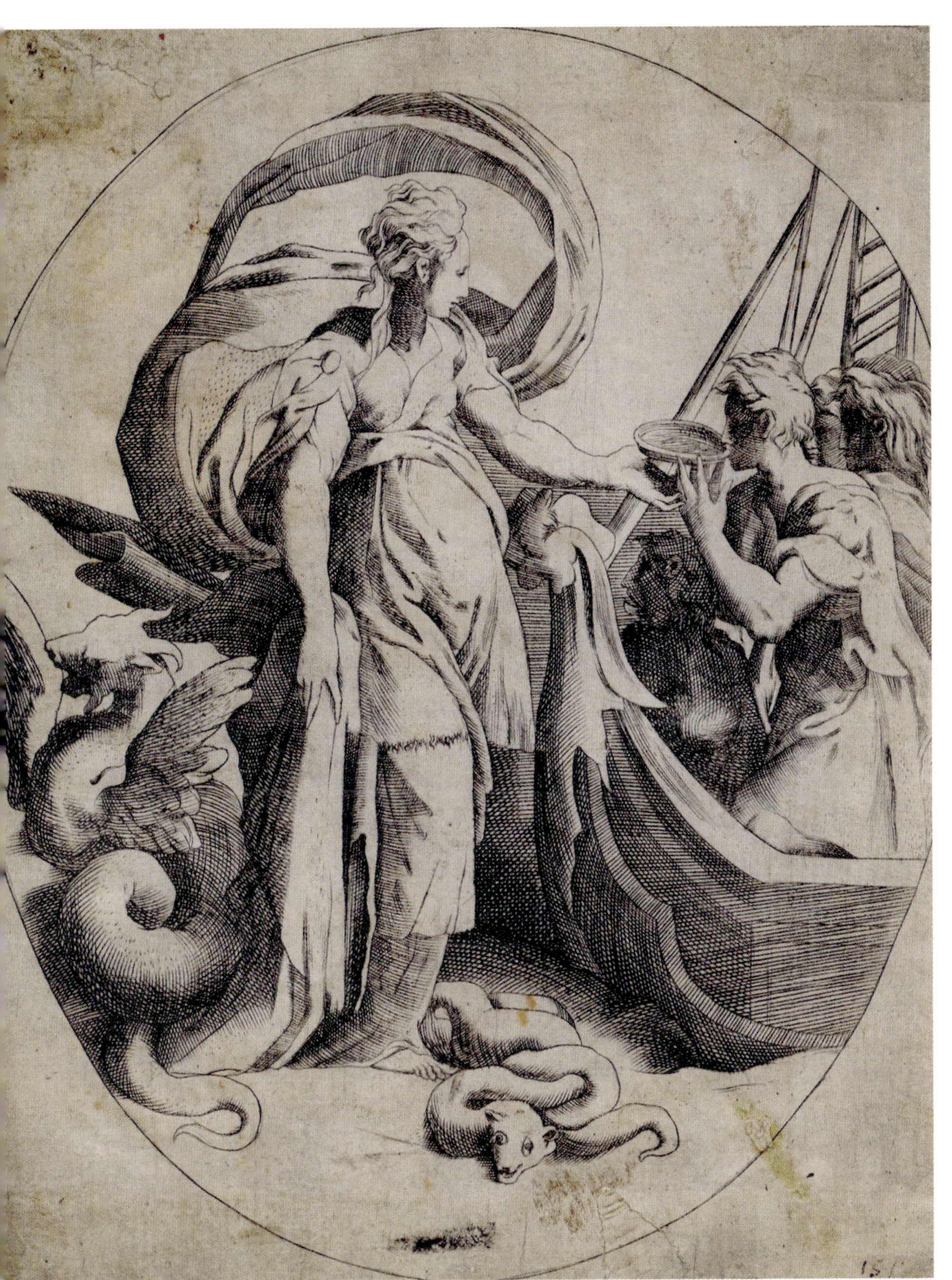

27

Terracotta female figurines

Pakistan
*c.*300 BCE–100 CE
Terracotta, *c.*5–15 cm

Archaeologists have found figures like these on several different sites in the North West Frontier plains of Pakistan. They all have more or less the same size and basic form. Their flattened bodies have an exaggerated, voluptuous bottom and a single tapering stump representing the legs.

They are easily recognised as a type, with their short, outstretched arms, pinched noses and split-pellet eyes that resemble cowries. When complete, the figures bear small pellets for breasts, which may be heightened with a hole to indicate the nipple. The separation of the legs is shown only by an incised line on the stump, with an incised triangle for the pudendum. Sometimes the navel is shown prominently by a small pellet, pierced or marked with an incised cross.

The artisans who made these figures expressed their creativity by giving them a wide variety of hairstyles and ornaments. Turbans, crowns, headbands, ear ornaments and pendants appear in various styles – some stippled, incised or stamped with circlets and others decorated with appliqué leaves, chakras, pellets and flowers. Some pieces are marked with a few lines painted in black pigment.

Excavations have demonstrated that these figures were made from the early third century BCE until the second century CE. The workshop in Charsada seems to have produced many of the more sophisticated figures, which have additional ornamentation and many different floral headdresses.

We don't know conclusively what these artefacts were for or why they have this peculiar shape in common. They can't stand up on their own, so we can assume they must have been stuck into the ground or some similar support, presumably for a ritualistic purpose. While it seems most likely that they would originally have been placed upright in the earth, it's possible they were held up in other ways. They might have been stuck in pots filled with grain, as is shown in terracotta plaques from the Ganga Valley dating after the Mauryan Empire (321–185 BCE). Perhaps they were held in the hand and decorated – puppets used as offerings. It's likely these figurines were used in popular cultic worship, the nature of which, with further discoveries, may eventually become clearer.

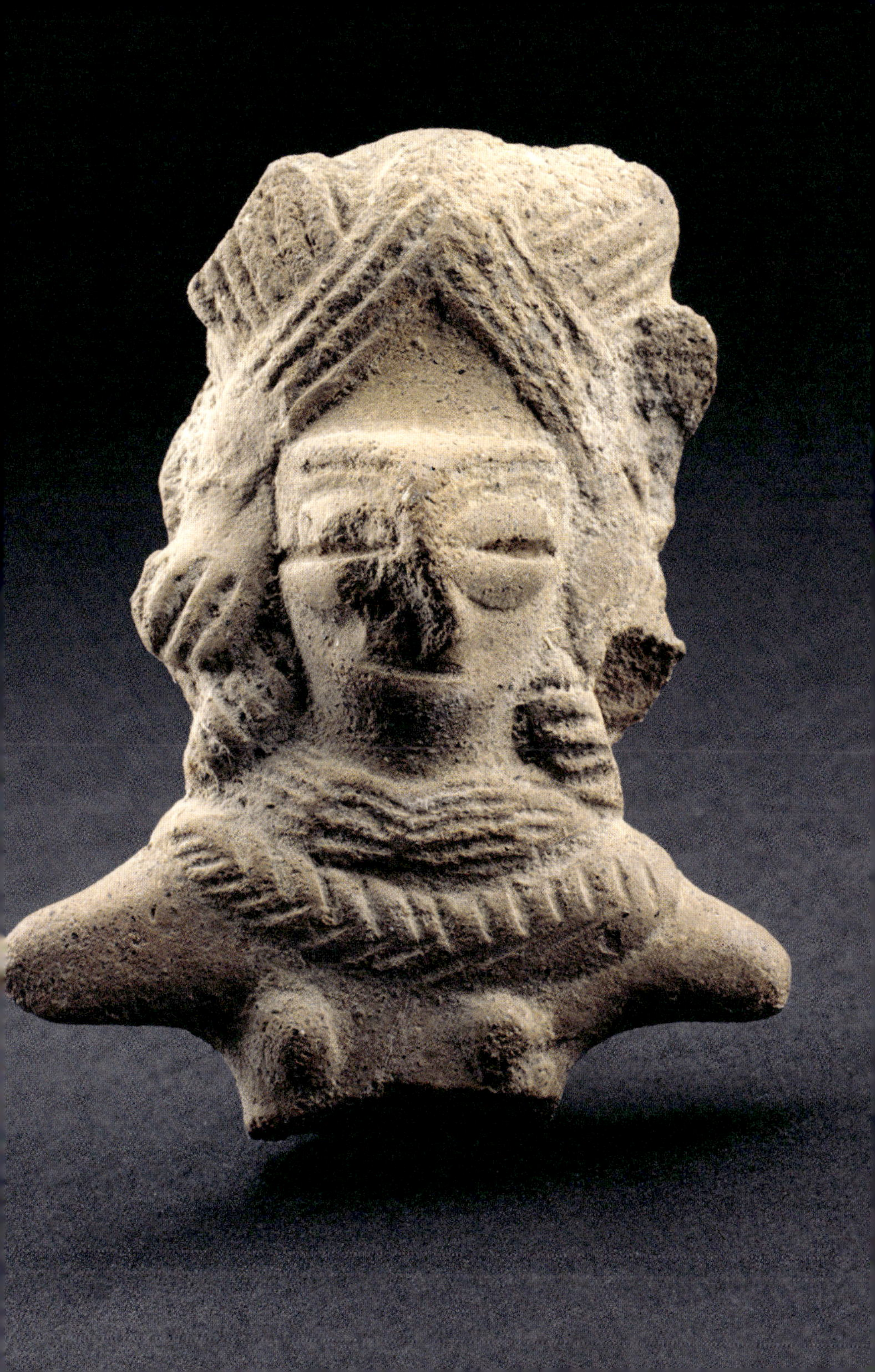

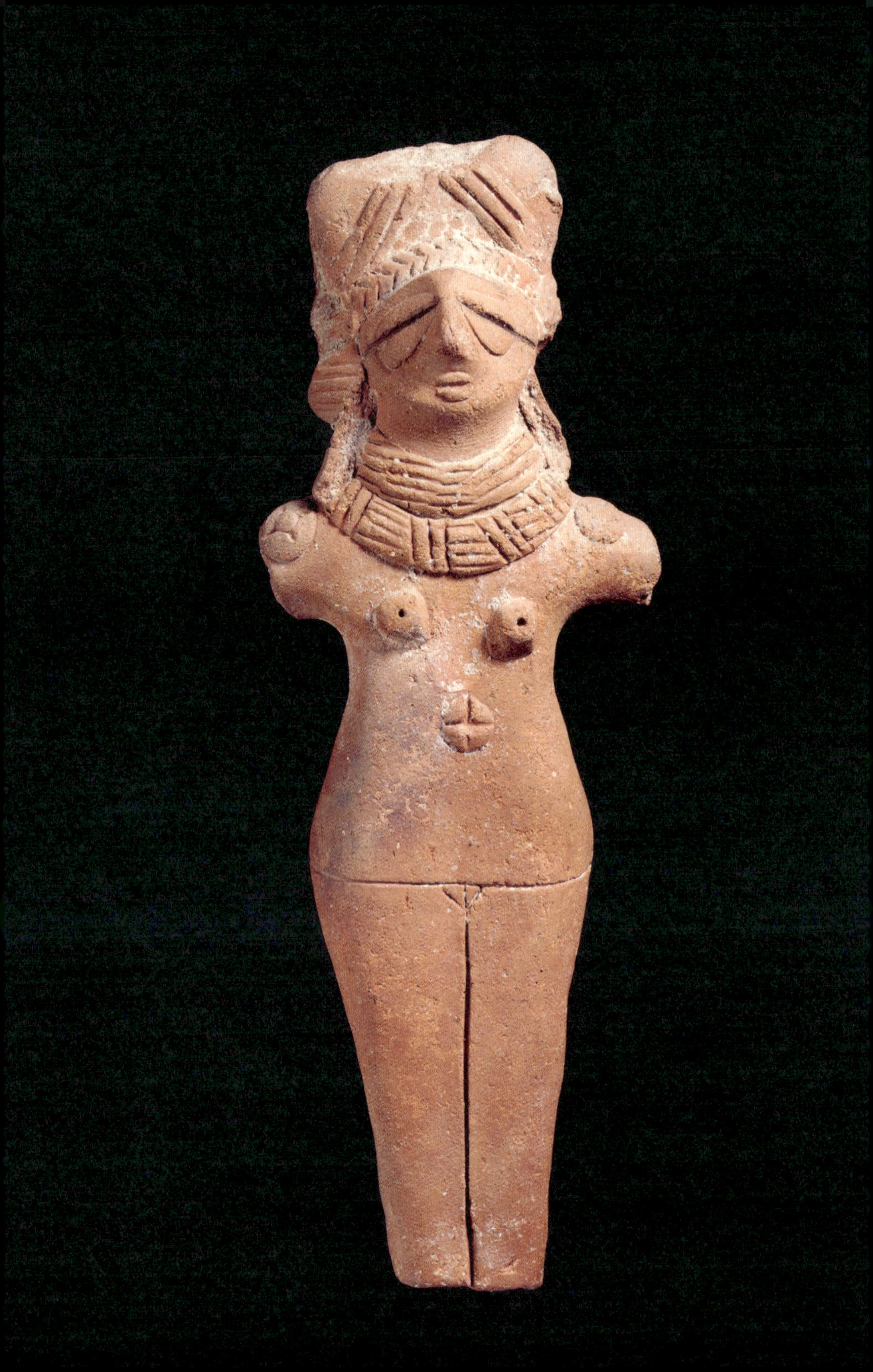

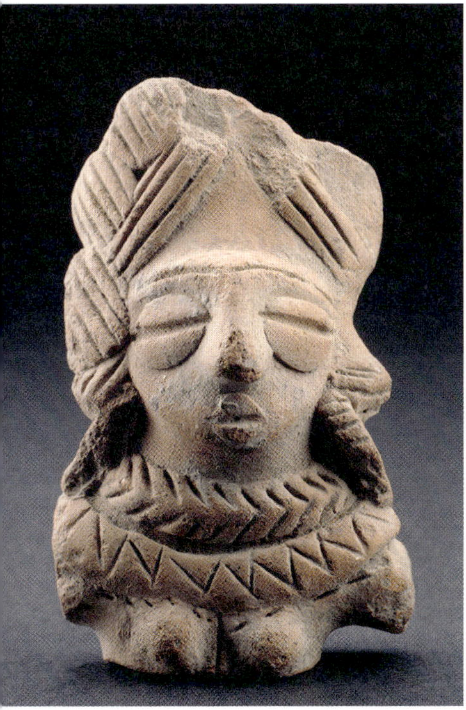

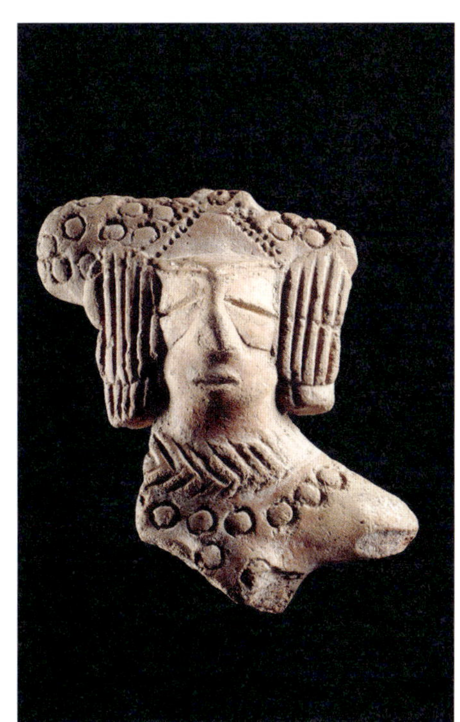

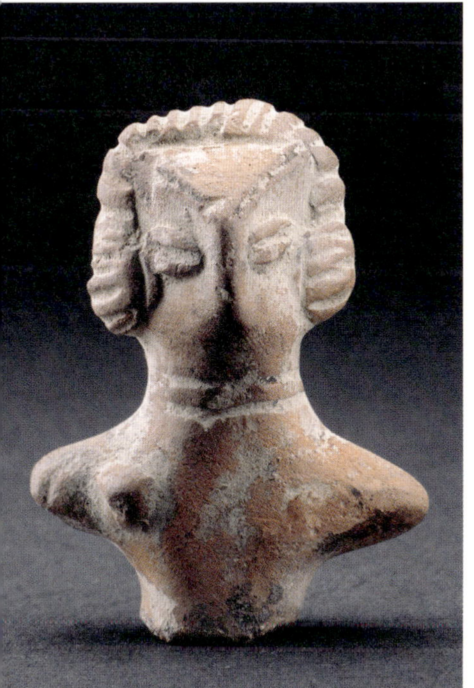

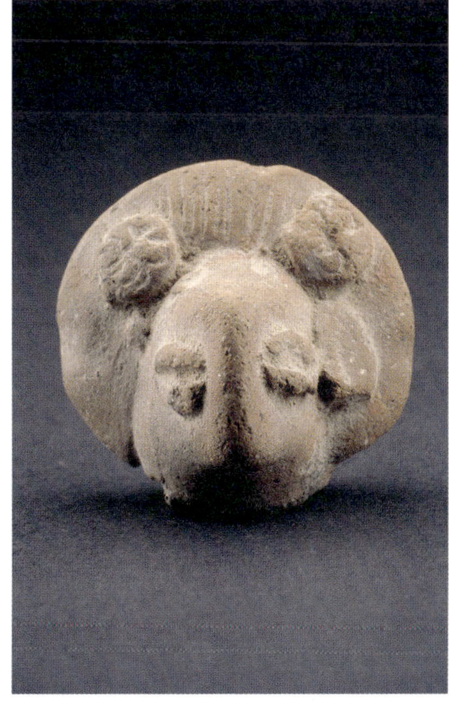

28

Plaque with *yakshi* (nature spirit) or mother goddess

India
200–101 BCE
Moulded terracotta, 21.7 cm

Terracotta has been an important and expressive medium for sculpture in India throughout the ages. It's been used, for example, to fashion religious icons, votive objects and toys. Many of the earliest sculptures represent standing women, which we think are probably mother goddesses who would have been venerated as auspicious fertility figures.

They vary widely in form, from simple hand-modelled figurines with pinch noses and split-pellet eyes to refined, elaborately detailed plaques like this one. This terracotta goddess was found by chance in 1883, emerging from a river bank at Tamluk, the ancient port of Tamralipti, a city in ancient Bengal. Often described as the "Oxford plaque", she was the first of her type to be discovered and is still the most famous of all early Indian terracottas. Her creator used a very finely detailed terracotta mould to impress her form in clay and added only a few additional details with the use of tools. The historic importance of this terracotta goddess, and the exceptional skill used in her creation, continue to set her apart from other known examples.

The goddess' smiling face is framed beneath a bejewelled headdress, one element of which displays five symbolic weapons worn as hairpins. Her elaborate jewellery includes huge disc earrings with pendant string of pearls, a massive collar and triple tubular bangles on each arm. These ornaments are patterned and minutely detailed, and include figures representing mythical animals such as the *makara* (aquatic monster) and pot-bellied dwarf *yakshas* (nature spirits or local godlings).

Various identities have been proposed for this mysterious female figure. Many others like her have come to light at archaeological sites, especially in Bengal. She may be an *apsaras*, or celestial being, a *yakshi*, or female nature spirit, or a forebear of the later, many-weaponed martial goddess Durga. She does have some of the qualities of the *yakshi* figures often found in early Indian stone sculpture, but it's more likely that she represents a mother goddess in the guise of a queen or high-born lady.

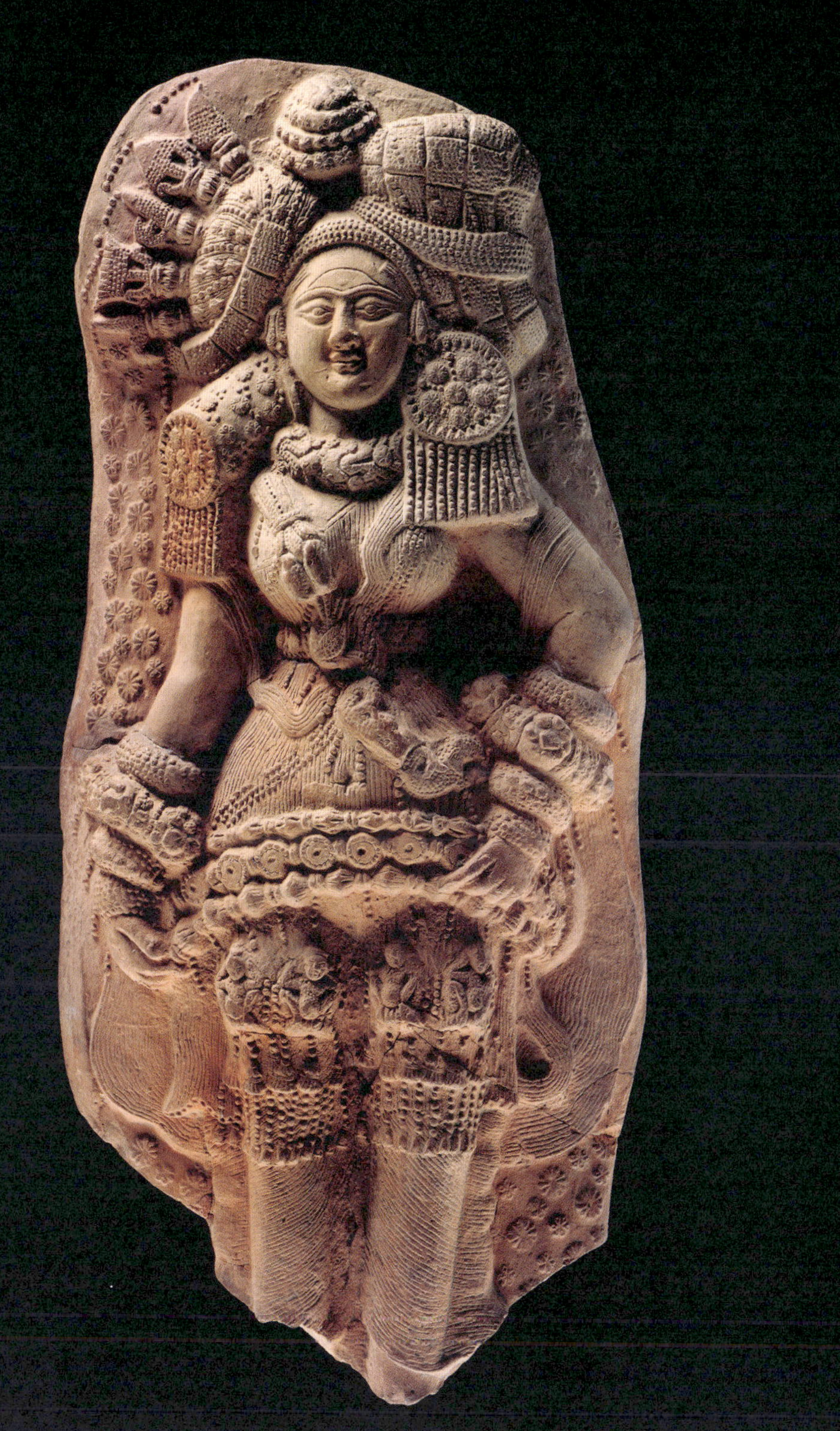

Reconstructed
Dead Sea scroll jar

Jordan
CE 1–70
Clay, 62.6 cm (restored)

One of the greatest archaeological discoveries of the twentieth century was made in 1947, on cliffs above the north-west shore of the Dead Sea. This jar is one of many that, 2,000 years ago, were carefully hidden in caves in the cliffs, and which were then preserved for centuries by the arid conditions of the Judaean Desert. The jars contained a literary treasure that was exceptionally important and rare.

The story begins with three Bedouin shepherds searching for a lost goat in the foothills of the cliffs around Qumran. One of the shepherds, just a teenager at the time, threw a stone inside a cave sealed with boulders and heard something shatter within. Later, he returned to the cave to discover ten cylindrical clay jars, sealed with lids. One of the jars contained three scrolls made of parchment: manuscripts from the Bible, the second oldest ever discovered, shedding light on the origins of Christianity.

Newspapers across the world announced the discovery, many suspicious of such a miraculous finding. The Bedouin were joined by archaeologists in the search for more caves. Between 1949 and 1956, hundreds of scrolls and thousands of scroll fragments were found in the caves around Qumran.

Few of the Dead Sea scroll jars have survived the passage of time and the excavation of the caves. In 1949, the Ashmolean asked to buy a jar from the Palestine Archaeological Museum in Jerusalem. Two years later, the Palestine Archaeological Museum sold the Ashmolean "a jar (made up and partly restored), a cover, and a specimen of linen" for £50. It was broken in transit and was unwrapped in shards in Oxford in 1951. In 2013, it was put back together, and in 2017, went on display for the first time.

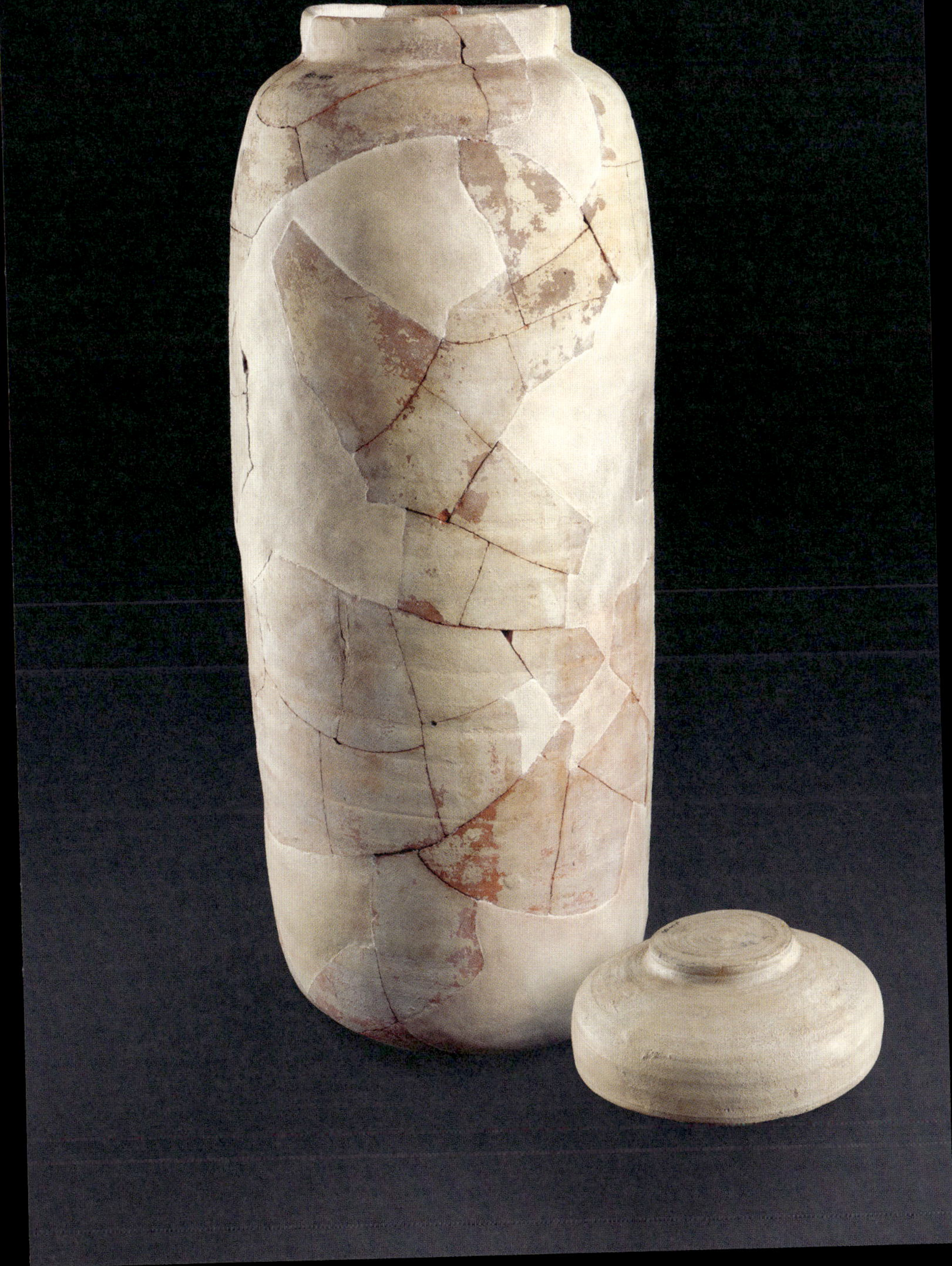

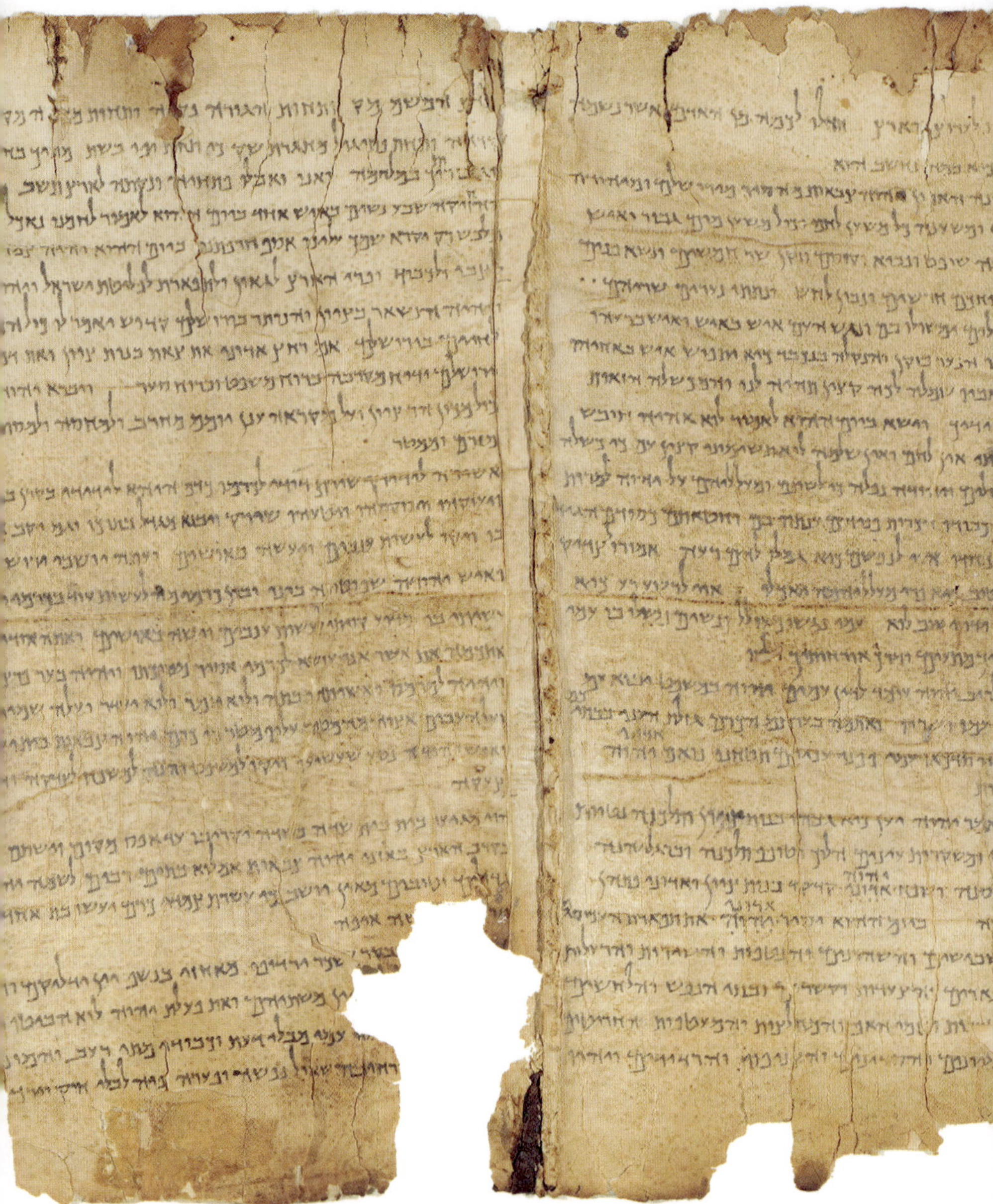

The Great Isaiah Scroll, columns 1-4.
Parchment, *c*.100 BCE. Part of the Dead Sea
scrolls, found in Qumran in 1947.

30

"Magic Bowl" with Aramaic script

Iraq
226–651 CE
Fired clay, fine buff ware, ink, 18.5 cm

This "magic bowl" is no ordinary piece of ceramic tableware that you would use for food or drink. The text that ripples out from its centre is a spell, the content of which is intended to protect against ghouls, demons and other evil spirits. This would have been the intention of the family of Babylonian Jews, living in Mesopotamia, who buried this piece beneath the floor of their house over 1,300 years ago.

The area where the bowl was found, in the early twentieth century, is now southern Iraq. Mesopotamia covered eastern Syria, south-eastern Turkey, and most of Iraq, and was the centre of the Persian Empire. At the time, all the cultures of the ancient Near East were invested in magic, demonology and exorcism, and these ideas were important in Jewish culture. Magic bowls have been found hidden under windows and doors, built into walls, and buried, as this one was, beneath a house's threshold. We think that the bowls were always buried upside down because they were designed to trap spirits with wicked intentions, who might be trying to get in.

The production of magic bowls was a trade and, as with most magic, the key concerns of customers seem to have been health, protection, childbirth, love and revenge, but the texts often also included stories and quotes from mythology and the Bible. The bowls that have been uncovered from places of concealment around a household often contain spells for protection and healing of the people who lived there. Some spells, however, intend harm to others outside the family. There is another Iraqi example in the British Museum, from *c*.600–800 CE (see image page 108), that combines both sinister intentions and passages from the Bible, albeit inverse ones. Commissioned by the client to inflict torture on an enemy through the means of black magic, the incantation instructs an evil spirit to:

> … blow up his belly like a bow and knead in him blood and thorns and sit like a slave on his heart … cast him in humiliation on his bed

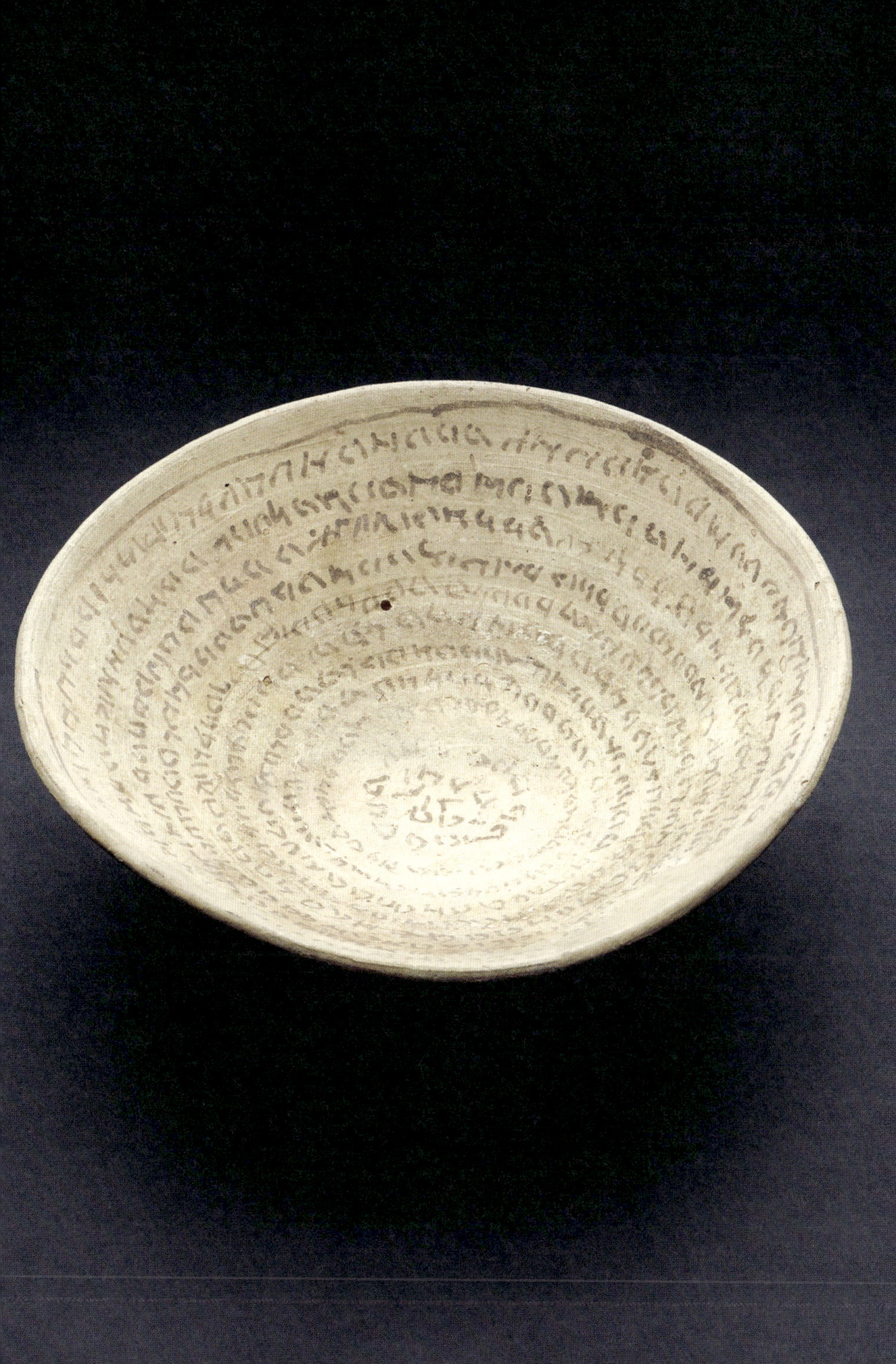

and do not give him either bread to eat or water to drink until he wails and screams and howls, until his fellow-creatures hate him … Cast down his power like the cast-off [afterbirth] of an ox of the field and draw out his spirit from the two hundred and forty eight limbs of his stature and kill him in anger and wrath and great fury.

Most magic bowls contain writing in Aramaic, in various regional forms. Bowls belonging to Christians use Mandaic; those belonging to pagans use Syriac, and a few bowls use Pahlavi, which was an early version of Persian. Jewish bowls like this one use Judeo-Aramaic, Aramaic written in the Hebrew alphabet. The incantations were applied in fine writing in brush and ink, the decorative texts spiralling upwards from the centre to the external rim, or radiating outwards in straight, ray-like lines.

The text in this particular bowl looks at first glance like a spell in Judeo-Aramaic, but is, in fact, nonsense. While knowledge of "demonic languages" was part of a magician's stock-in-trade, it's also possible that this spell is a fake. The painter may have persuaded his (most likely illiterate) customer that he was a real scribe, then sold them a bowl that didn't contain any actual magic at all.

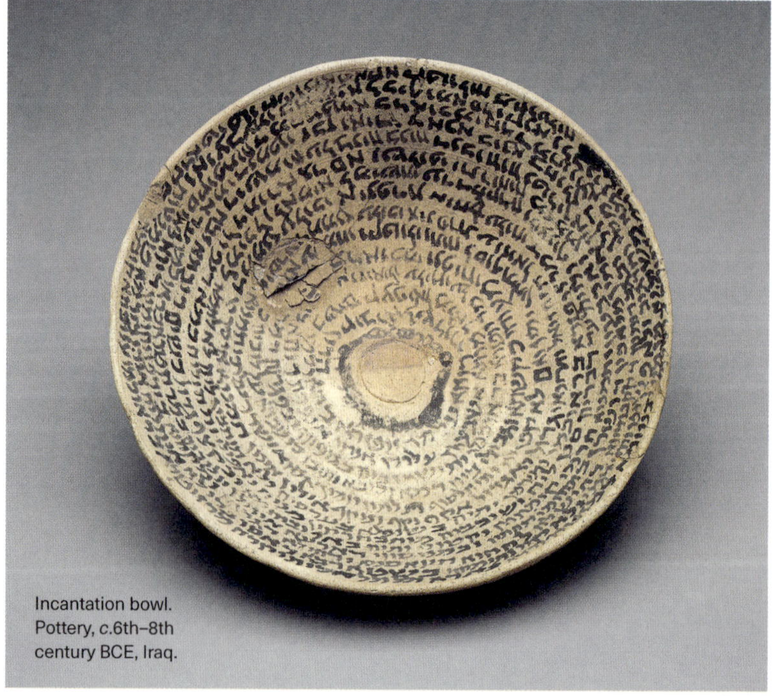

Incantation bowl.
Pottery, c.6th–8th
century BCE, Iraq.

Jar with heavily moulded decoration in Central Asian style

China
500–600 CE
Stoneware, glaze, 38.3 cm

"Greenware" like this jar was the mainstay of high-fired Chinese ceramics for 3,000 years. The term refers to pieces with an iron-rich glaze that have been fired in a reducing atmosphere (that is, without oxygen) to produce a green surface.

This technique resulted in pieces that range in colour from a yellowish-olive hue to a delicate green-blue. Greenware first appeared around 1500 BCE and potters continued to make them until midway through the Ming dynasty, some 3,000 years later.

Some wonderful greenware was made in north China during the eleventh and twelfth centuries, but the heartland of production was always in the east, in the modern province of Zhejiang. The Yue kilns of northern Zhejiang began large-scale production around 300 CE and were working at their peak during the later part of the Tang dynasty. At this time, a particularly impressive type of greenware called "secret colour Yue" was used by the emperor and written about by poets.

This vase dates from shortly before the Tang dynasty (618–907 CE). It's very similar to a vase excavated from a tomb dated 581 CE in the north at Jingxian, Hebei province. It was unusual for potters to apply moulded ornament to high-fired ceramics.

The motifs on this vase reveal connections with areas beyond China's borders. When it was made, the lotus leaf was fairly well established as an ornamental motif in China; it had first appeared in Chinese decoration with the arrival of Buddhism during the Han dynasty (206 BCE–220 CE), associated with the birth of Buddha. At first, it was used on temple architecture but soon became an ornamental motif on small items as well. The beaded roundel was originally a Sassanian motif (from the Neo-Persian Empire) and the half-palmette ornament was ultimately a Hellenistic motif (from Greece). So, the decorative scheme of the vase links a quintessentially Chinese type of ceramic with the cultures of Central Asia, South Asia and southern Europe.

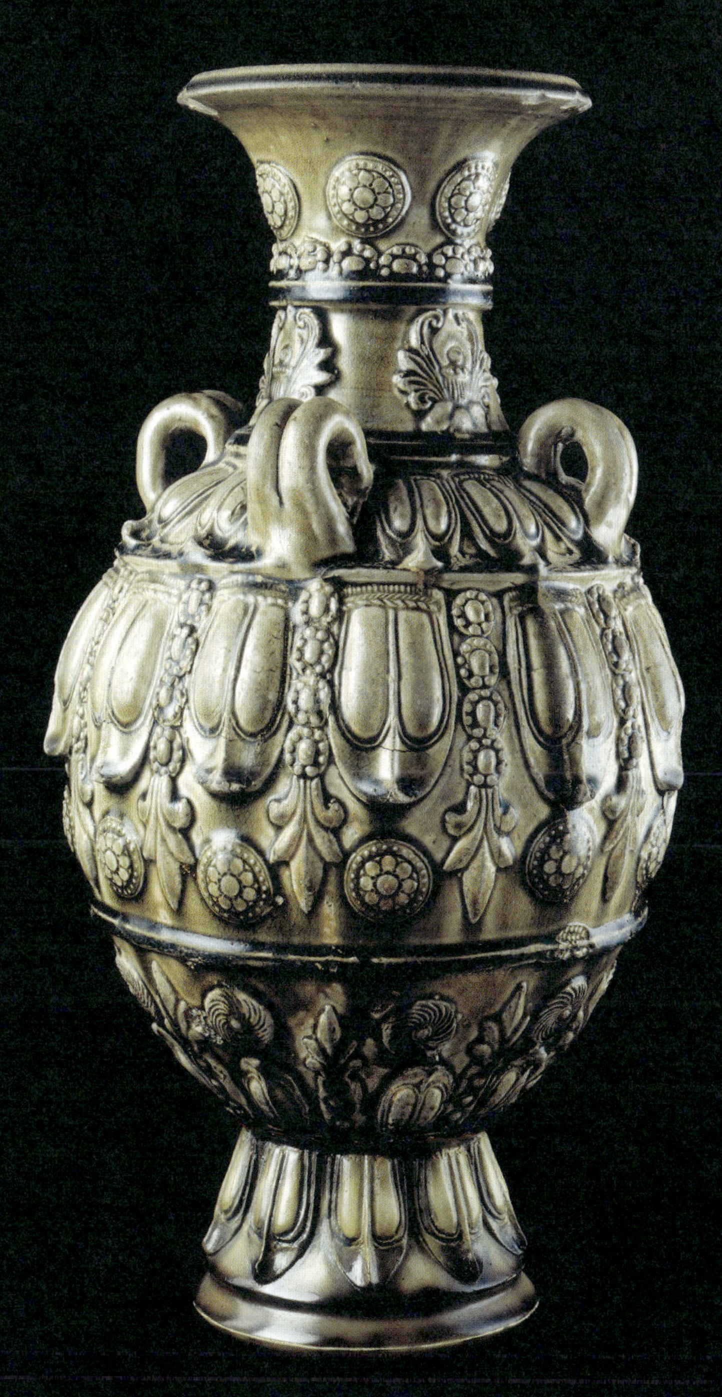

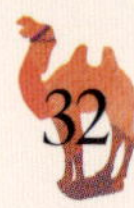

Earthenware figure of a camel

China
618–907 CE
Earthenware, glaze, 52 x 36 x 19.5 cm

This pottery camel was made from flat sheets of earthenware clay. Its creator pressed the clay into moulds and decorated the result with white clay slip and coloured glazes. The modeller carefully depicted the different features of the camel – its beautifully textured coat with colour variation on the mane, the open mouth and upright humps showing the creature is in rude health – the model is full of life and detail. The camel stands 52 centimetres high, and thermoluminescence dating confirms that it was made in Tang dynasty China.

Camels were not only mere working animals, but played an essential role in the movement of goods. This model represents the enormous wealth and opportunity there was to be found at this time by those trading on the Silk Road. This huge network of trade routes between the Far East and Europe, via the Middle East, made China fabulously rich, opening up the opportunity to trade ceramics, textiles, leather, fruit and vegetables, precious stones, metals, paper and gunpowder. Goods were bought and sold but also exchanged, stored and distributed via the routes – which also provided a marketplace for the exchange of language, religion, science and culture.

Bactrian camels – those with two humps, rather than the single-humped dromedary – were the ideal beasts of burden on these tough and often hazardous routes; they could withstand the hugely varied weather conditions, from intense desert heat to freezing mountain cold; cover 30 miles a day, carrying heavy loads; and go without water for several days.

However, this camel was created with a different journey in mind. It is a *mingqi*, or spirit object, created to journey with its owner to the afterlife. It was buried in the tomb of an elite member of Chinese society in the tenth century. In Tang dynasty China (618–907 CE), the tombs of the rich and powerful were filled with ceramic models of people, horses, camels and mythical beasts. The quantity and durability of *mingqi* that have been excavated mean that these objects are vital to our understanding of Chinese culture. During the Tang dynasty, *mingqi* were part of the elaborate funerary practices that

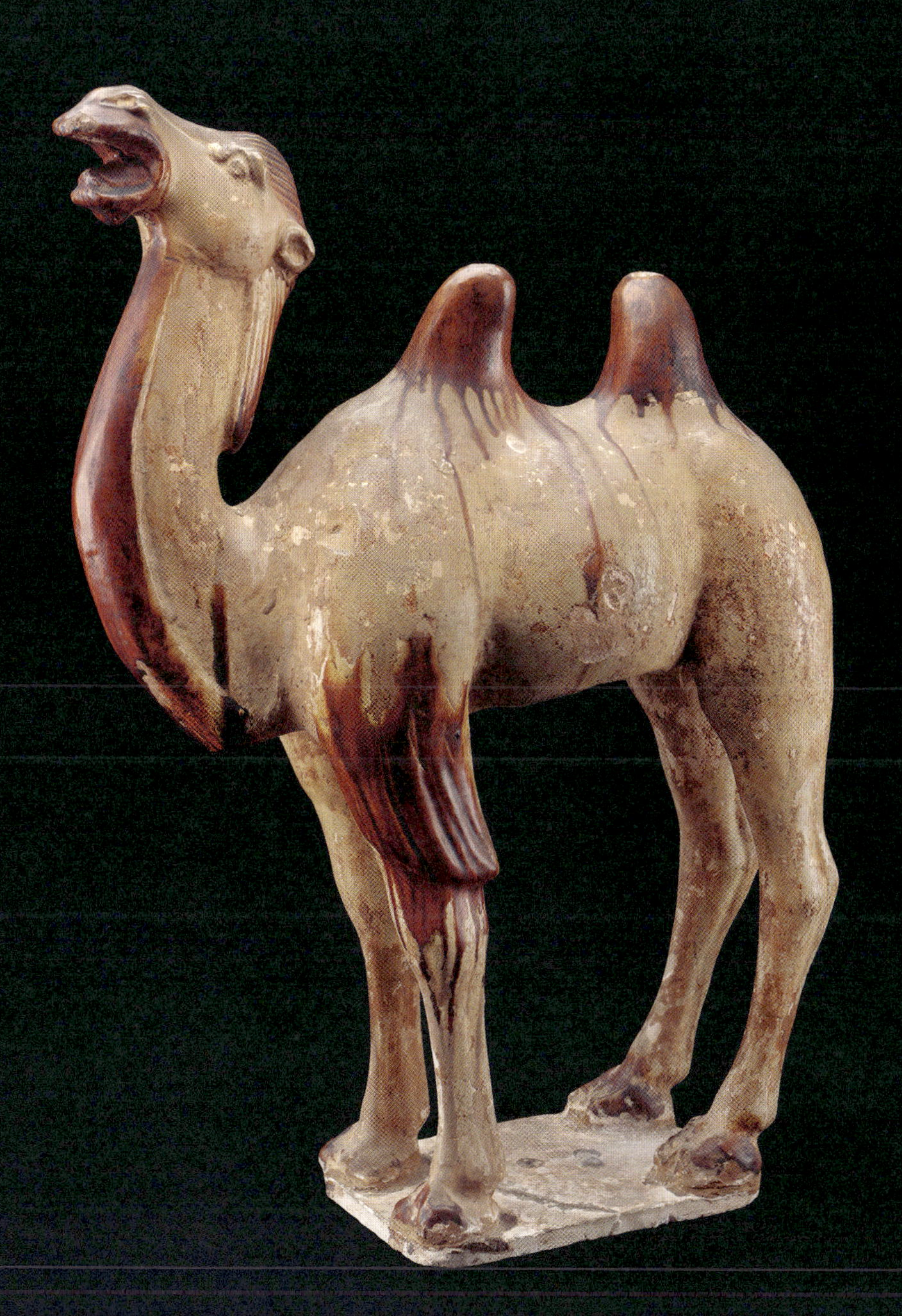

would demonstrate the status of the deceased and their family. In addition to models of animals, the great and good might be accompanied by dancers and musicians to entertain them in the afterlife, along with servants to serve them.

After being excavated, this camel probably made its way to Europe as a Far Eastern antique. It was discovered, by Katharina Ulmschneider and Sally Crawford at the Oxford University Department of Archaeology, that the eminent archaeologist Paul Jacobsthal and the camel had both made long and perilous journeys to Oxford. Jacobsthal, who in 1912 had been appointed Professor of Classical Archaeology at the University of Marburg, Germany, bought the camel at an auction in the town in 1924. In 1933, the Nazis came to power in Germany and, in 1935, Jacobsthal was forced to resign from his job, like many other academics, simply because he was Jewish. Many German Jews, in fear for their lives, faced insurmountable difficulty finding safety in Europe, as countries swiftly closed their borders.

In a stroke of luck for Jacobsthal, he counted John Beazley among his friends and colleagues. The Professor of Classical Archaeology and Art at the University of Oxford arranged for a position to be created for Jacobsthal and, a few months later, he escaped to England, eventually being appointed a Fellow of Christ Church in 1936. Although he had found refuge in Oxford, in 1939, Jacobsthal, like thousands of other foreigners, was arrested as an "enemy alien" and sent to an internment camp, first at Warth Mills in Lancashire and then on the Isle of Man. After the war, he chose to remain at Oxford, and remained there until his death in 1957.

The earthenware camel sat on the windowsill by his desk in his North Oxford house from 1937 until the end of his life. It had escaped Nazi Germany with him, and even became a code word in his letters back to Germany to refer to himself and the other Jewish academics still trapped there. It had become a symbol of a journey that must be taken in order to survive. The camel was donated to the Ashmolean in 2012, and eventually Katharina Ulmschneider and Sally Crawford uncovered the full story of its journey, which traversed 11 centuries, thousands of miles, and the bridge between death and life.

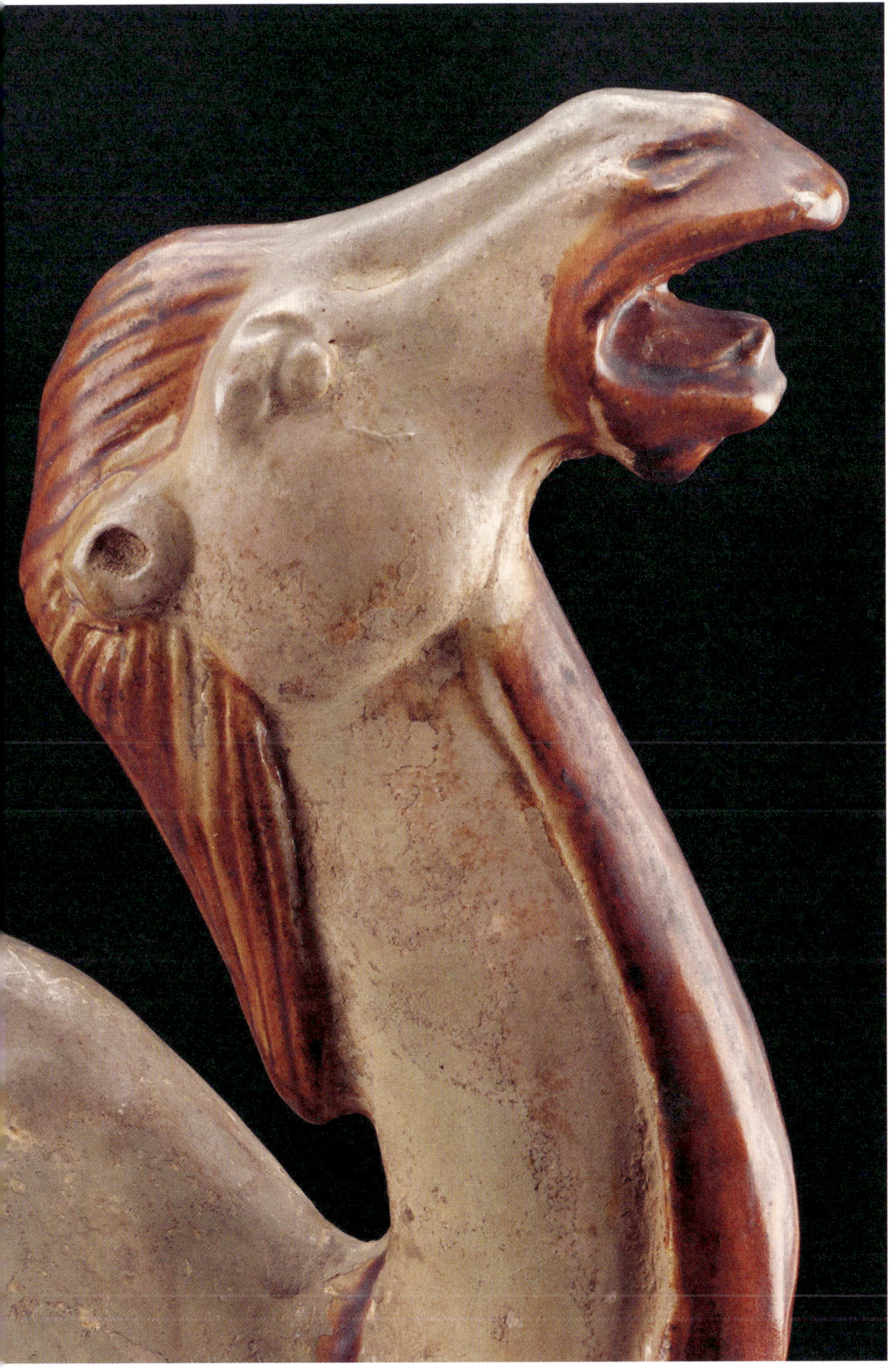

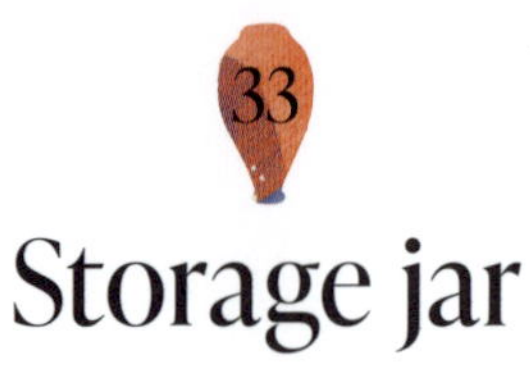

33

Storage jar

Iraq
*c.*700–800 CE
Earthenware, applied decoration, turquoise glaze, 77 cm

This striking storage jar is one of the earliest objects in the Ashmolean's Islamic collection. It's particularly impressive because of its size and beauty. At 77cm tall, it's a prodigious feat of pottery. Thrown on a wheel in two sections and joined in the middle, it was glazed and fired along with dozens of similar pieces.

This jar would have been one of thousands turned out by a veritable industry. The rise of Islam was accompanied by a major expansion in manufacturing and trade. These vessels were often used to store and transport goods such as dates and honey. Though it has been decorated, this vase is still a functional object, and it was almost certainly valued for its contents rather than its form.

This kind of glazed ware existed even before Islam. In fact, it was the only kind that was made in the Middle East prior to the rise of the religion. More than 1,000 years old, this pot is an example of the fundamental type of ceramics upon which more inventive and creative glazing techniques were developed, and proliferated, across the Islamic world. It's part of a tradition of glazed wares in the Middle East which stretches back in time to the Bronze Age and continues into the present day. In Iran, you can still buy your yogurt in turquoise-glazed earthenware.

There are only a few dozen jars like this that have survived down the centuries without being smashed into fragments. This is the most elegant of them all. Umpteen sherds of this type of jar have been found along the Asian trade routes: around the Persian Gulf, down the East African coast, around the coast of India and Southeast Asia, and as far as China and Japan. We can see that the jars were used in long-distance commerce, a glimpse of the extraordinary interconnectedness of the ancient world.

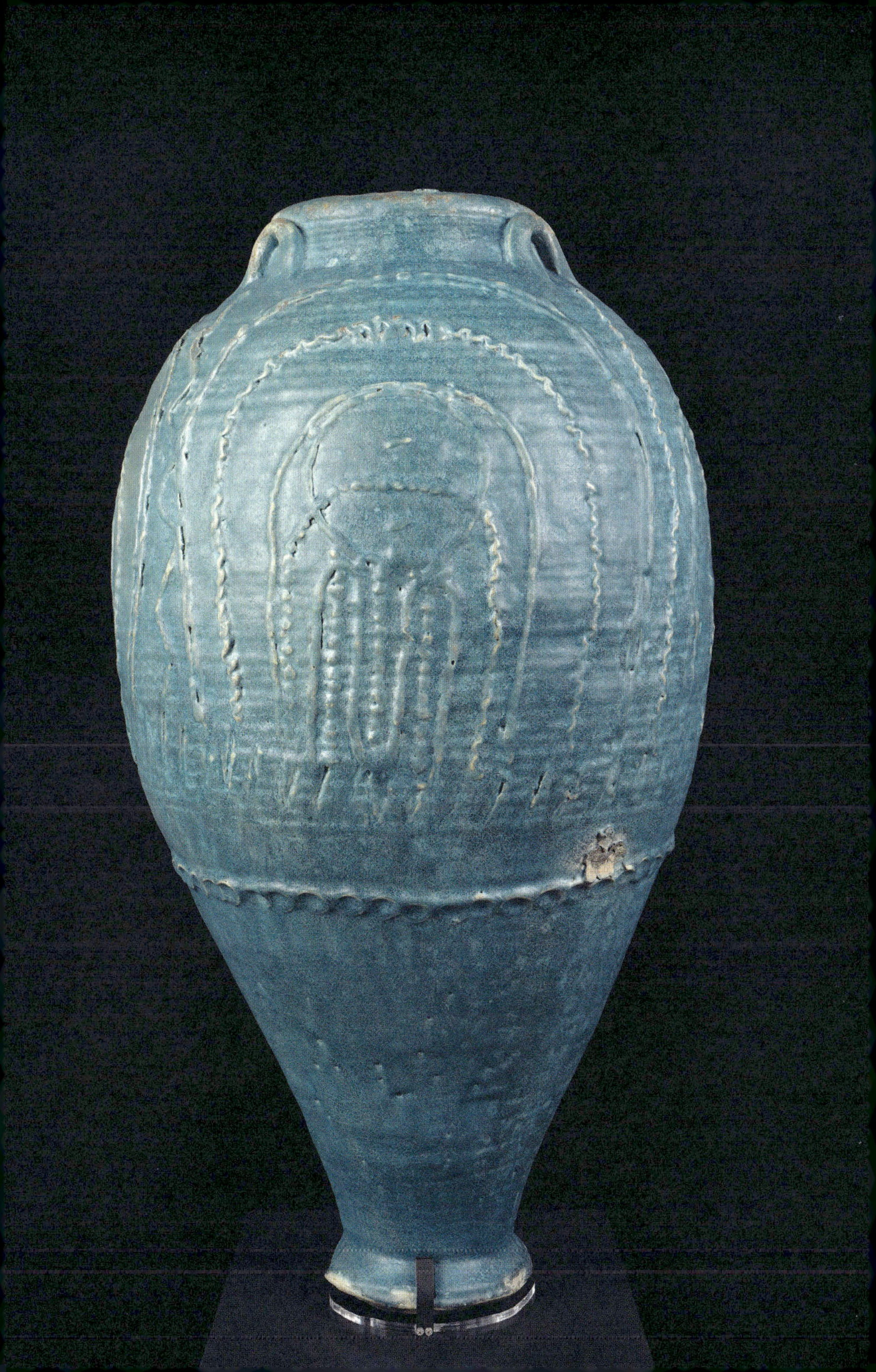

34

Two tin-glazed bowls

Iraq
801–900 CE
Earthenware, inglaze painting, glaze,
6 x 20.2 cm & 9.5 x 13 cm

The global history of ceramics has been heavily influenced by Islamic potters. They were the first to use white tin glaze as a background for ceramic decoration. They were also the first to use cobalt blue on that white ground, inventing blue and white pottery centuries before the colour scheme became the trademark of Chinese export porcelain and the hallmark of European taste. Furthermore, they developed lustre, a luxurious metallic sheen that seems to transform clay objects into gold.

The rise of decorated ceramics in Europe owes its original inspiration and technical knowledge to Islamic countries situated around the Mediterranean, such as Muslim Spain. Without the ceramic tradition of Islam, we would not have the iconic ware that later emerged in Europe: maiolica and Delft, Lambeth and Nevers. The Chinese originally developed their own decorative porcelains for the Islamic market. From this tradition emerged willow pattern and the many other chinoiserie designs that decorate contemporary European china.

Islamic artistry is marked by the love of colour, a taste for geometry and pattern, and the ability to adopt and adapt. Islamic potters took Chinese models and suited them to their own taste: down the centuries, we see enterprising potters first copying, then making their own versions of Chinese ideas. When Chinese white wares and high-fired porcelains were imported to Iraq in the ninth century, local Arab potters tried to reproduce the whiteness of the Chinese originals in their own low-fired pottery.

The Iraqi pottery fired to a yellowish buff colour, and one of the ways the potters made this look white was to cover it with a tin-glaze – a glaze which is made opaque by the addition of tin oxide. The use of this tin oxide indicates the eastern trade links of early Islamic Iraq: the metal would have had to be imported by sea from Southern Burma and Malaysia. Sometimes, the Iraqi potters left their ware completely plain, like the Chinese work they copied, but more often, they decorated it, using the white as a ground for blue and green designs made from cobalt and copper.

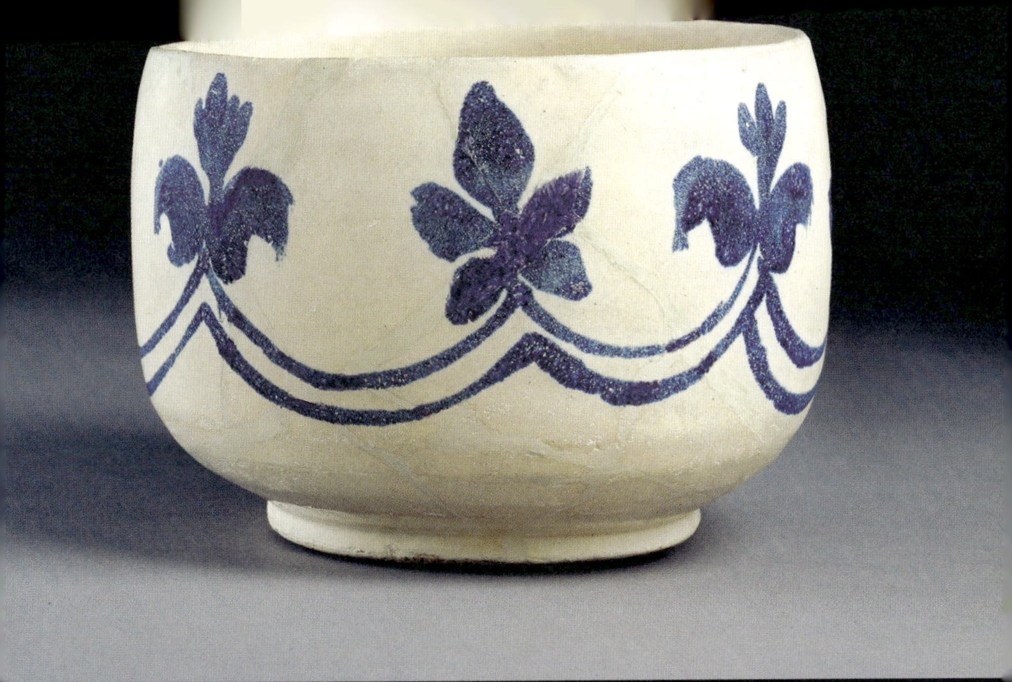

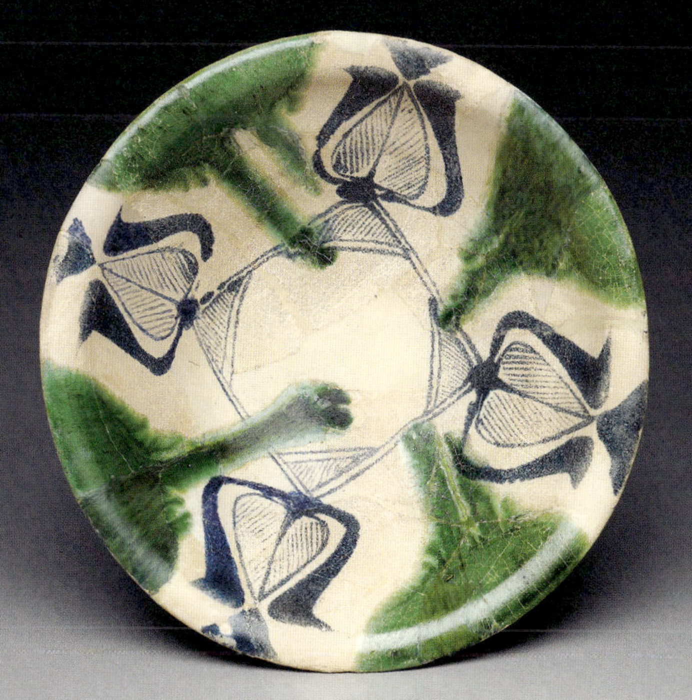

These two bowls illustrate four decorative characteristics which are typical of early Islamic pottery. The first, the palmette, appears on both pieces. Traditionally, this is a five-petalled blossom, which later was adapted into the "arabesque" theme. The second is geometry: the central part of the design of the second bowl is made up of a square inside a diamond. The third is stylisation: in the second bowl, the highly stylised palmette is about as far from its natural parent as it could be, illustrating a taste which also shapes the Islamic painting tradition. Finally, the second bowl is ornamented with green splashes, a common decoration on less expensive pottery during early Islamic times, across an area stretching from Egypt to Iran.

Another technique common to Islamic pottery is *sgraffito*, a term that comes from the Italian *sgraffire*, "to scratch". An object is covered with a thin clay slip, and a design is incised through the slip before the glaze is applied. If, for example, the body of the pot is buff, the slip is white and the glaze is colourless, the incised lines will be buff on a white ground. *Sgraffito* wares came into fashion in Iraq and Iran in the tenth century. They were less technically demanding than ceramics painted with cobalt blue designs or lustre and were used by many less affluent town and village dwellers. This second class of ceramics was in use for centuries.

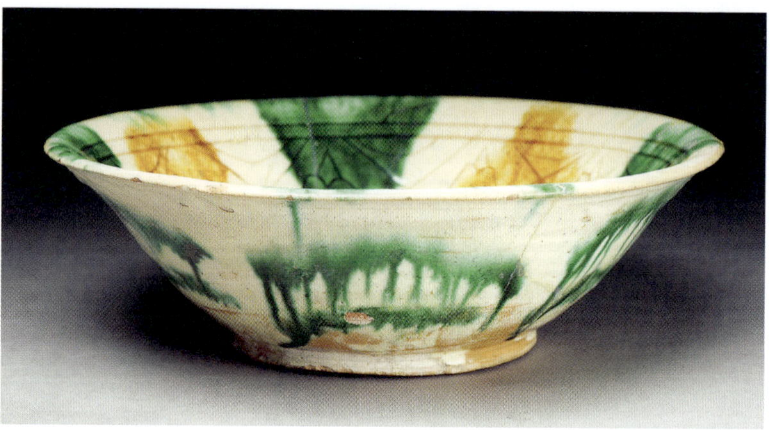

Earthenware bowl with *sgraffito* decoration, with decoration incised through a white slip and colours painted in the glaze. Iran, 901–1000 CE, 22.5 cm.

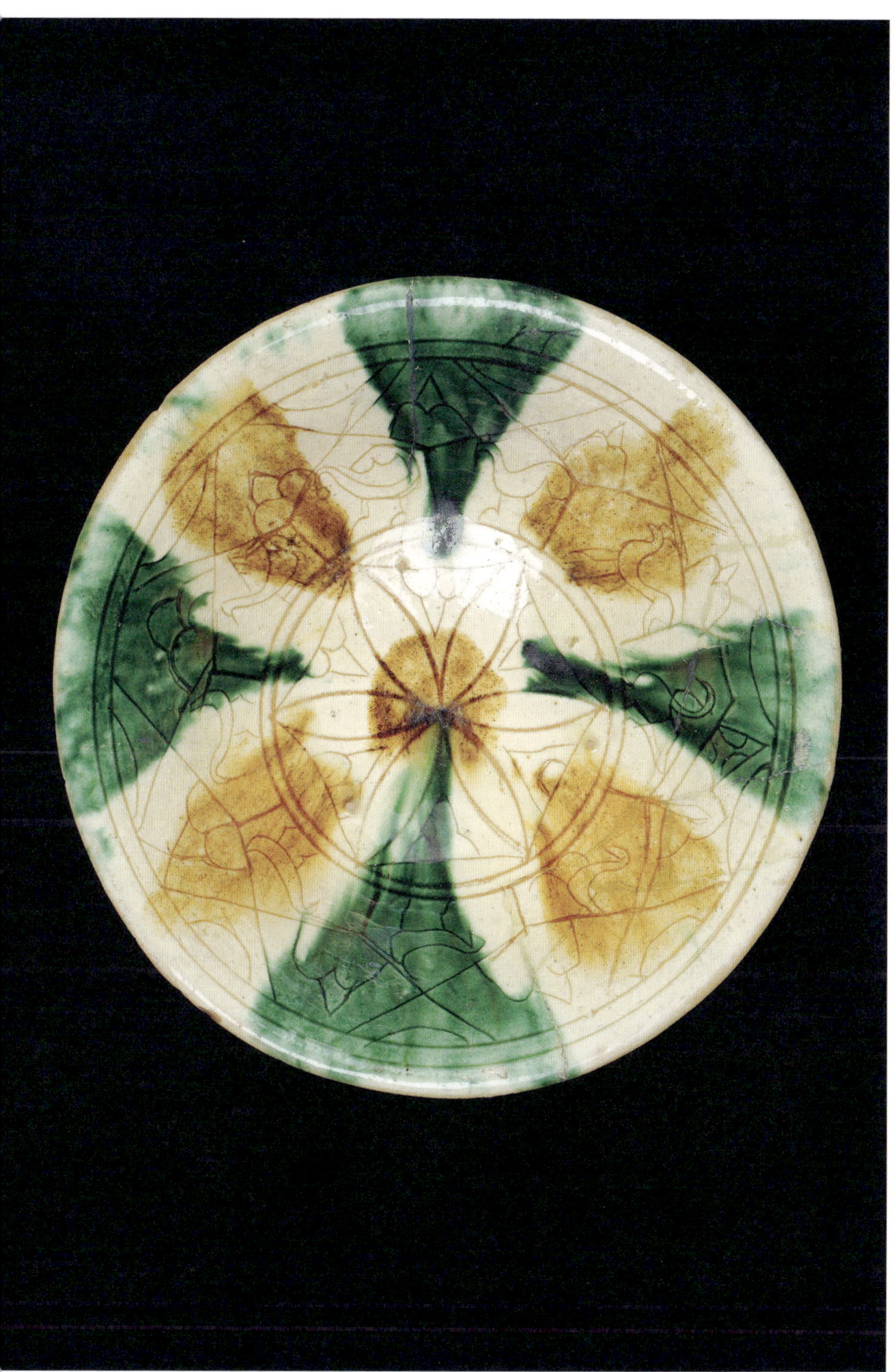

Bowl with animals and plants

Iran
Tenth century CE
Earthenware, underglaze painting
in black slip and colours, 20.4 cm

Four ibexes surround a hare in the curve of this lively bowl. Peacocks and unidentified quadrupeds frolic around them, surrounded by vegetal forms and stylised epigraphic motifs, covering all available space. This vessel is a piece of buff ware, a type of multicoloured ceramics from eastern Iran, characterised by the vigour and density of its decoration.

Buff ware was first brought to light during the American excavations of the medieval city of Nishapur in the 1930s (modern day Iraq), which, at the time of this bowl's production, was among the 10 largest cities on Earth. Looking at a variety of this type of ware, we can follow the massive progress in glazing techniques achieved in the Islamic world due to rising demand for luxury tableware during the tenth century. The makers' use of a bright yellow pigment – lead stannate – for its distinctive palette could be a reference to yellow-glazed wares produced in Syria in the eighth and ninth centuries.

Buff ware – which unusually highlights the somewhat hidden earthenware fabric of these vessels rather than their eye-catching surface decoration – was produced at the same time as the more sober and elegant epigraphic ware (see pages 126 and 136). But the humbler materials used to create buff ware, and the rather naïve quality of its decoration, do not necessarily suggest that the people who used it were less affluent. The fact that both types of ceramics were available in this prosperous region of the medieval Islamic world gives us a picture of the range of interests and consumers that were served by potters at this time.

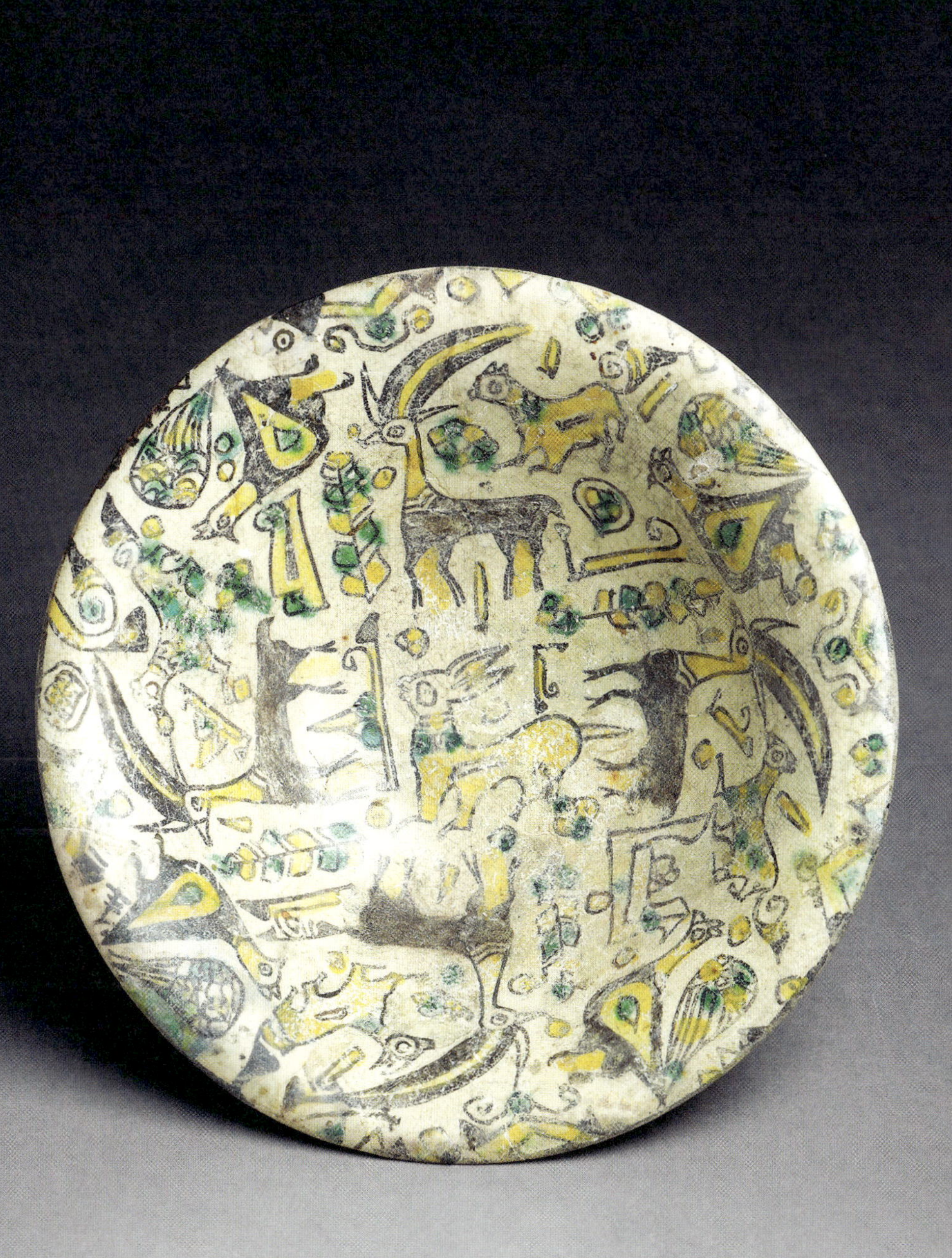

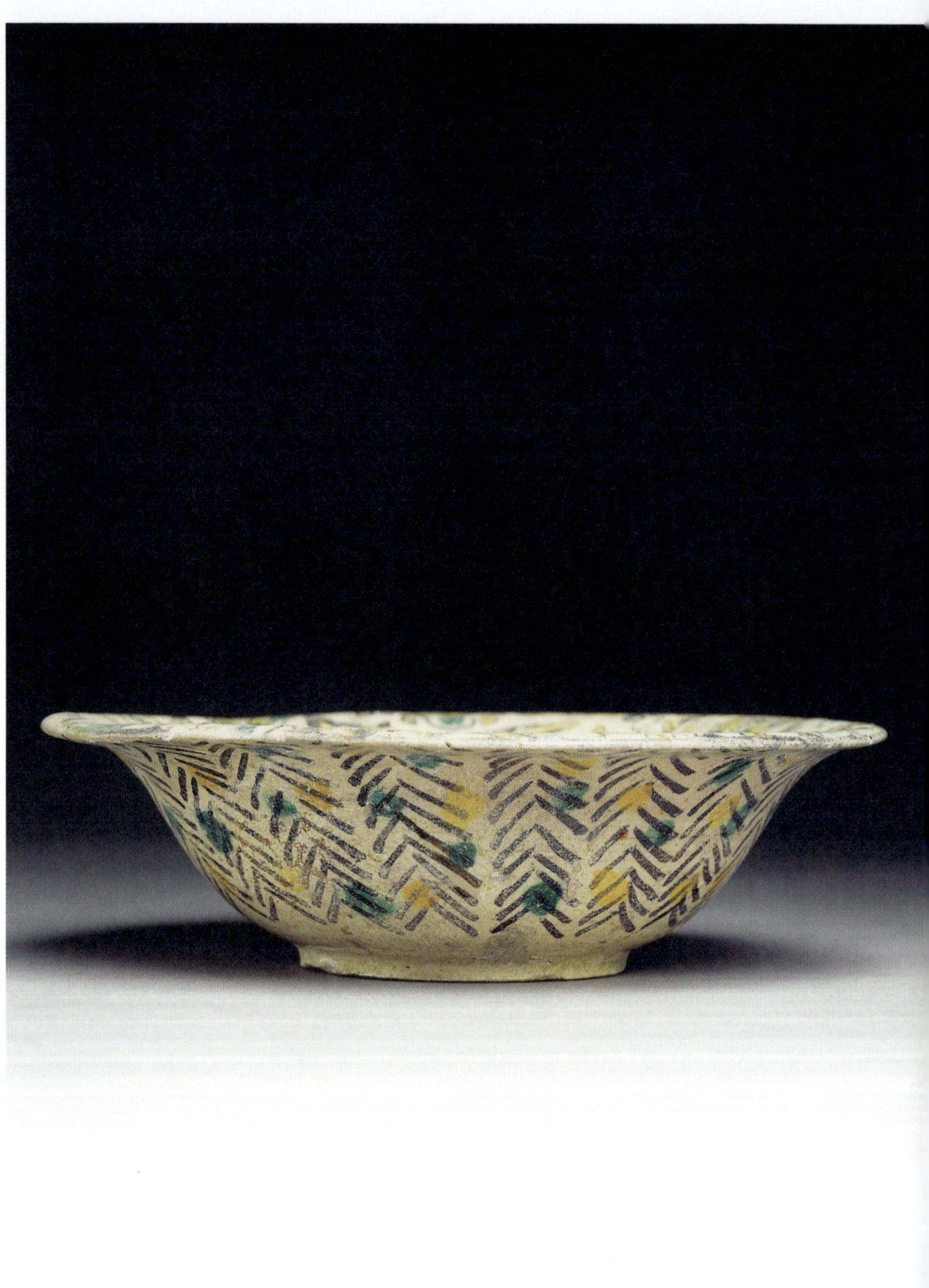

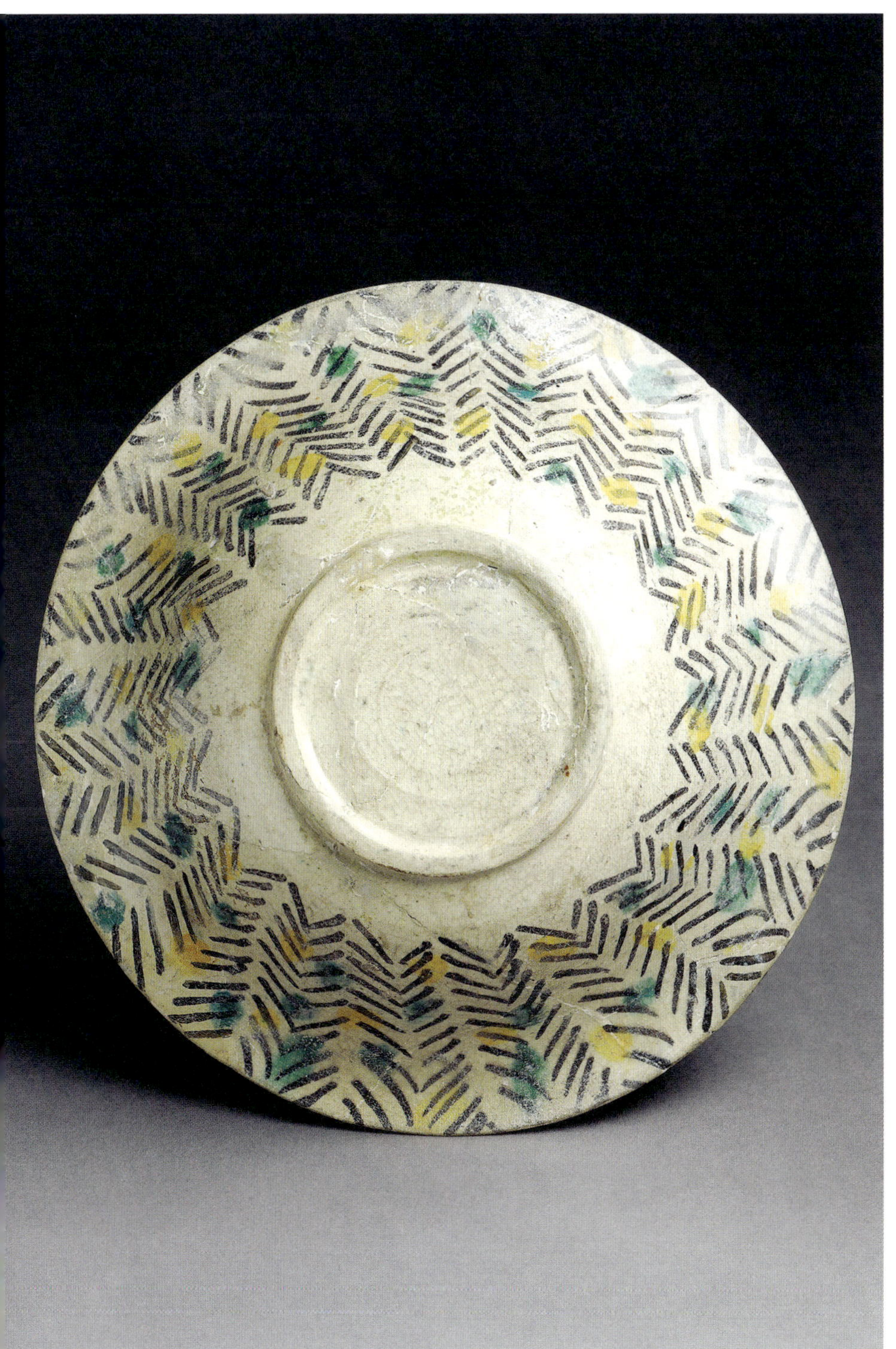

36

Bowl with epigraphic decoration

Nishapur (Iran) or Samarqand (Uzbekistan)
*c.*10th–11th centuries
Earthenware, slip, transparent glaze, 21.3 cm

Only a few words survive of the Arabic inscription on this bowl. It reads: *"alaniyya al-faqr"* ("[greed is] a sign of poverty"). The elegant epigraphic band unfurls along the bowl's rim, against an opaque, milky ground.

Some of the letters stand high and proud, springing from a continuous black baseline. They pivot upon themselves to form knobbly protuberances or conclude in curls and ornamental motifs resembling the leaves of palm trees. Horizontal strokes stretch to join decorative loops, used by the potter to fill awkward empty spaces.

It's no strange thing that a vessel like this should carry a moralising message. This bowl is an example of a type of fine slip-painted ware produced between Nishapur (modern-day Iran) and Samarqand (modern-day Uzbekistan) under the rule of the Samanids (819–999 CE). This ware was typically decorated with calligraphy. As they cleaned their plates, diners uncovered Arabic proverbs celebrating virtues like generosity, modesty and the pursuit of knowledge. These gracefully painted, sophisticated aphorisms suggest that vessels like this one were used by well-educated people. These diners would recognise the precepts as extracts from the ethical teachings of pious men.

Scholars have suggested that this type of dish would have been a conversation piece, starting discussions at the dinner table. The vessels offered diners a timely reminder that when they attended social gatherings, they were expected to behave with moderation, kindness and conviviality.

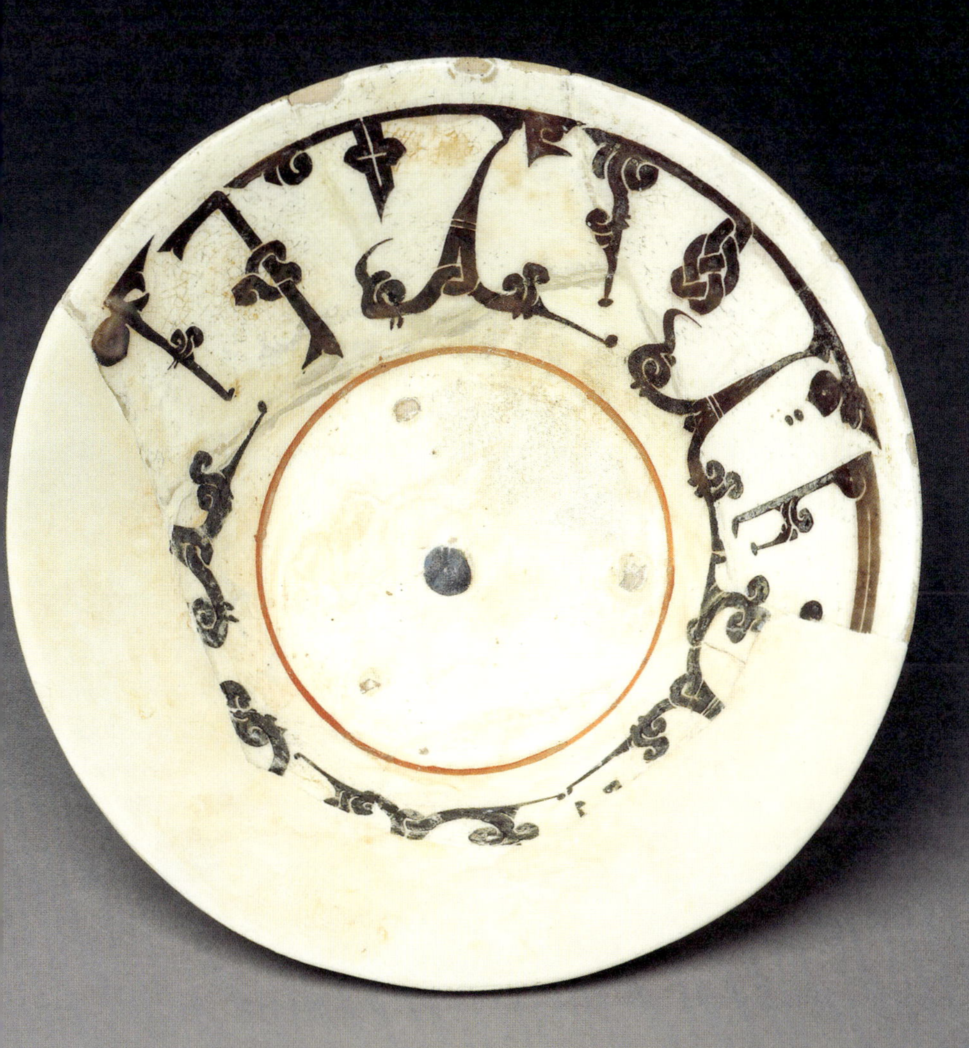

37

Black ware jar
with white stripes

China
960–1279 CE
Stoneware, slip, brown iron glaze, 20.9 cm

This jar was made in north China, in either Henan or Hebei province, and belongs to a type known as "Cizhou". Cizhou is, in fact, a village in Hebei. However, as a ceramic category, it refers to a range of wares boldly decorated in black or clear glazes and black or white slip by means of painting, incising and *sgraffito*, and produced in kilns across four provinces of north China.

With its utilitarian forms and wide distribution, Cizhou ware underpinned a Chinese ceramics industry which was unprecedented and unmatched for the variety and quality of its products. The black glazes and slips had quite a high iron content, 6 per cent, and were fired in an oxygen-rich atmosphere to produce the dark, glossy colours.

Large quantities of Song ceramics were exported around Asia and as far away as Africa. But we could argue that their most enduring influence was due not to the transportation of pots but to people travelling to China. The Japanese monks who visited their Chan (Zen) Buddhist colleagues in China particularly admired the black-glazed ceramics they encountered and took them back to Japanese monasteries. Eventually, black-glazed ceramics were also made in Japan.

The Ashmolean's jar offers many clues to potters anywhere about the techniques used in its creation. The horizontal ribbing on the interior shows that the jar was thrown on a wheel. The protrusion of the white ribs demonstrates that they were applied to the pot, while the thin transparent glaze that covers them shows that this was done before glazing. The thin glaze layer near the foot is yellow, its colour revealing it to be a single coating of the same iron glaze that is black when thick. The scarring of the glaze of the interior base was left by a smaller jar fired inside this one. These marks remind us that this vessel was created not by an artist craftsman but as part of an economic enterprise for which kiln space had to be maximised, at the end of a production line.

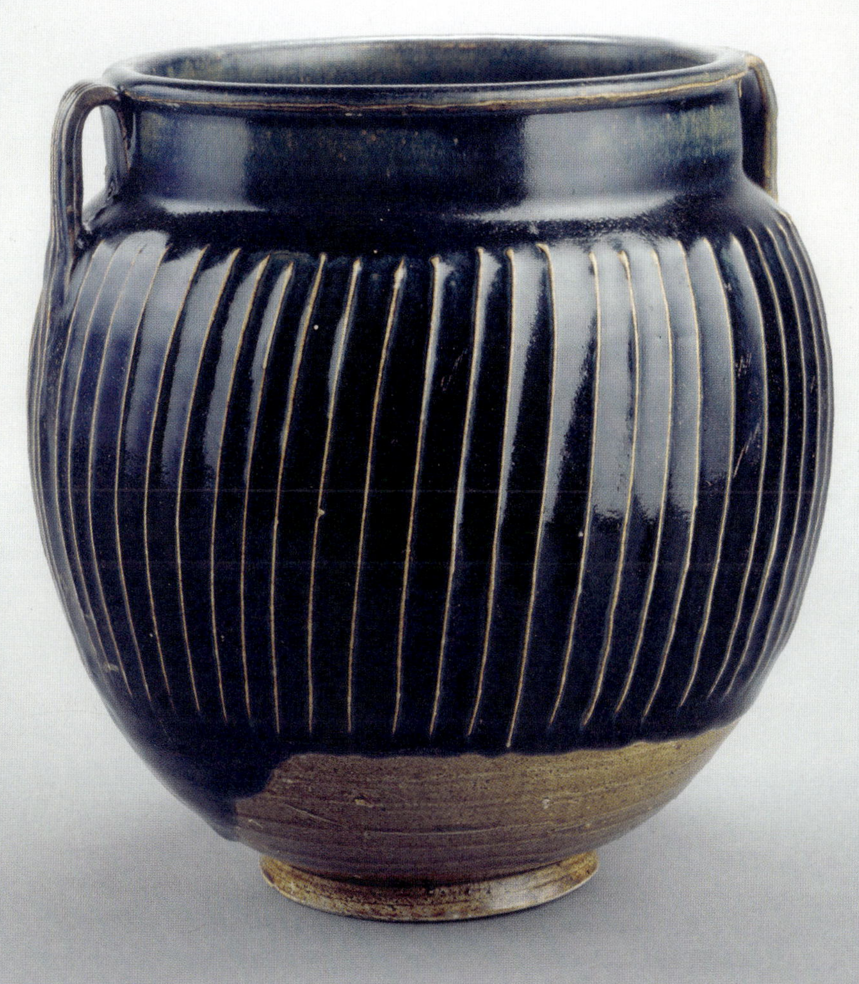

Dish with purple splash

China
960–1279 CE
Stoneware, glaze, 17.1 cm

This dish was created at the Jun kilns, at Linru in the province of Henan in central north China. "Jun ware" pieces are some of the best known ceramics from the period between the tenth and thirteenth centuries. At this time, the high-fired ceramic traditions established in north China and in the eastern coastal provinces were at the height of their technological development.

China's economy was flourishing and lots of new kilns were being established near major kiln complexes or export routes. Overseeing this success was the great Song dynasty (960–1279 CE) – though in the 1120s, the Jin invaded the north, ruling it until 1234 CE while the Song dynasty continued in the south, with a new capital at Hangzhou.

Jun ware is famous for its easily recognisable glazes, an attractive light blue often ornamented with splashes of purple. Occasionally – as on this piece – the purple glaze is controlled to make a clear pattern. The dense colour of Jun wares is due to a quality they possess which is unique among Chinese glazes – they are opaque. All other glazes are transparent, but the composition of Jun glazes means that their components separate when the ceramics are fired, resulting in a milky or opaque effect. The dish's creator used a copper colourant to achieve the purple splashes.

Jun ware is typically heavy and dense, and the bases are generally unglazed. The sedimentary clay from which they're made needs to be fired at a high temperature for a long time. The complete firing cycle of Jun ware, from building up the temperature, through saturation (the maintaining of the peak temperature) and cooling could take up to 14 days. Northern kilns were designed to manage high temperatures. They were known as *mantou*, or "bread-bun", kilns, after the distinctive shape of a typical northern food. They were tall and narrow, to make best use of the short flame length of the coal fuel with which they were fired.

Jun ware was first made during the Northern Song period (906–1127 CE) and the copper splashing was introduced in the late eleventh century. The heavy potting and simple shapes mean that Jun ware has been understood to

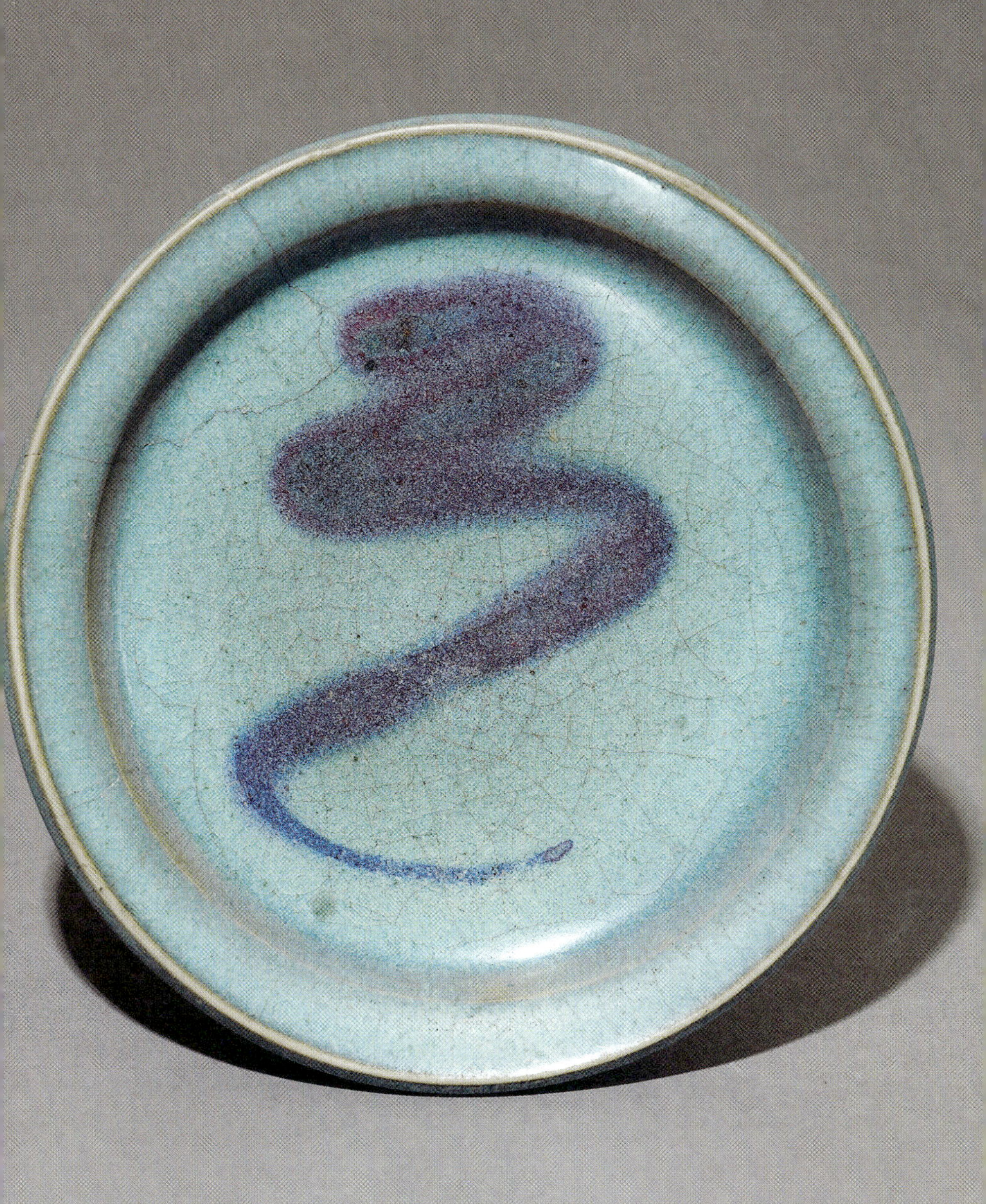

have been created for domestic use. The ware is not cited in Song dynasty records as one of the highly admired ceramic types. However, later connoisseurs appreciated its colours and shapes. Collectors began to seek it out as an example of the period often referred to as the "classical" era of Chinese ceramics, the Song Dynasty.

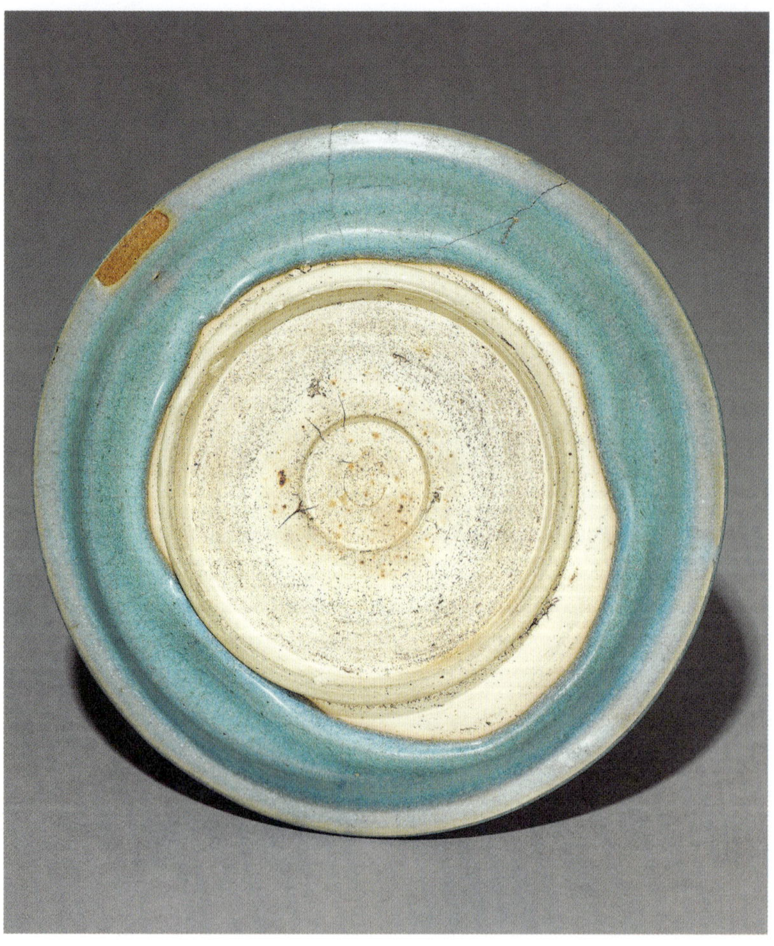

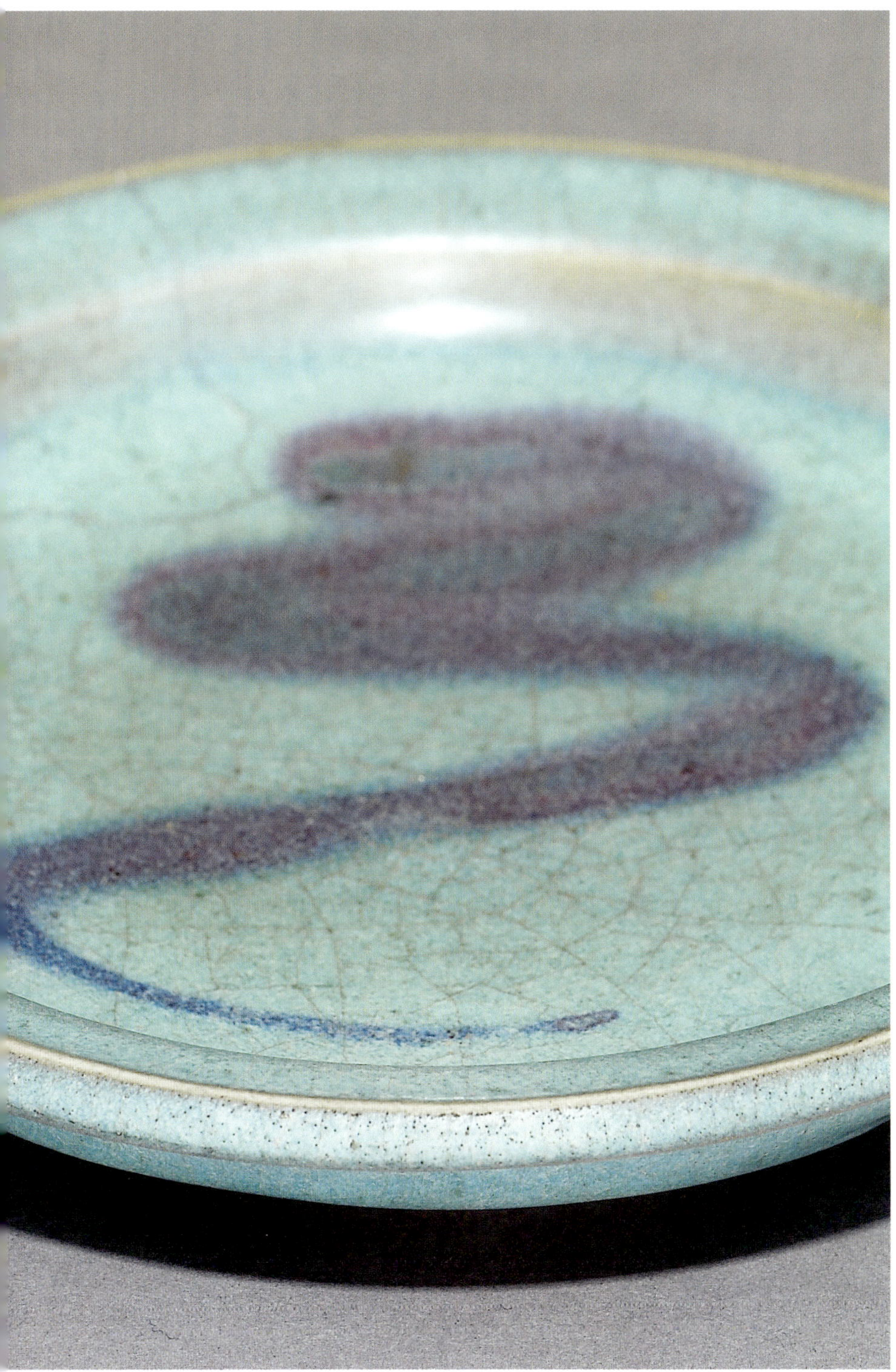

39

White ware dish
with incised lotus decoration

China
1100–1200 CE
Stoneware, incised decoration, glaze, copper rim, 21 cm

It was one of the most creative and innovative periods in Chinese history. Experts regard the Song dynasty as the era in which the foundations of modern China were established. Where ceramics were concerned, it was a high point of the industry, with kilns all across the country producing huge quantities of wares: these were destined both for markets within China and for export to East and Southeast Asia, the Middle East and even as far away as Africa.

It was also a point of transition. At the beginning of the dynasty, the most inspired and experimental potters were based in the north, but porcelain kilns dominated in the south by the end of the era.

This dish was made at the Ding kilns in Hebei province. The kilns were fuelled with coal, which gives an oxygen-rich firing atmosphere, resulting in warm tones in the glaze colours. Imitations of Ding wares, produced at the wood-fired porcelain kilns of the south, have a cooler, bluish tone. Coal has a short flame length, so kilns had to be high and narrow, the wares stacked for maximum efficiency. At the Ding kilns, the wares were piled inside earthenware saggars, or firing boxes, which could contain more pieces if they were fired upside down, resting on the rim. This is why the rim of this dish is unglazed. It has subsequently been bound in copper for smoothness and protection. The potter incised the lotus decoration by hand, probably with a sharpened bamboo tool. Later in the twelfth century, decoration was applied by moulding: the clay was pressed over a dish- or bowl-shaped mould already carved with the ornament.

Ding wares have been excavated in East Africa and in modern Iraq. They were among the most highly regarded ceramics in Song China. Some examples are incised with the character for "official" on the base; we know that Ding wares were presented as tribute at the Song imperial court. They are among the "five great wares" (the others being Ru, Jun, Guan and Ge) identified by a fourteenth-century writer and have been highly sought after by collectors.

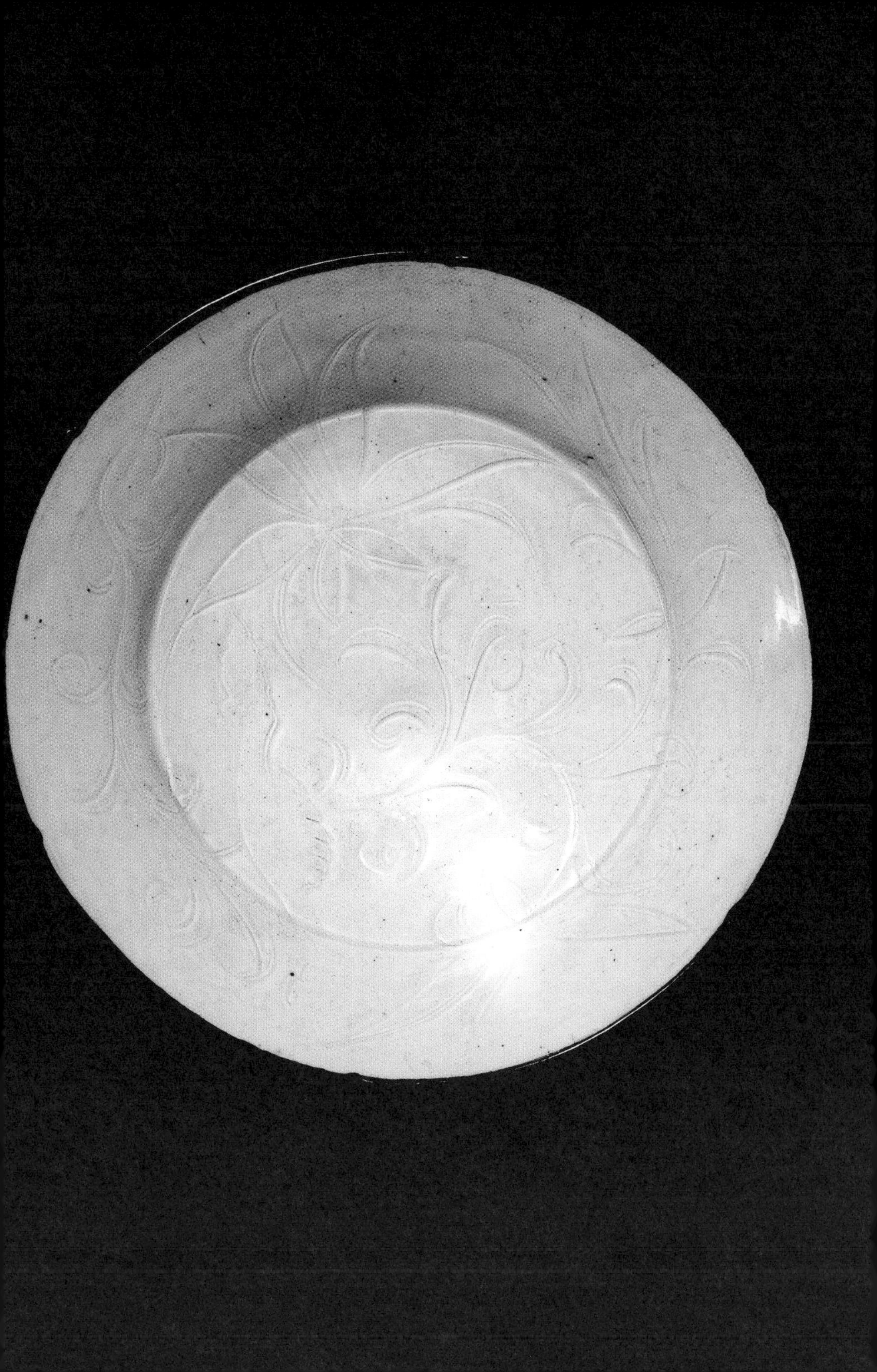

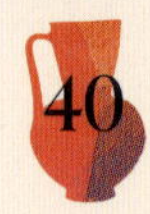

Jug with epigraphic decoration

Iran
1151–1220 CE
Fritware, moulded decoration, glaze, 21.6 cm

"Fritware", also known as stone-paste or faience, is a mixture of clay and ground glass originating in the Medieval Islamic world. The Islamic potter who made this jug began by creating a stone-paste body to imitate white Chinese porcelains. But the next thing was to add colour. First, a single colour was added to the glaze – using cobalt for blue, manganese for purple or copper for turquoise (as seen here).

In a lead glaze, copper turns green, but in the alkaline glazes used on stone-paste, it turns turquoise. This turquoise has been the hallmark of Persian ceramics and tiled domes ever since artisans started using it in the eleventh or twelfth century.

The jug's design features stylised *kufic* script, an elegant, angular script that is one of the oldest forms of Arabic calligraphy. Calligraphy was far more important in Islamic culture than in the Christian west. Muslims understood the Arabic text of the Koran to be a transcription of an eternal Koran, written in Arabic, kept in heaven. The words and letters themselves were holy. Islamic artists were strongly opposed to the use of images in a strictly religious context. So they fully explored the decorative possibilities of the Arabic script, and calligraphic designs were popular even in media, like ceramics, that were less suited to them. On this jug, it looks as though the Arabic words for "glory" and "prosperity" have been designed by an illiterate craftsman. Despite this limitation, his use of Arabic script would have appealed to his customers. Even if they, too, were illiterate, they would have understood the script to symbolise the good wishes that the potter had intended.

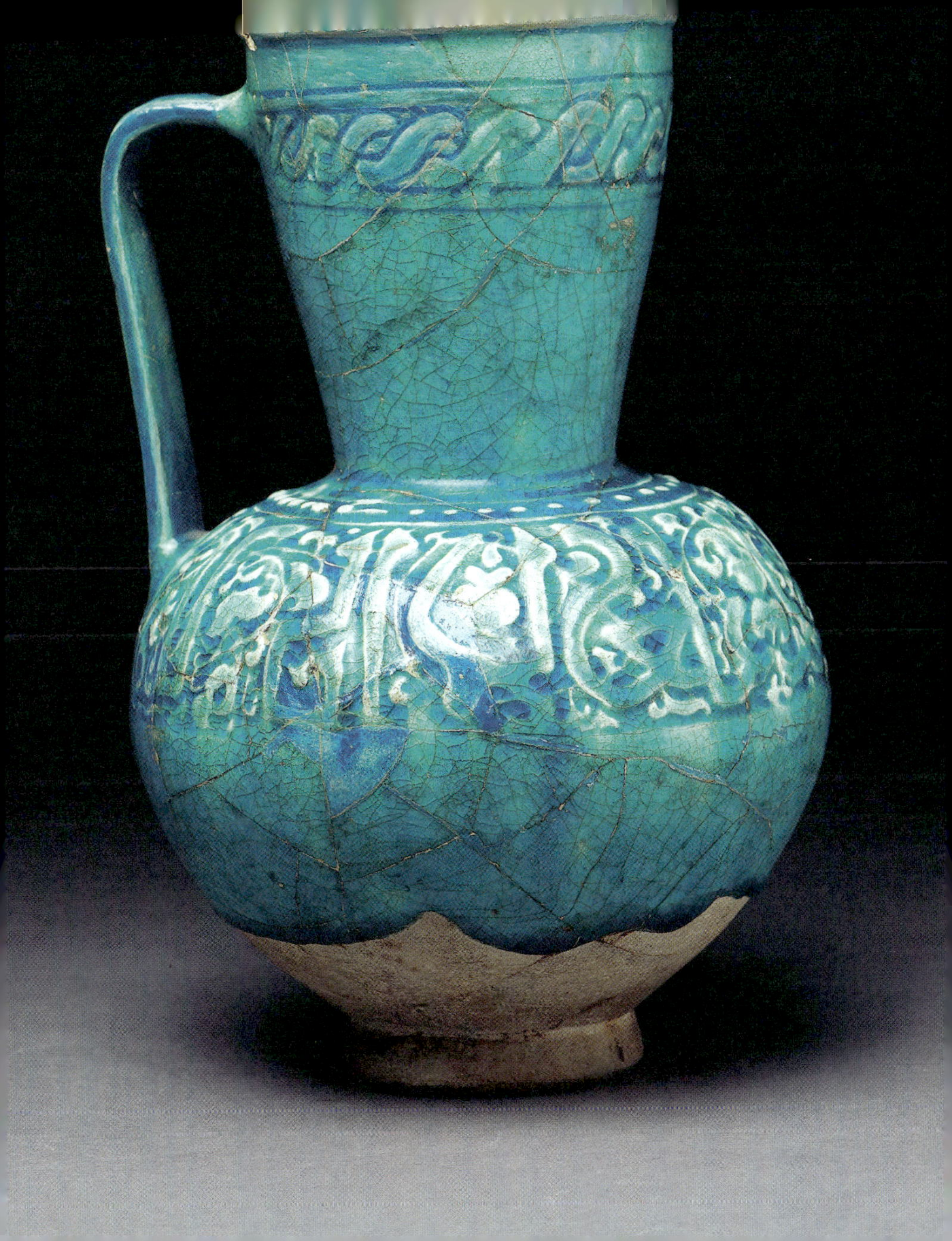

Nabeshima porcelain cup

Japan
*c.*1660 CE
Porcelain, overglaze enamel decoration, 6.5 x 6 x 5.5 cm

Some of the most elegant Japanese porcelain in existence was originally made exclusively for the Nabeshima clan. They were the governors of Hizen province in Kyushu during the seventeenth and eighteenth centuries CE.

We think that Nabeshima porcelain emerged when the production of Chinese porcelain was disrupted after the fall of the Ming dynasty in 1644. The Nabeshima lords were accustomed to presenting official gifts of fine imported Chinese porcelain to the ruling Tokugawa shoguns. Unable to get hold of it, they had to find local substitutes.

Excavations and documentary evidence suggest that the Nabeshima lords began to produce porcelain around 1650, at the Iwayagawachi kiln in Arita. This town was the centre of Japanese porcelain manufacture. In the 1660s, the lords moved their kiln to Ōkawachi, in the mountains, a few kilometres north of Arita. The idea was to separate the Nabeshima kiln from other local kilns in Arita, so that it could keep secret its techniques and designs. This porcelain cup was made at the new site.

The Nabeshima lords could afford the best materials and to hire the most skilled artisans. The kiln produced first-rate porcelain to be used by the Nabeshima family and presented to the lords of other provinces and the shogun. Nabeshima ware was not sold on the general market and other kilns were not allowed to imitate Nabeshima designs. Examples of this porcelain can be recognised by their smooth, elegant bodies, uniform shapes and meticulously painted designs in underglaze blue and overglaze enamel colours. A lot of porcelain made in Arita drew inspiration from Chinese designs, but Nabeshima porcelain used traditional Japanese motifs in striking, highly inventive decoration. These motifs were often inspired by textile or lacquer patterns of the period.

Nabeshima production was at its peak during the late seventeenth to early eighteenth century. As part of a raft of financial reforms in the 1720s, Shogun Tokugawa Yoshimune stipulated that gifts of Nabeshima porcelain should be less extravagant, with no overglaze enamel colours. After this, the designs became increasingly restrained and formalised.

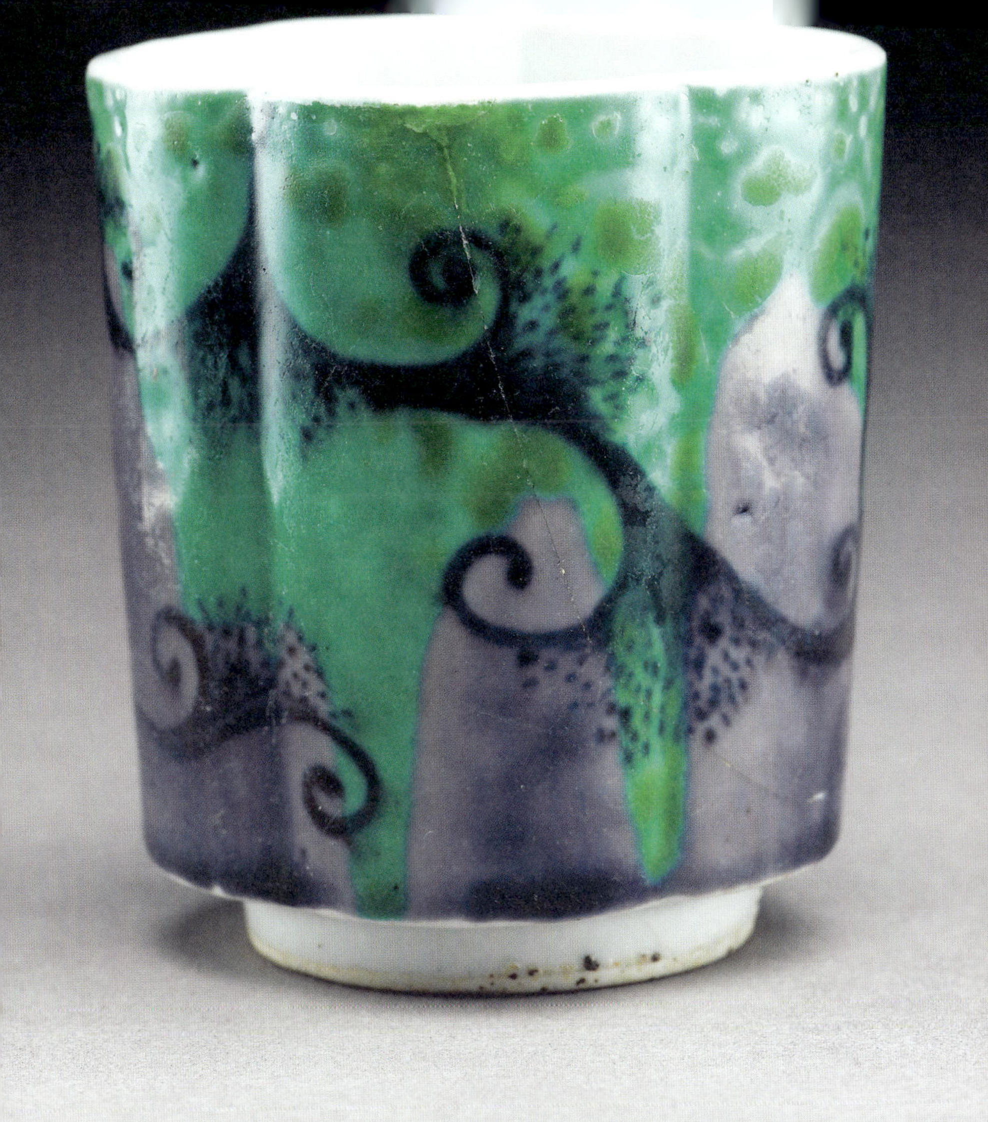

Kintsugi bulb bowl with purple and blue glazes

China
Song or Yuan dynasty,
13th–14th centuries
Stoneware, opaque glazes, gold lacquer repair, 20 cm

Glowing in gorgeous bright blue and purple hues, this small Chinese bowl is an example of Jun ware, known for the rare opacity of its blue glazes. It's a bulb bowl, used in the imperial palaces to hold plants such as narcissi. Jun ware bulb bowls were particularly admired during the Ming and Qing dynasties, and their creators also made large planters and jardinières.

This bowl demonstrates the art of *kintsugi*, an ancient "upcycling" process which originated in Japan. Examples have also been found in Vietnam, Korea and China, and the practice was not limited to Asia. *Kintsugi* means "joining together with gold", and this attractive method of repair arguably makes the piece even more appealing and precious than it was before. The shining tracery contrasts with the blue and purple glazes to striking effect.

In repairing a broken object, *kintsugi* also celebrates its history. The owner of the bowl has decided not to throw it away or attempt to hide the cracks. Instead, the golden join links past and present, connecting the moment in which we admire the bowl with the moment in which it was damaged. It indicates to us that this was a treasured object that the owner has brought back lovingly to a state of wholeness. However, we are left with something different from what the potter originally envisioned. The mend is a creative addition to the bowl's life story, both a mark of age and a symbol of recovery.

In Japan, the art of *kintsugi* is part of the Zen Buddhist concept of *wabi sabi*. This is both a world view and an aesthetic in which beauty is found in things that are simple, imperfect or impermanent. The word *wabi* means "an aesthetic ideal that finds beauty and significance in what is humble or commonplace and appears natural or artless". *Sabi* refers to "the beauty of materials or spaces which have been worn down over time to become withered and aged". These days, the philosophy means not only accepting but actively honouring change and the flaws and imperfections it brings. When the idea

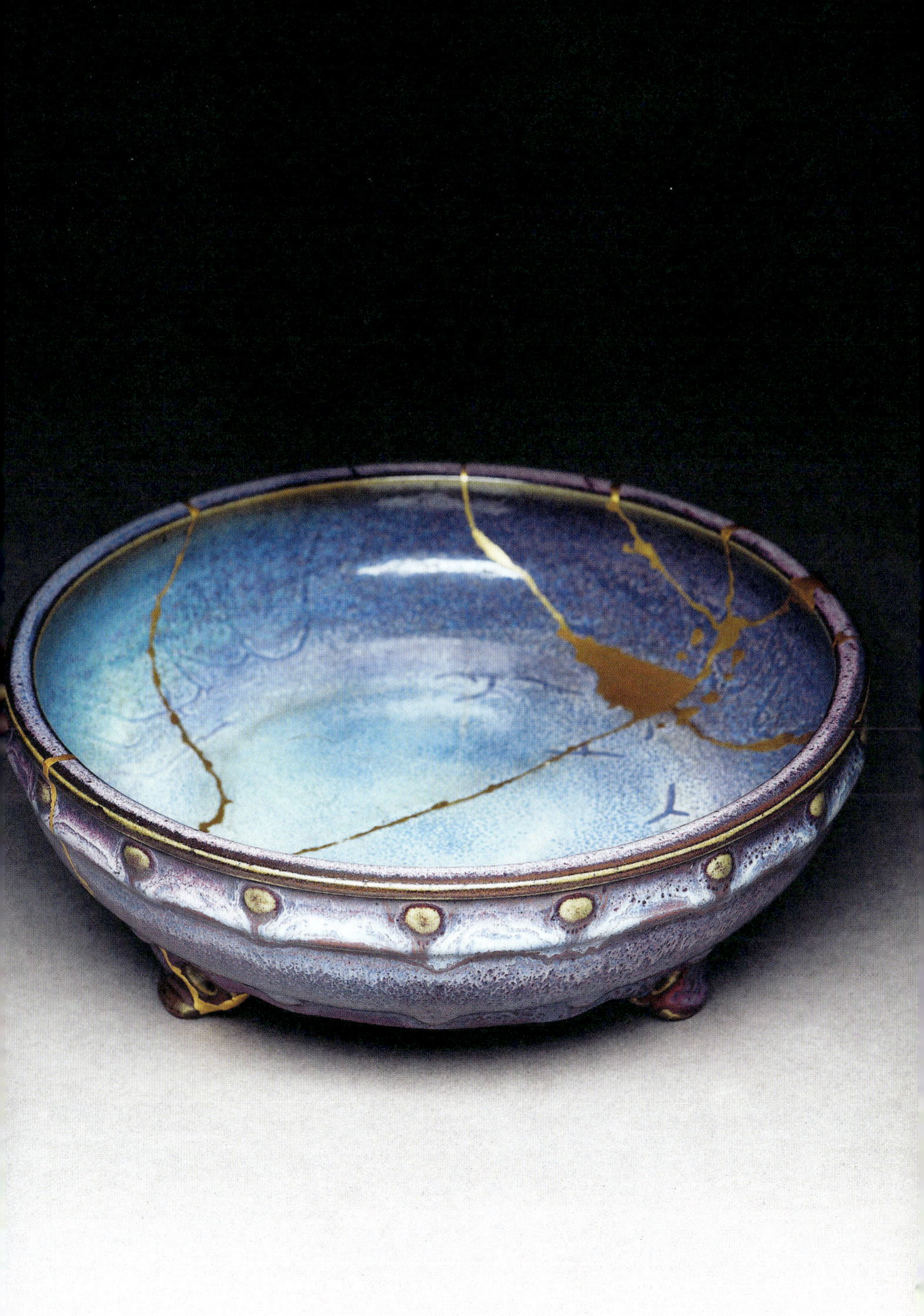

is applied to people, for example, you could say that experience makes someone more resilient and distinctive.

Kintsugi continues to be popular with artists and designers today. There has been a resurgence of interest in the practice since the COVID-19 pandemic, perhaps growing in popularity in a time when many of us sought to rebuild our lives after experiencing losses. It's easy to find *kintsugi* kits for sale, including instruction manuals and bowls to practise on before you attempt to revive your own treasures. In the context of the climate crisis, *kintsugi* seems likely to grow ever more relevant as people try to be less wasteful and more aware of the impact of human activity on our planet.

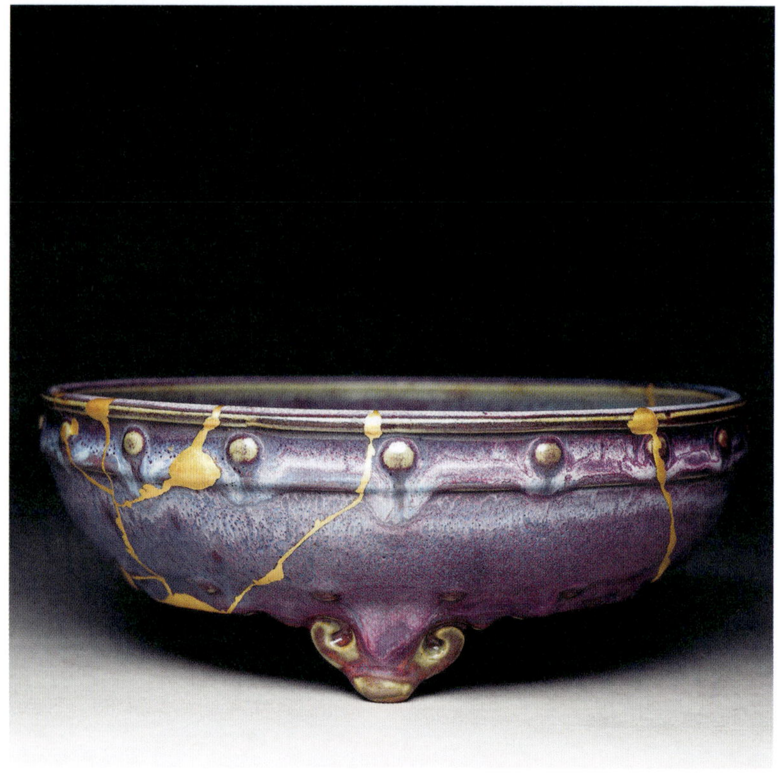

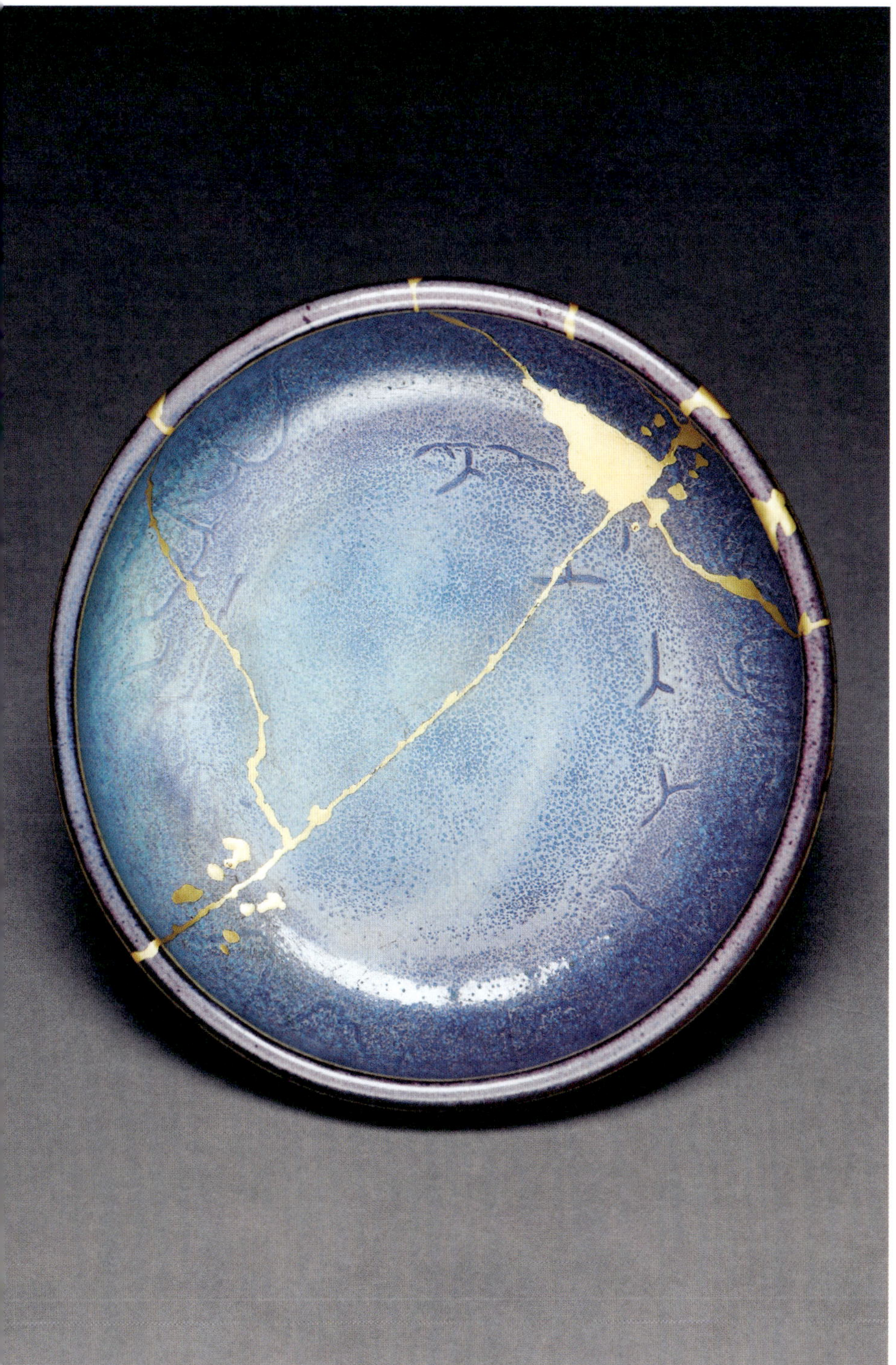

Bowl with seated figures
by a stream

Iran
1211–1212 CE
Fritware, overglaze painting in lustre, 20.5 cm

One of the most significant contributions made by Muslim craftsmen to ceramics around the world? Lustre. Initially used to decorate glass vessels, it was first applied to pottery in ninth-century Iraq. In the tenth century, the technique reached North Africa and Egypt. It then travelled back to Syria and Iran, spreading to Spain in the twelfth and thirteenth centuries. Those responsible for its spread were migrating artisans, on the hunt for new markets and job opportunities.

They were highly skilled: lustre is not only an expensive technique, but a difficult one to master. The potters diluted metal-based pigments (and other substances) and applied them carefully to pre-glazed pots, creating elaborate decorations which featured figurative, abstract and calligraphic motifs. The vessels were then placed in kilns with a reduced atmosphere to be fired a second time. During this process, carbon monoxide triggered a chemical reaction that permanently fixed the metallic oxides onto the object's surface, resulting in the distinctive sheen that characterises lustreware.

This radiant bowl, which dates to the year 608 of the Islamic calendar, is an example of the best of early thirteenth-century lustre production in Iran. Its design and execution reveal the hand of an experienced artisan. Elements of its imagery are clues to its likely origins in Kashan, a city which emerged as the most prominent centre of lustreware between the late twelfth and early fourteenth centuries. Moon-faced figures with sumptuous costumes, portrayed singly or in groups, like the ones we can see here, are typical of designs associated with this city. The figures often appear in outdoor settings, surrounded by a group of stylised plants, rivers, birds and fish which suggest a landscape. Another clue that a vessel might have been produced in Kashan is a single or double band of inscriptions on its inner and outer surfaces. These inscriptions are generally poetic excerpts from the classical Persian literary repertoire.

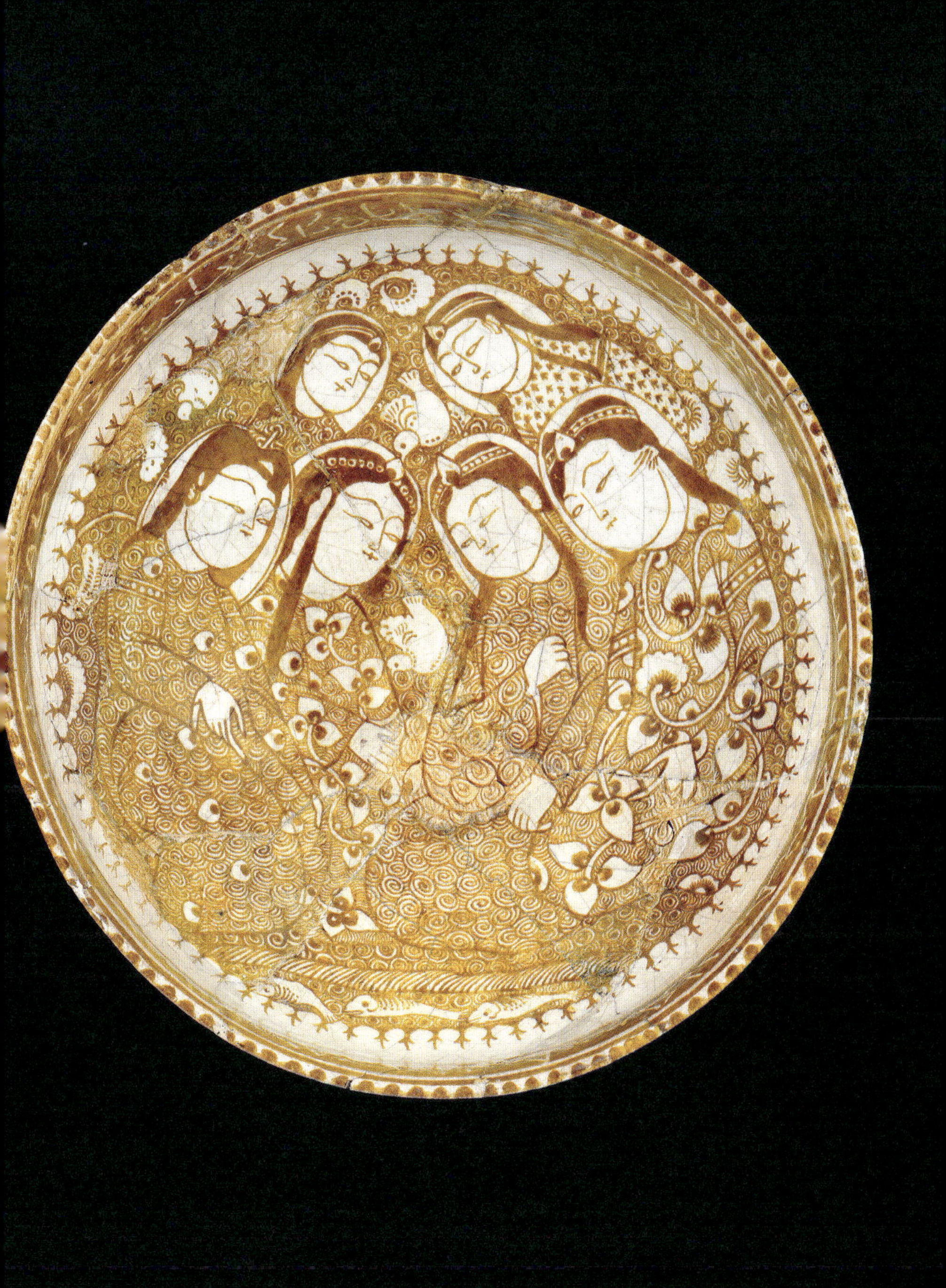

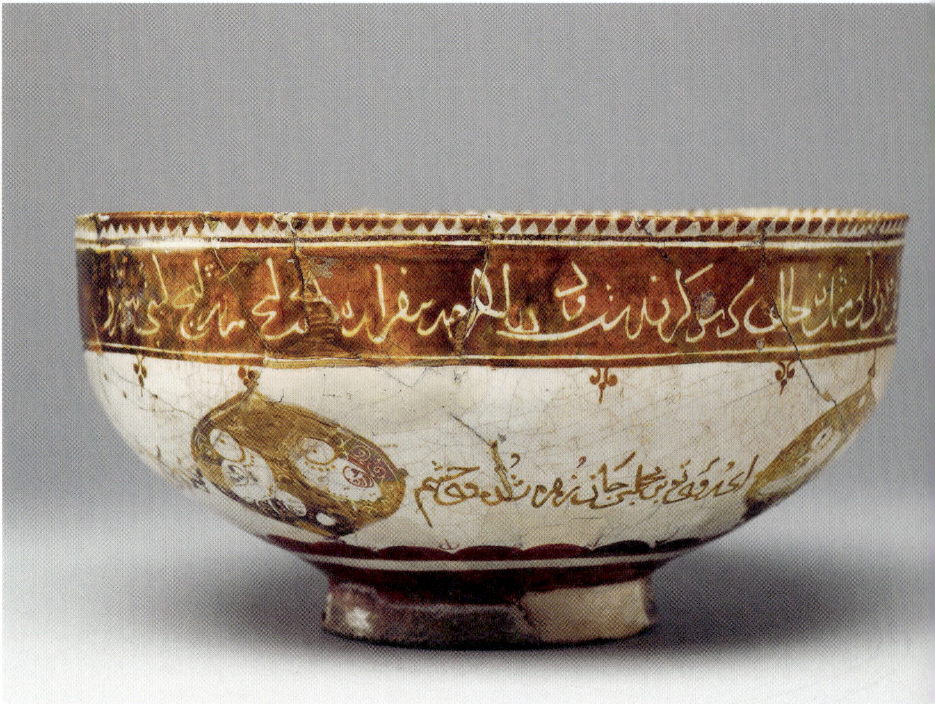

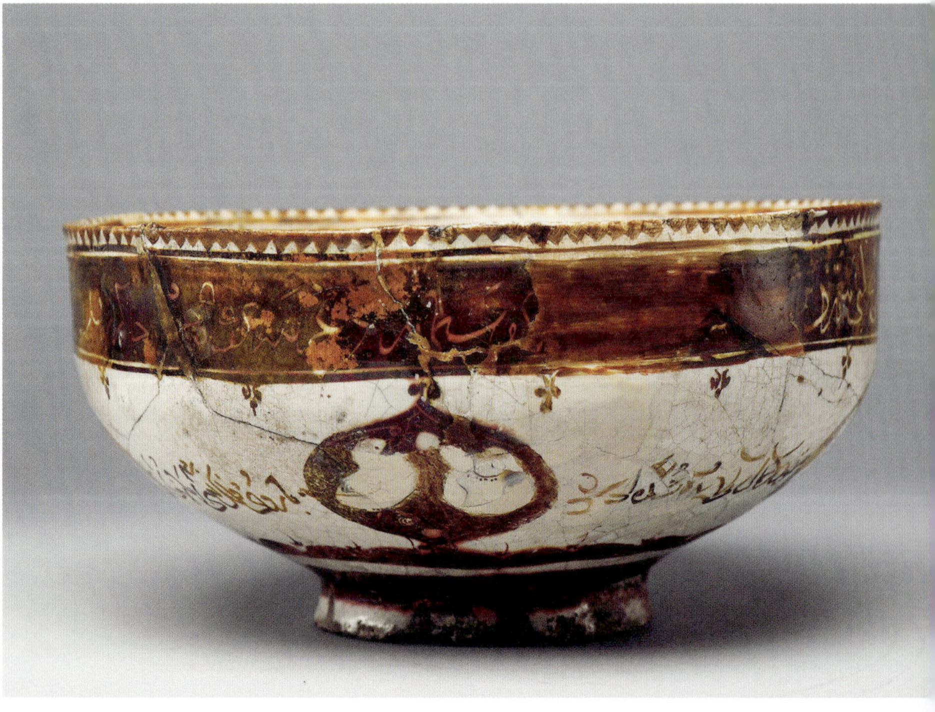

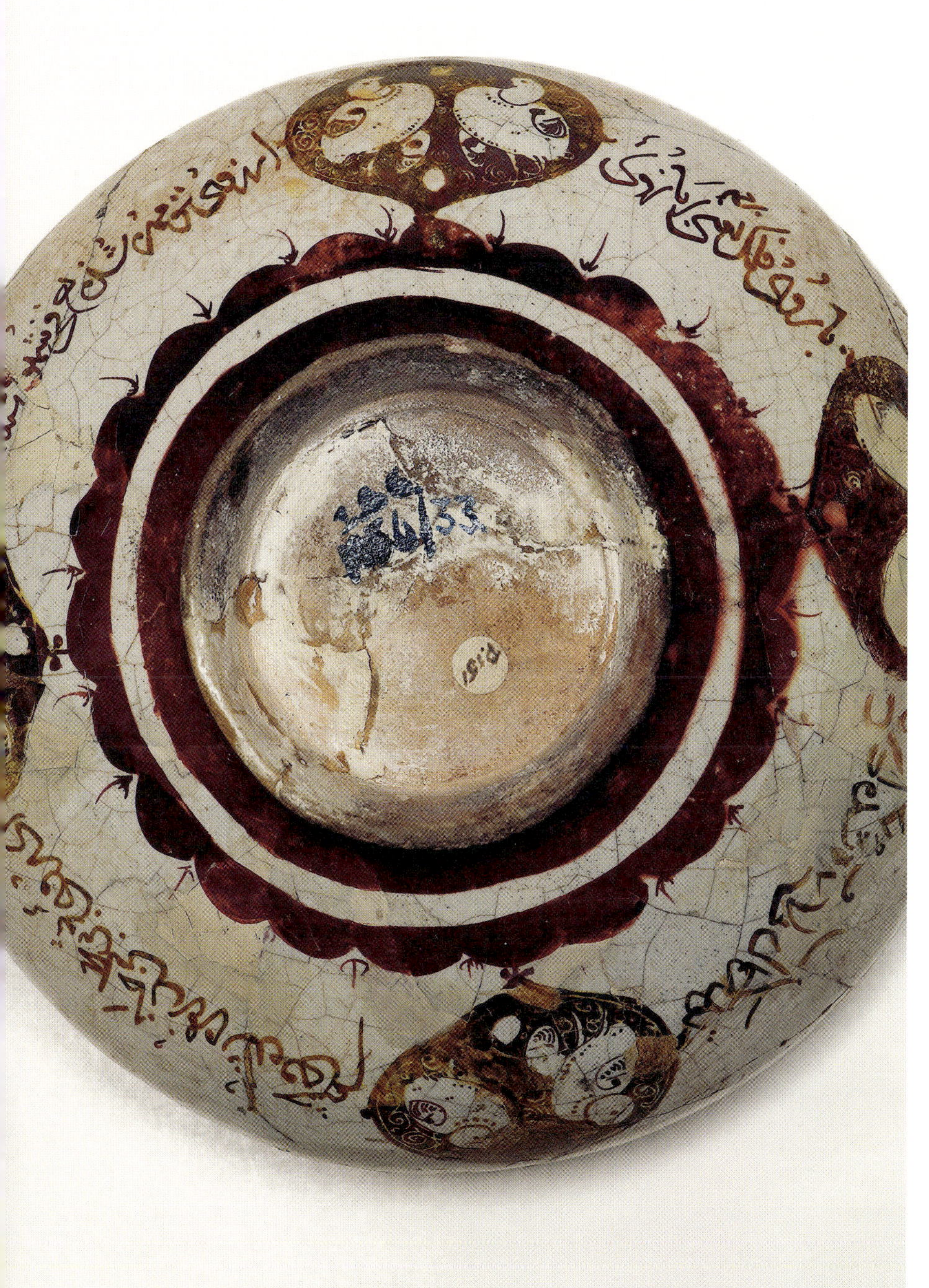

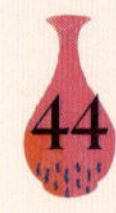

White ware vase with floral decoration

China
1279–1368 CE
Porcelain, slip-trailed and beaded decoration under
a bluish-white glaze (*qingbai* ware), 28.2 x 15.5 cm

In the mid-fourteenth century, some years before their conquest of China was complete, the Mongols established a porcelain bureau to regulate the production of ceramics in the town of Jingdezhen in Jiangxi province, south-east China. The region first produced porcelain in the tenth century and the town (literally "town of Jingde") takes its name from the Jingde reign period (1004–07) in which it was founded.

During the Ming dynasty, Jingdezhen became the site of the imperial kilns, which remained there until the end of the empire in the early twentieth century. Even now, Jingdezhen Ceramic Research Institute is a leading centre of innovation in porcelain.

In the eleventh century, during the Song dynasty (960–1279), an expanding class of scholar officials pursued painting, poetry, calligraphy and antiquarianism. Eleventh-century China enjoyed a vigorous economy, including a flourishing ceramics industry. The centres of innovation were in the north of the country. However, high-quality ceramics were produced throughout China. Consumer goods were widely circulated, raw materials were available and pottery technology was understood empirically. The artisans of Jiangxi province were one group that made exquisite pottery. They often employed techniques and styles that were inspired by the Ding ware of the north, which were used at court. At first glance, it's hard to see why the remote town of Jingdezhen, surrounded by hills, was a centre of pottery production. However, good-quality porcelain stone was available in the region, the result of ancient volcanic ash deposits across East Asia that extend across southern China and into Southeast Asia. Wood for the kilns could be obtained from forests in the hills surrounding Jingdezhen, and when the pots were finished, they could be transported across the country on the river Chang, north of the city. These are probably the reasons why the Jiangxi porcelain industry was established in such a seemingly unlikely place.

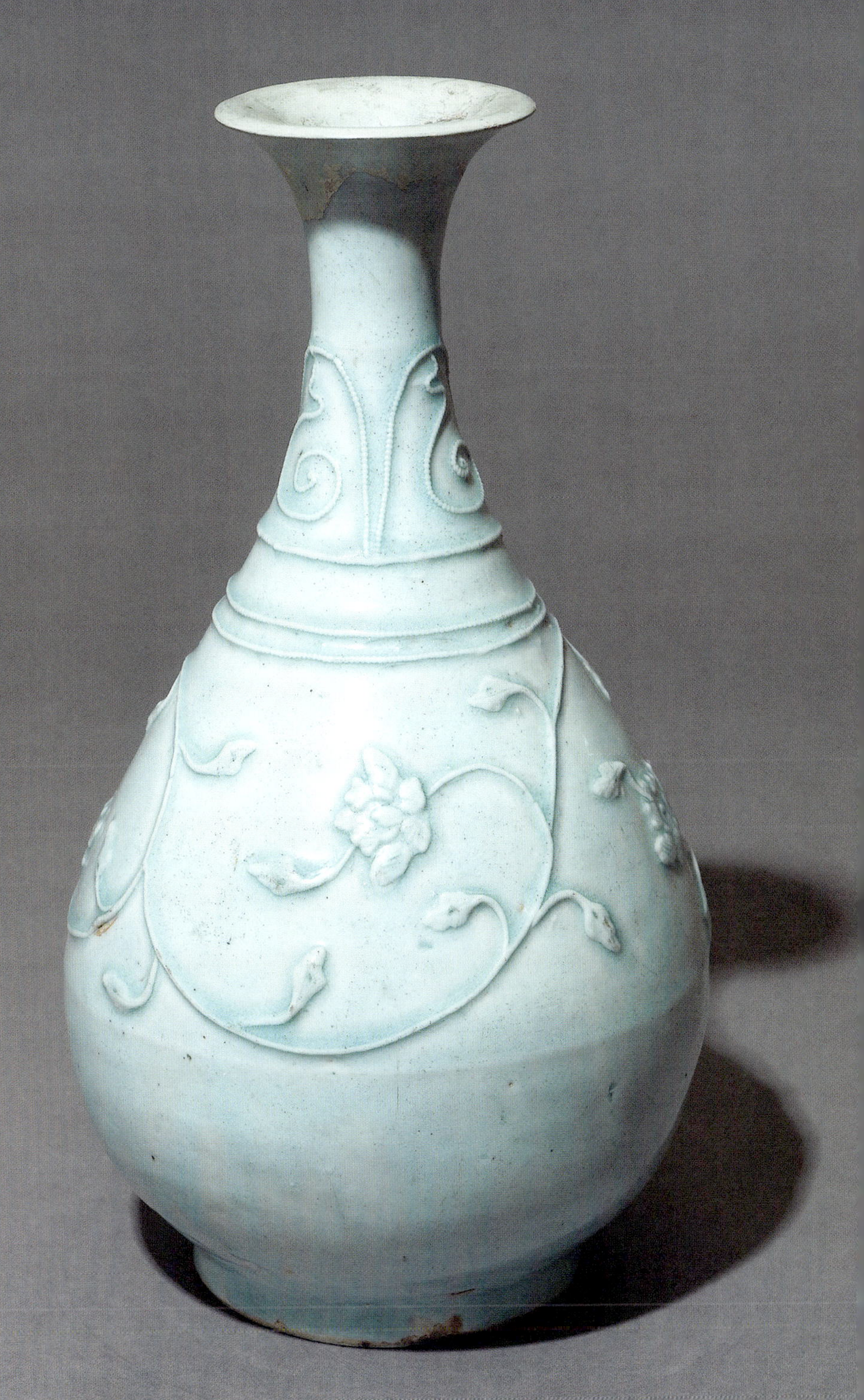

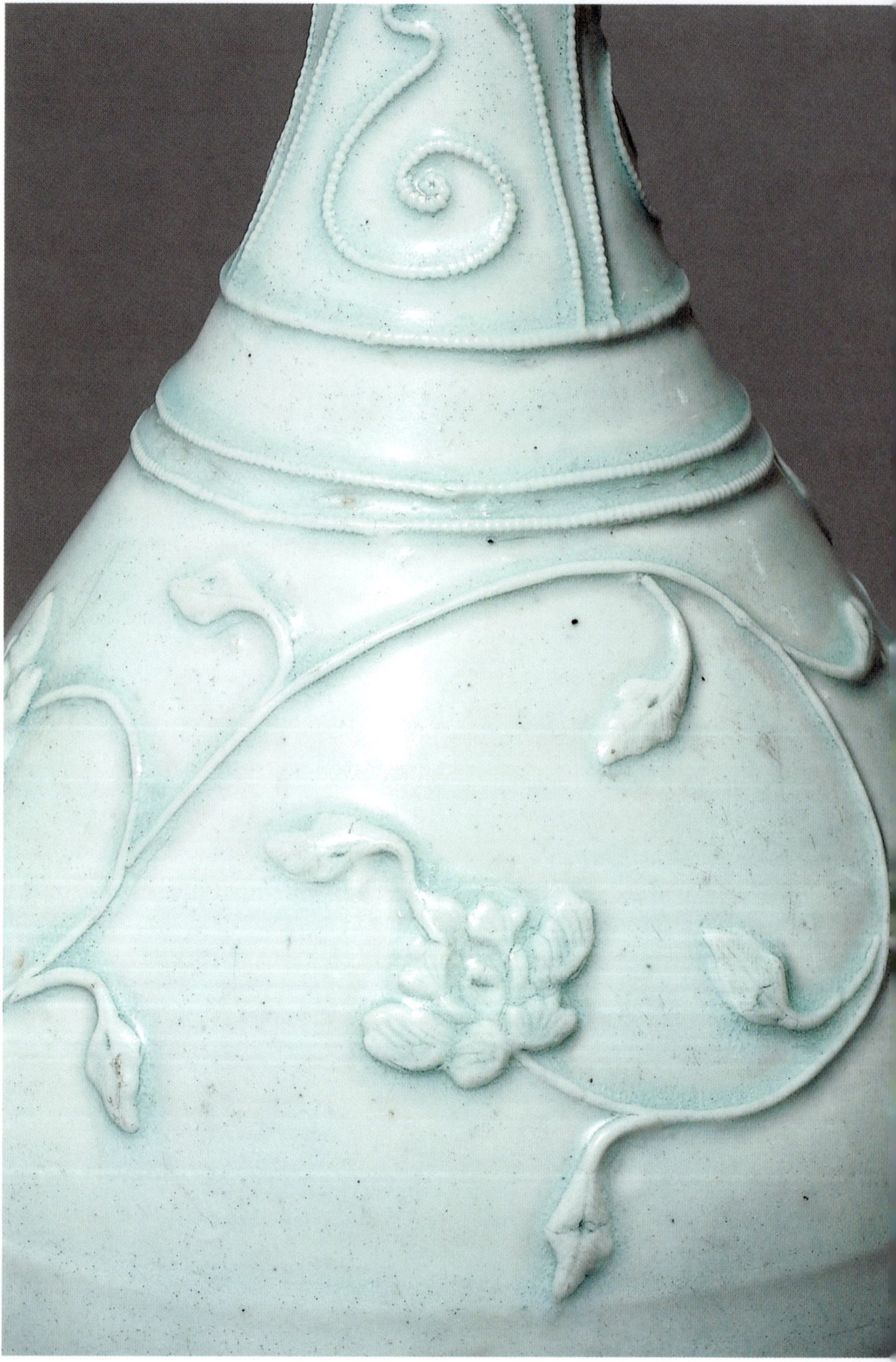

Wood fuel creates a reducing atmosphere in the kiln, resulting in cool-coloured glazes with a bluish tinge, in contrast to the warm-toned, coal-fired white wares that were produced in the north of China. Ceramics with this chilly colouring are known as *qingbai* or "bluish-white" wares. *Qingbai* wares were made for everyday use and though they're now highly regarded and collectable, at the time when they were produced, they seem not to have been particularly cherished. Even in the Southern Song (1127–1269), Yuan (1279–1368) and Ming (1368–1644) dynasties, when connoisseurs collected and often wrote about antique Song ceramics, *qingbai* wares are not mentioned.

During the fourteenth century, designs that had previously been incised under the pale blue glazes were instead painted onto the unfired body in cobalt blue and then covered with a translucent or pale blue glaze. Some decades later, pieces made in this way were decorated further, with enamel colours – which were fixed by firing a second time at a lower temperature. From that point on, improvements in the quality of Chinese porcelain were more a matter of refinement than innovation, apart from experiments with glazing in the eighteenth century. The exquisite imperial wares of the Ming and Qing dynasties had the finest bodies and the purest glazes, but the raw materials from which they were created were very close to those used in Song *qingbai* pieces. From the sixteenth century onwards, the quantities of blue and white and enamelled porcelain that informed the image of China in the West were another aspect of a monolithic industry centred in one kiln complex, which has now been active for more than 1,000 years.

Spouted bowl with two queens flanking a pillar

Italy
c.1275–1375 CE
Earthenware, unglazed, 9.8 x 29 cm

This medieval spouted bowl is originally from Orvieto in Italy. Dating to *c*.1275–1375, it's one of the earliest pieces of maiolica in the Ashmolean's collection. Its creator covered the earthenware with a glaze made opaque by the addition of tin oxide and painting – an approach which was introduced into Italy from the Islamic world around 1200.

This object, decorated with a simple palette of copper green and manganese purple-brown, is one of the earliest pieces of its kind. It belongs to a group of so-called "archaic maiolica" ceramics in the Ashmolean.

This type of pottery rarely survived above ground or intact. When this particular bowl is looked at closely by eye, or under a microscope, the individual sherds vary in surface colour and texture, which generally indicates slight differences in burial conditions. This would seem to suggest that the bowl was probably thrown away because it had been broken.

The bowl was acquired by the Ashmolean in 1954, when it underwent extensive conservation treatment. The conservators' methods reflect the materials and techniques available at the time, and a typical approach to conservation when the profession was still in its infancy. Few records exist within the museum from this period but advances in conservation materials enabled fine restoration work to be carried out and pieces were restored as closely as possible to their original appearance. An X-ray image showed the original break lines and further areas of fill, both clearly visible as patches of white. It could be seen that lost fragments, including the missing rim sections, were reconstructed in plaster. The repairs were then overpainted in the style of the original design. Although it's possible to spot the fill areas and over-painting on very close examination with the naked eye, they become immediately obvious when exposed to UV light as they fluoresce brightly compared to the ceramic. When the bowl is compared to the X-ray image, it is also clear that the overpainting or varnish extends beyond the fill area.

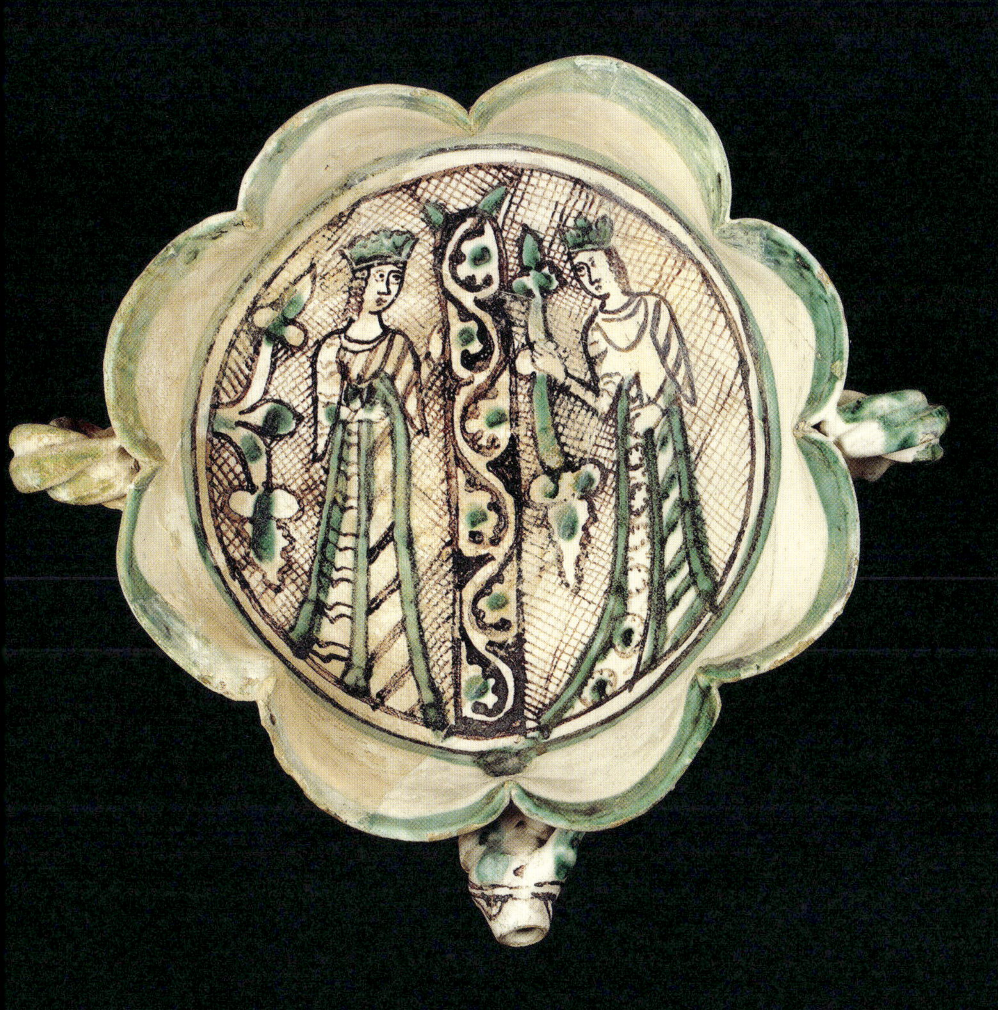

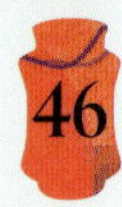

Albarello, or storage jar

Syria
1301–1400 CE
Fritware, underglaze painting in blue and black,
24.3 x 15.1 cm

This elongated cylindrical jar belongs to a group of storage vessels now known by their Italian name of *albarelli* (the singular form of the word is *albarello*). But these jars have been produced since the ninth century and their origins are in the eastern part of the Islamic world, where they may have been designed to hold pharmaceuticals or to be used in medicine.

Their popularity spread rapidly across the rest of the Islamic world before they became a sought-after export across the Mediterranean. Syrian and Egyptian earthenware drug jars are documented in Italy as early as the fourteenth century. They're recorded among the imports of the Datini firm in Prato (Tuscany), beginning in 1384, and listed in the inventories of Florentine estates. Later accounts confirm that by the middle of the fifteenth century, Damascene wares including *albarelli* were sold in Florentine shops. So, by this point, they were a popular purchase and a desirable object to have in your home. These ergonomically designed containers also frequently appeared on the shelves of Italian pharmacies or *spezierie* where they were used to hold a variety of items and liquids, as their different shapes and proportions suggest.

Islamic *albarelli* are usually decorated using a number of different underglaze techniques, but their ornamentation does not tell the viewer anything about their use or contents. As we can see in this example, the motifs instead emphasise the vessel's flaring walls, which in this case are accentuated by vertical bands of stylised floral motifs. The shoulder and neck of the jar are filled with friezes of abstract motifs and incomplete Arabic words, which echo commonly used well-wishing phrases. However, when Italian potters began to make *albarelli*, producing a variety of the containers at some of the most famous hubs of Renaissance ceramics, they eventually began to add inscriptions labelling the substances that the jars were designed to hold. The Ashmolean's collection contains a Venetian *albarello* marked with an inscription which reads "bear fat" (the Italian is "*g.d.orso*"). We know that in the sixteenth century, bear fat was used both as an ointment and for culinary purposes.

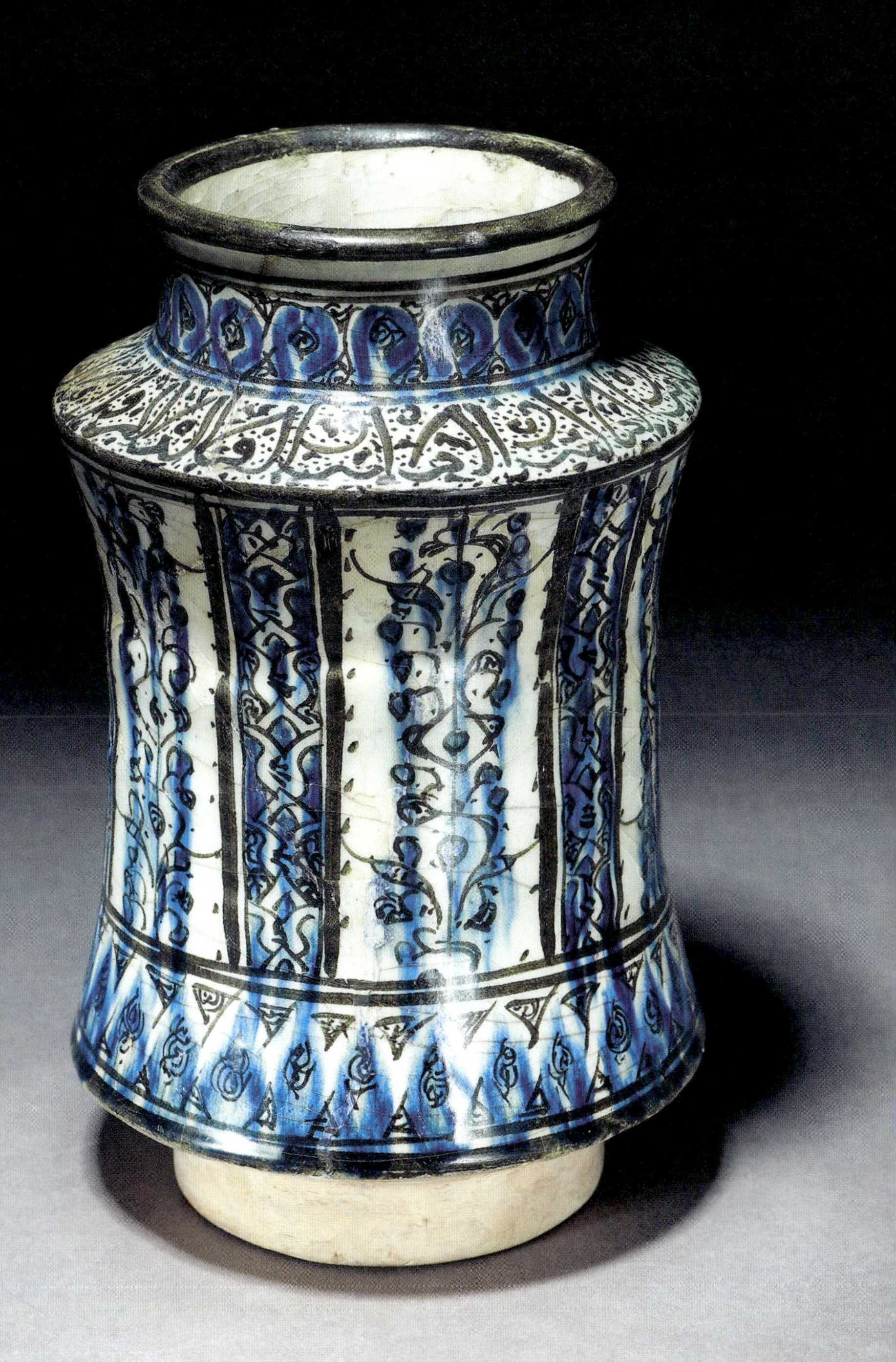

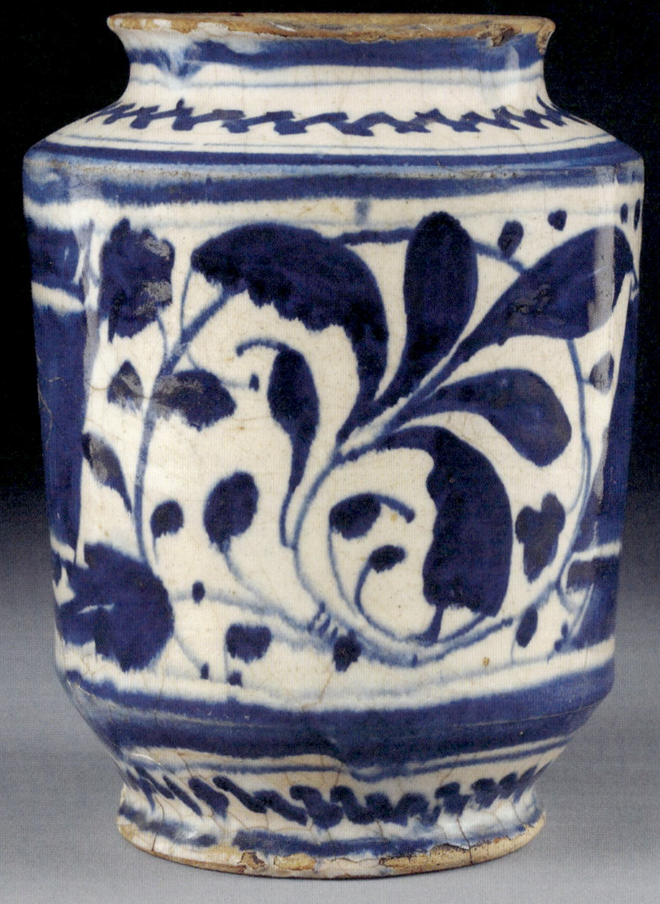

Albarello, or pharmacy jar, straight-sided, flanged at rim. Painted in blue with plant motifs and a scroll. Earthenware, tin-glazed (maiolica), blue and white, Italy, *c*.1500–60 CE, 17.4cm.

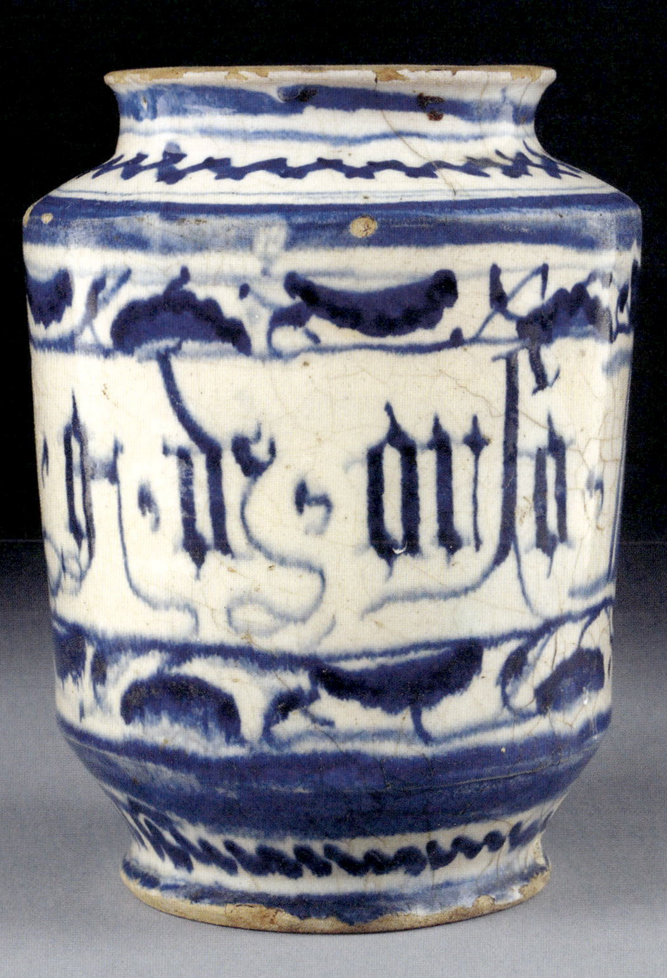

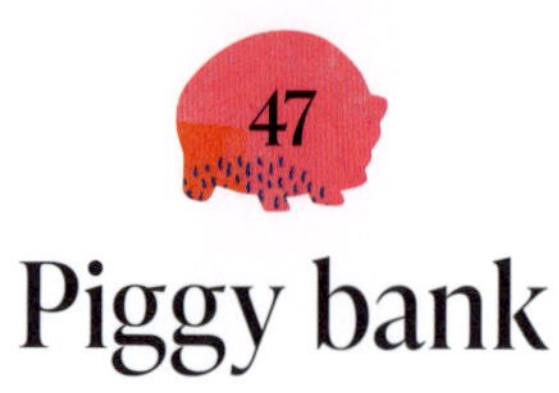

Piggy bank

Java, 15th century
Terracotta, 8.3 x 11.3 x 7.4 cm

In the Middle Ages, people stored their money in earthenware pots and other containers known as "pygg" pots. Pygg was an abundant and affordable type of clay which was commonly used for making household objects and storage vessels, including pots to store money. Over time, the spelling and pronunciation of "pygg" changed to "pig", and the shape followed suit, possibly as a play on the word by the potters.

The earliest known pig-shaped money pots were found in the twelfth century on the Indonesian island of Java. This one is made of terracotta and dates back to the fifteenth century. In Indonesia, boars and pigs were symbols of power and plenty, and so became associated with saving money. This piggy bank has the characteristic slot in the top into which money is dropped. However, it does not have the hole and bung in the bottom to get the money back out again. The only way of retrieving the contents was to literally "break the bank". It is because of this that so few intact examples of complete piggy banks survive today, making this example rare and valuable (albeit empty).

People tended to use these containers precisely so that they could not easily access the money once it had been deposited within. Its function was therefore similar to – and a forerunner of – that of a bank, and the form of a piggy bank is now universally recognised as a symbol for savings for a rainy day. Retrieving the money was an irreversible act, so required the owner to make a definite and decisive choice. Savings in the fifteenth century were, therefore, a very firm commitment.

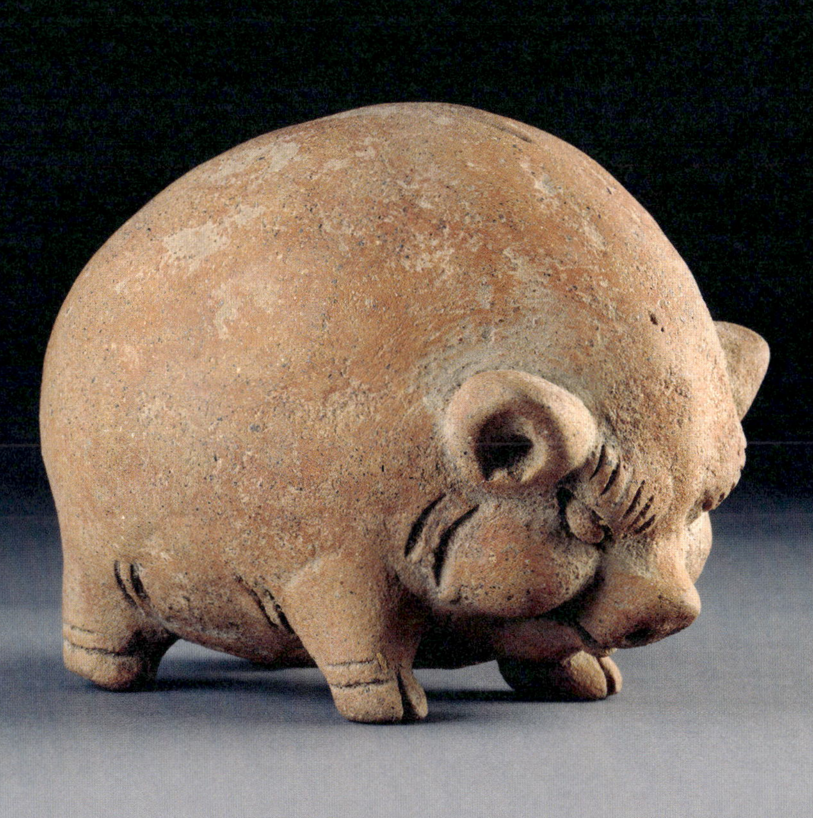

48

Plat de la Passion dish

France, 1511
Earthenware, lead-glazed, 37 cm

This exceptionally rare dish is one of a very small group of "Plats de la Passion" that were made in Beauvais, in northern France, in the early sixteenth century. Six other examples are known, all of them in French museum collections and all bearing the same dated inscription. Fragments of dishes from the same moulds were excavated in Beauvais during the 1840s and again in Nantes, in the north-west, during post-war reconstruction in the 1940s. Before the Ashmolean acquired this example, there was no comparable piece in a UK public collection.

The dish is crisply moulded around the rim with the stages of the Passion of Christ. Details of the different scenes are interspersed by crowned shields bearing arms including those of France, the Dauphin or heir apparent, and the queen consort Anne of Brittany. Moving clockwise from the actual crucifixion, we can see next an image of the pillar at which Christ was scourged, with a cat-o'-nine-tails-type whip, a birch-type bundle of twigs and the cock that crowed after Peter's third denial of Christ. In the following image, we can see the ladder used for the deposition of Christ's body, together with a hammer, a pair of pliers, a roll of linen used as his shroud and a flask for the myrrh used to anoint his body. Next, we see Christ's seamless garment for which the soldiers diced, and then the armed troop that arrested Christ in the Garden of Gethsemane, together with a lantern and the ear of Malchus, the High Priest's servant, attached to the sword with which Peter cut it off. The following image is of Judas's 30 pieces of silver, together with the lance that pierced Christ's side and the sponge on a hyssop stick used to moisten his lips with vinegar.

The central boss features the IHS monogram encircled by tongues of flame. This monogram represents the first three letters of Jesus's name in the Greek alphabet, transcribed into Latin: *iota* (the Greek letter for I, which was not differentiated from J when written in Latin at this time), *eta* (the Greek letter for E, which resembles capital H in the Roman alphabet) and *sigma* (the Greek S). Around this raised decoration are smaller shields bearing the monogram of Charles (Karolus) VIII (King of France from 1483–98) and

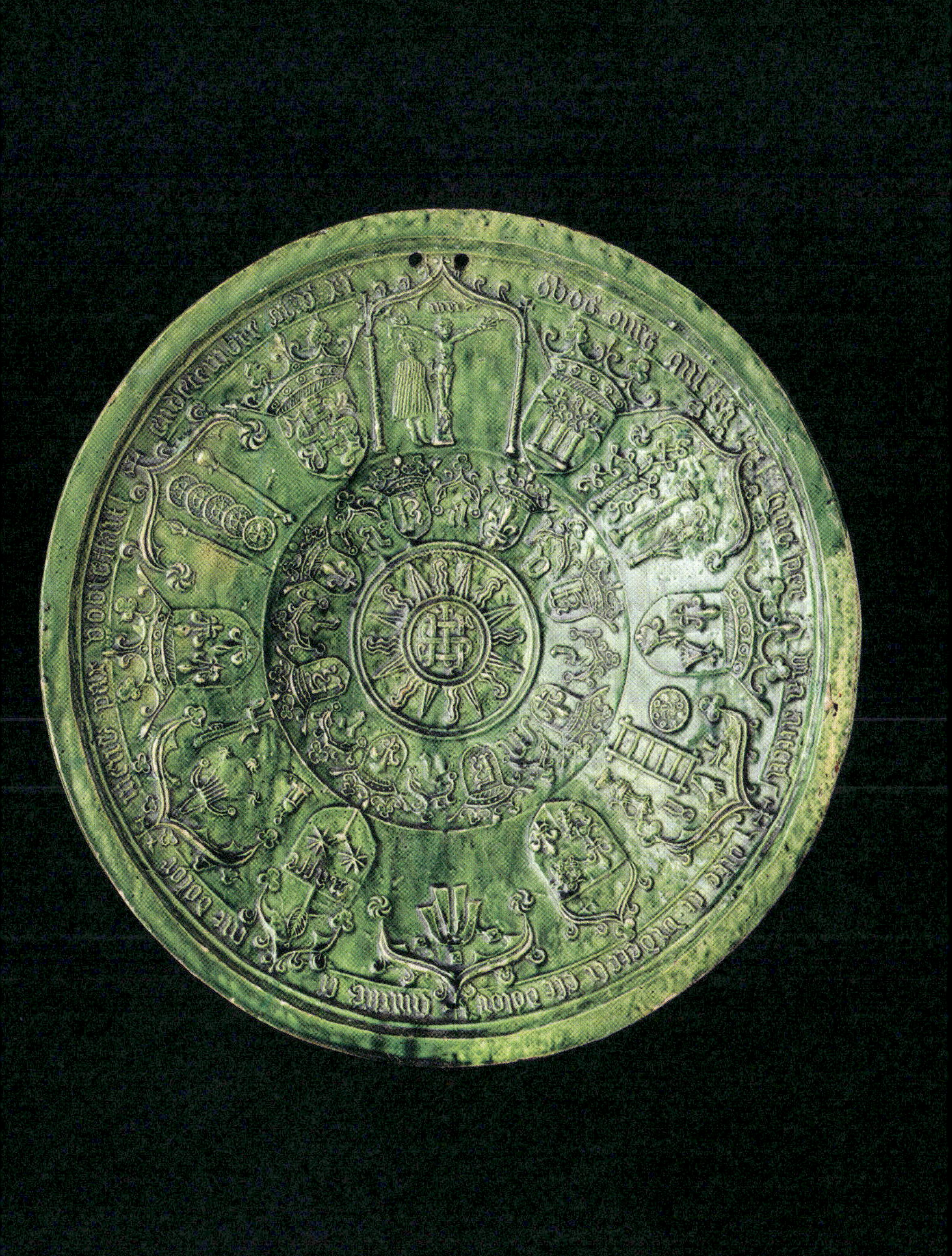

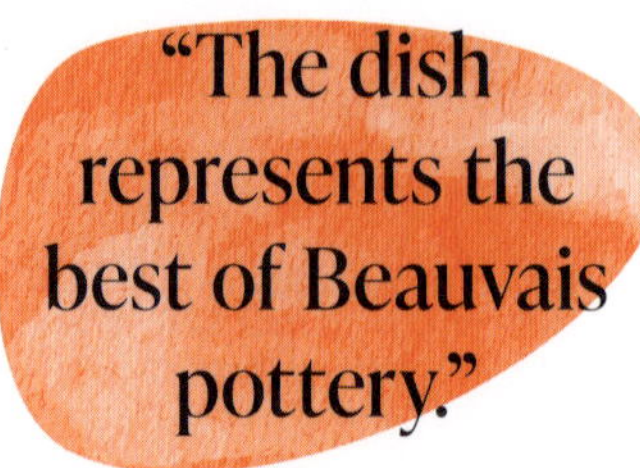

fleur-de-lys divided by the letters spelling out *AVE MARIA* ["Hail, Mary, the mother of Jesus"]. The outer rim bears an inscription in Latin and French, adapted from Lamentations 1, verse 12: "*O vos omnes qui transitis per viam attendite et videte si est dolor similis sicut dolor meus pax vobis. Fait en décembre MVCXI*" ["O all you who pass by the way, attend and see if there is any sorrow like my sorrow. Made in December 1511"].

There are suspension holes in the back of the dish, so we can tell that it was made to be hung on a wall, as an object of veneration. The arms of Anne of Brittany (1477–1514) and the cipher of Charles VIII (1470–98), together with the dated inscription, suggest that the dish may have been commissioned to commemorate what would have been the royal couple's twentieth wedding anniversary. They had married in December 1491, initially to Anne's serious reluctance. The marriage contract provided that the spouse who outlived the other would retain possession of Brittany; however, it also stipulated that if Charles VIII died without male heirs, Anne would marry his successor in order to secure Brittany for the French crown. Following Charles's death in 1498, Anne therefore remarried Louis XII after his marriage to Joan, sister of Charles VIII, had been annulled. An important patron of the arts, Anne was admired for her intelligence and shrewdness. She died aged 36, worn out by 14 pregnancies, from which only two children survived.

The dish represents the best of Beauvais pottery in the sixteenth century. The quality of clay from the Pays de Bray region and the wide range of objects produced from it made the Beauvais area one of the greatest ceramics centres in France. The iconography, heraldry and detailed inscriptions on this piece make it a fantastic teaching and research resource.

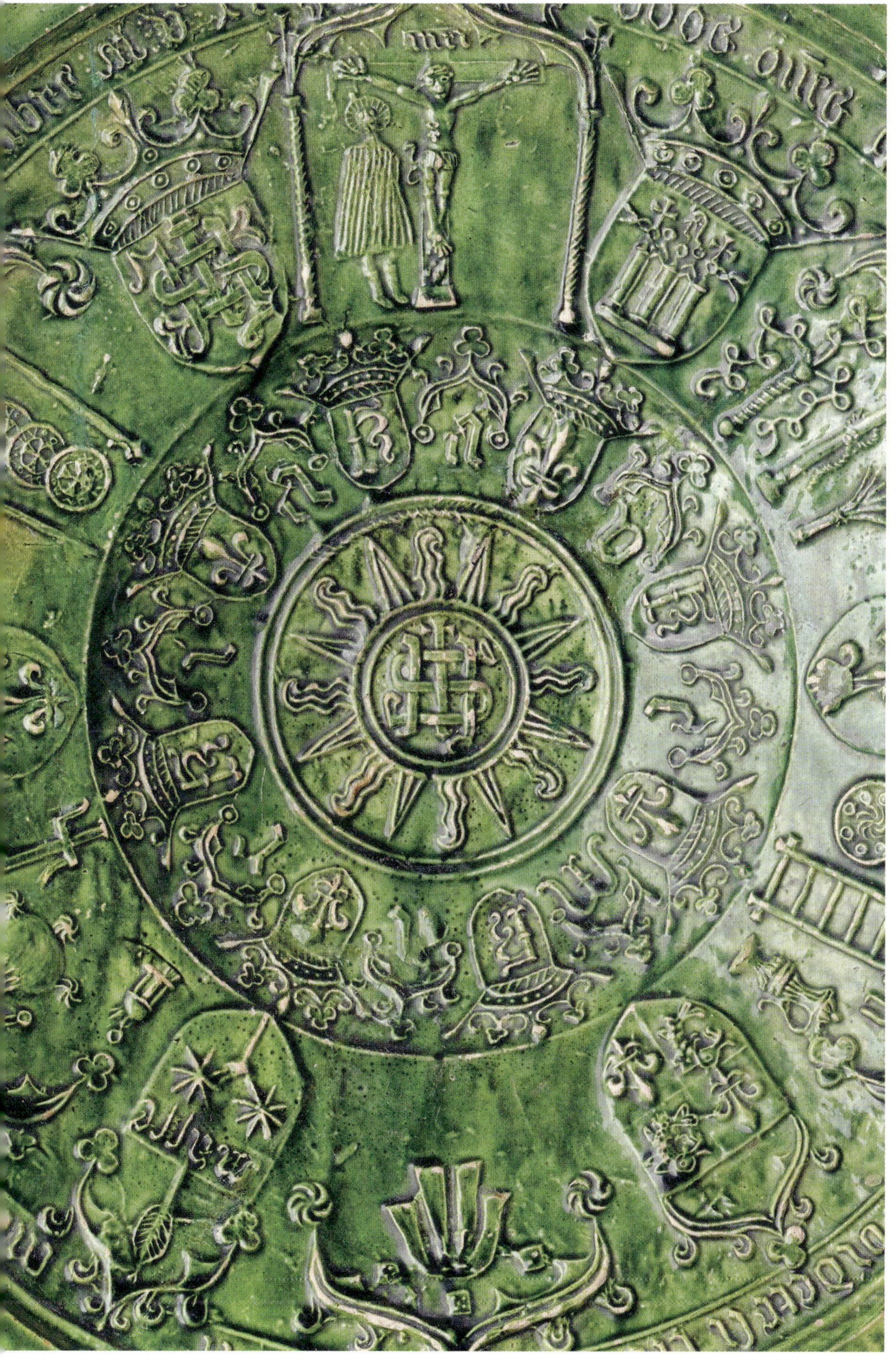

49

Dish with a composite head of penises

Italy
1536 CE
Francesco Urbini (active 1530–36)
Earthenware, tin-glazed (maiolica), 23.2 cm

This somewhat notorious Italian maiolica plate has been widely exhibited and discussed since the Ashmolean bought it, after a public appeal, in 2003. It's one of the most extraordinary images in any medium of Renaissance painting.

The plate is, first and foremost, a joke, but it's a jest which makes reference to a broader artistic context. The overall design of the piece is a parody of the Renaissance *bella donna* portraits of women, or their rarer male equivalents. The phrase *testa de' cazzi*, very similar to the modern expression "dickhead", is applied to the person represented by the head, who we may assume was the subject of an affectionate joke thought up by friends.

The surface of this piece of maiolica – a type of tin-glazed earthenware – bears a skilfully painted head made up of interlaced penises, one of them sporting a ring. On the *banderole* (long scroll) are the words: "*OGNI HOMO ME GUARDA COME FOSSE UNA TESTA DE CAZI*" ("Every man looks at me as I were a head of dicks"). On the reverse of the plate is written the date – 1536 – and "*El breve dentro voi legerite Come I guide se intender el vorite*" ("If you want to understand the meaning, you will be able to read the text like the Jews do"). This hint refers to the fact that the inscription on the scroll is written right to left, as Hebrew is.

The maiolica painter Francesco Urbini worked in Gubbio and Deruta, in central Italy, in the 1530s. This extremely rare piece is a parody of maiolica dishes bearing portrait heads of named girls – an established maiolica genre sometimes made for young men to give to the objects of their affections. It's a Renaissance joke, reminding us of both the sexual explicitness found in the writing of the satirist Pietro Aretino and the artistic ideas and interests of Leonardo da Vinci. The bowl resembles, but predates, the heads composed of vegetables and other objects which the painter Giuseppe Arcimboldo made popular at the court of Rudolf II in Prague, from 1562 onwards.

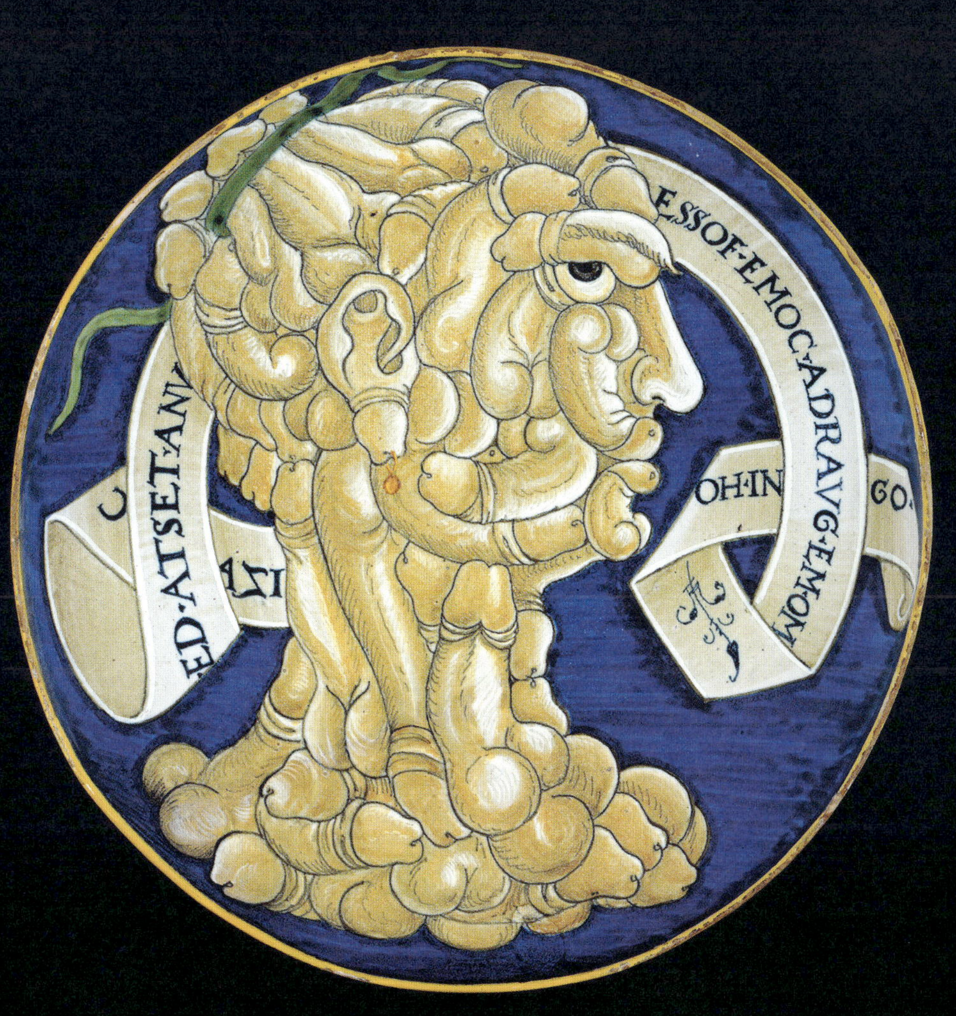

ESSOF·EMOC·ADRAVG·EMOM
OH·IN
GO·
ED·ATSETANI

Childbirth bowl and cover

Italy
*c.*1570 CE
Francesco Patanazzi (active 1560–93)
Earthenware, tin-glazed (maiolica), 15.5 x 8.4 cm

This bowl and cover are from a characteristic type of sixteenth-century maiolica: a stacking set designed as a present for a new mother. Within the bowl is a scene of a mother seated beneath a canopy with her baby son on her knee, reaching out to her. He appears to have small wings, suggesting that the figures represent Venus and Cupid.

Around the image is a triple rope-pattern border. On the inner sides of the bowl are grotesque figures made up of birds, vases and winged monsters. On its outer sides are grotesques incorporating winged monsters and un-winged herms in striking headgear (herms are male figures made up of a head and upper torso on a column). There are leaping lions, and animals that may be intended to represent lionesses (though they also resemble dogs). The foot is painted with a band of linked discs.

The bowl's cover has a figure at the centre of its upper surface: Cupid, reclining on a rock, wearing a quiver of arrows. Circling this central roundel are grotesques – two pairs of dogs; fantastic female half-figures holding up swags; and insect-like creatures. Around this, outside the "flange" or projecting rim, is a broad band of rope ornament flanked by two bands of simpler ornament on the inside and two on the outside. On the lower surface is a central roundel painted in orange-brown grisaille, a style of painting that imitates sculpture. Winged Cupid stands in a rocky landscape, holding a bow and arrow. Outside this centrepiece are grotesques: herm-creatures with exotic headgear, fantastical birds and fine scrollwork.

In Renaissance Italian society, childbirth was celebrated with rituals and the production of a wide range of material goods in honour of the occasion. These were often consumable delicacies, but they could be presented in or on objects that had an afterlife and might be treasured as family possessions. In the fifteenth century, painted wooden trays were a popular gift. After 1500, as the sophistication of maiolica painting developed, a fashion arose across Italy

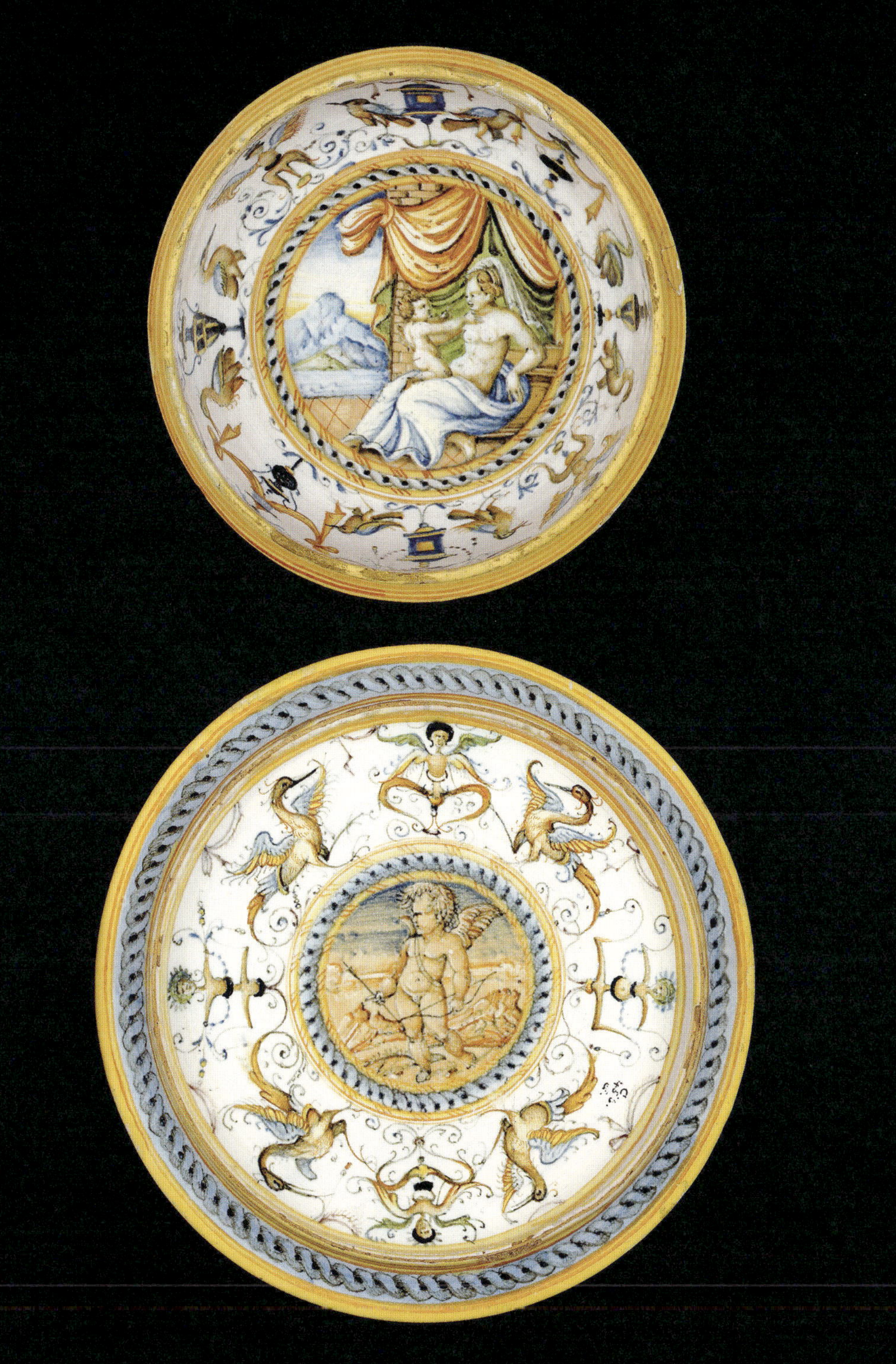

for a particular kind of childbirth gift made of maiolica: a set of bowls, covers and stands which could be stacked up on top of each other. No complete Renaissance set has survived. One of the main production centres for childbirth sets of this sort was Urbino. It became customary in Urbino potteries for childbirth sets to be painted, on one or more surfaces, with scenes specifically of childbirth. Sometimes these scenes show the actual moment of childbirth, with the mother supported and encouraged by a group of other women; on other examples, there are images of family life with the new baby swaddled and cared for. The scenes are celebrations of womanhood and of the nursery – fathers are rarely present.

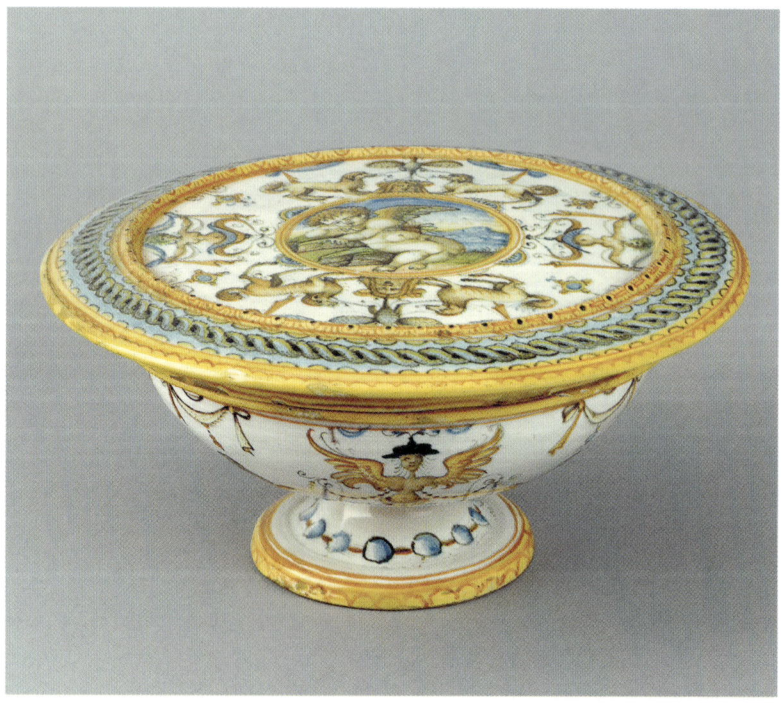

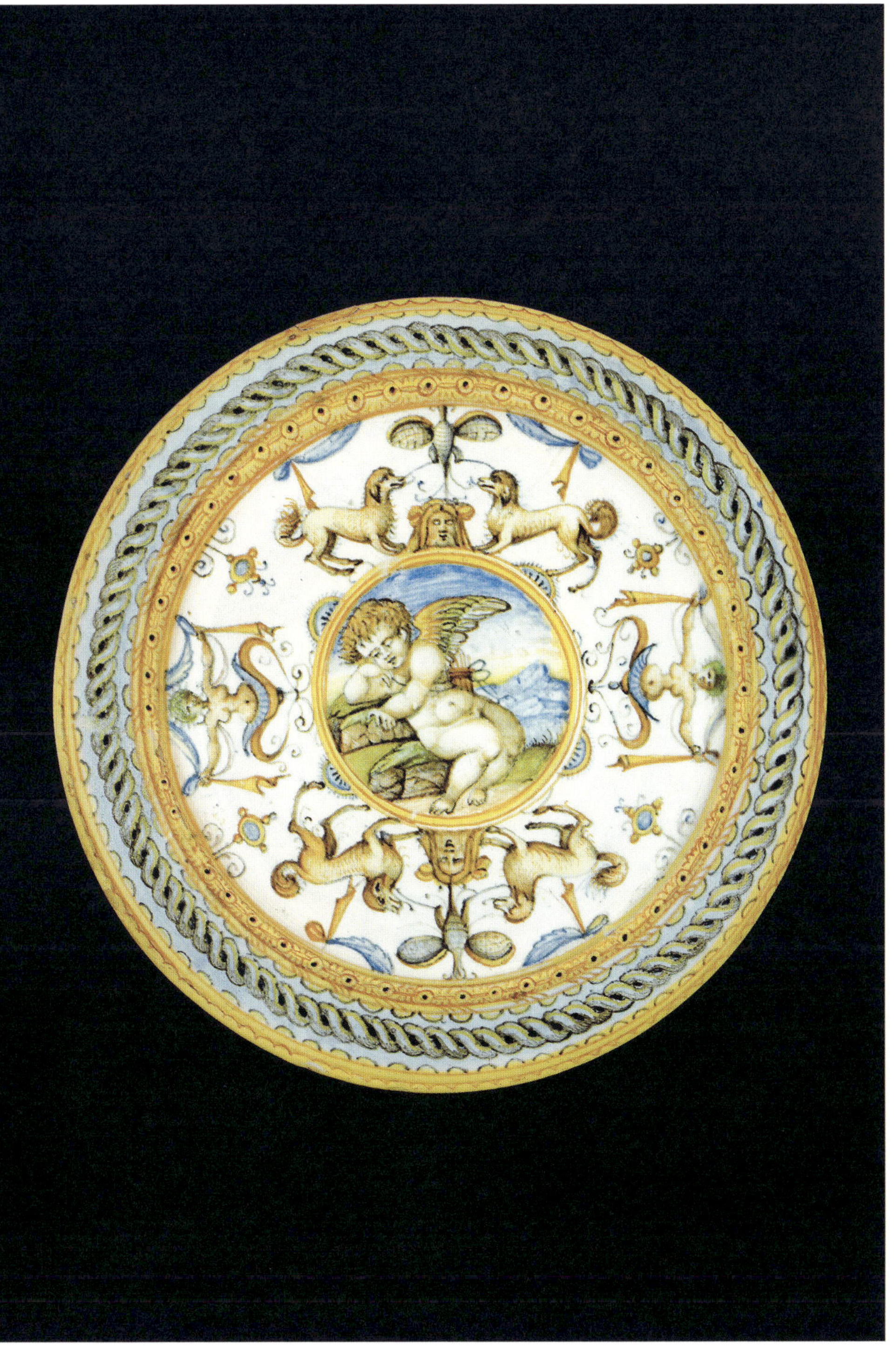

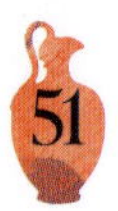

Porcelain ewer

Italy
*c.*1575–87 CE
Soft-paste porcelain, moulded with underglaze,
18.5 cm

This ewer is one of only 70 or so examples of Medici porcelain known to have survived the march of history. Medici porcelain isn't truly porcelain, but a material created in an attempt to copy Chinese ceramics before Europeans discovered the secret of how they were made. Since vessels like this were rediscovered in the 1850s, Medici porcelain has become one of the most admired categories of European ceramics.

In medieval Europe, Chinese porcelain was treasured because it was so rare. But, by 1500, it was finding its way through Islamic trade routes into Italy, where artisans were impressed by its hardness and translucency. Records show early efforts to recreate porcelain taking place in Venice, and further experiments are documented in Lodi, the duchy of Urbino, and Turin. Europeans probably achieved the first successful imitations of Chinese porcelain in Ferrara, around 1561, under the patronage of Duke Alfonso II.

However, the earliest surviving examples of European "porcelain" were made in Florence, in the workshop of Francesco de' Medici, Grand Duke of Tuscany. De' Medici was an enthusiastic experimenter across science, alchemy and art. In 1575, the Venetian ambassador in Florence recorded that: "Grand Duke Francesco de' Medici has found the way of making Indian porcelain, and in his experiments has succeeded in equalling its quality – its transparency, hardness, lightness and delicacy."

It wasn't until the early eighteenth century that Europeans discovered how to make real, hard-paste porcelain in the Chinese manner. To make real porcelain, you need kaolinitic clays and kilns capable of very high temperatures, and these were not available in sixteenth-century Florence. The ceramic body of this flask combines elements from Islamic pottery and Italian maiolica. Medici porcelain was painted under a transparent glaze, unlike maiolica, which was painted on top of the glaze. But the main inspiration for the decoration of the flask was Chinese porcelain, of which the Medici had a large collection. It was also influenced by Islamic blue and white pottery, itself imitating Chinese models.

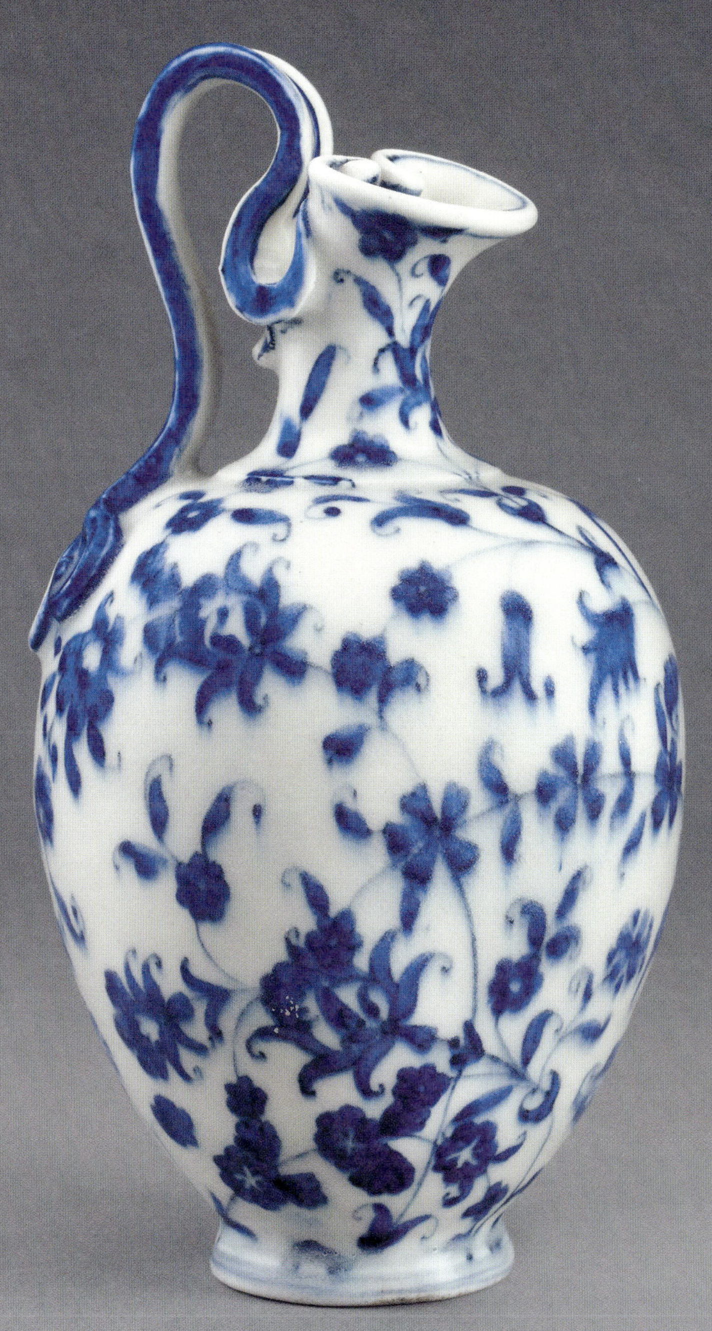

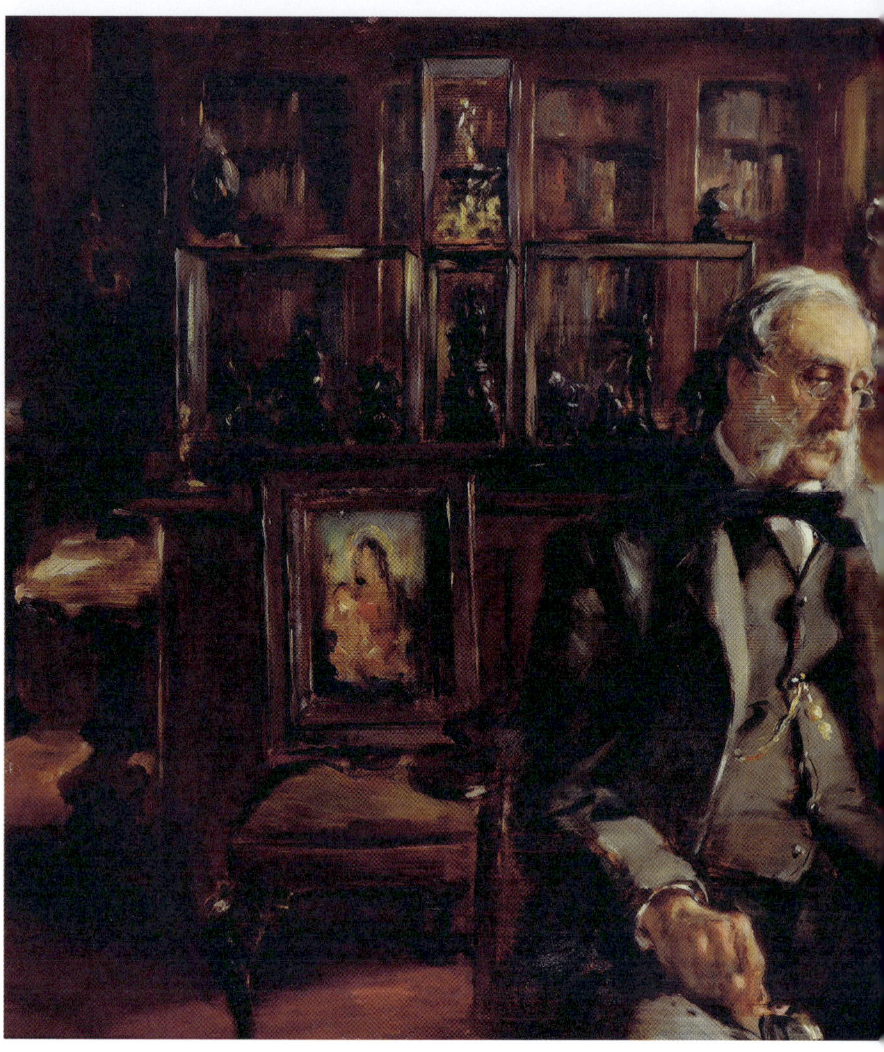

The shape of this piece, and others like it, does not imitate a Chinese model. It owes more to maiolica and to the goldsmiths, hardstone workers and glassmakers who worked alongside the potters in Francesco's workshops. The temperatures needed to fire the vessels meant that the potters were working at the limits of Italian kiln technology. Many of the surviving pieces show firing defects. However, this piece, apart from a slight running of the blue, is virtually flawless. Most of the surviving pieces that have been identified, including this flask, bear a mark consisting of a cupola and the letter F, for "Florence" (or, just possibly, for "Francesco"). The cupola represents the

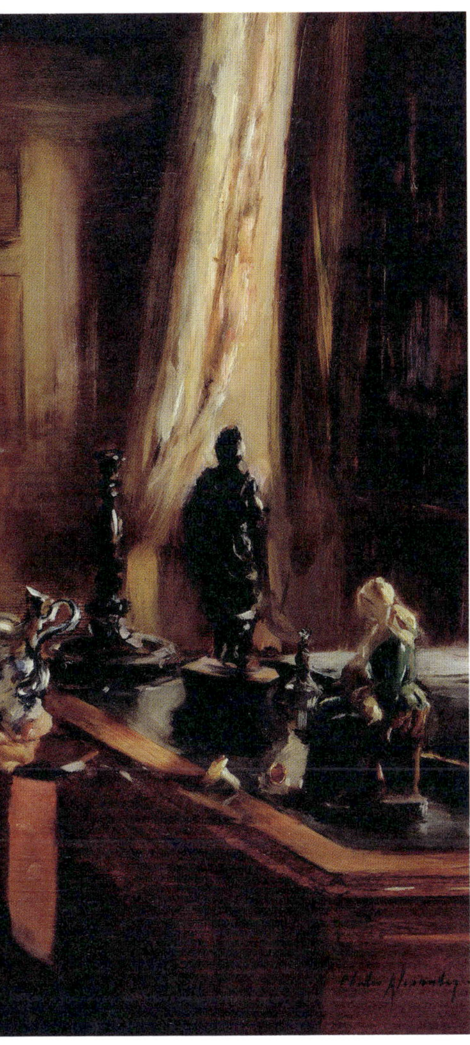

dome of Florence Cathedral. In the end, porcelain production in Florence petered out after Francesco's death in 1587. His project doesn't seem to have influenced the direction of later European porcelain.

In the late nineteenth century, Medici porcelain was intensively sought after by collectors. This flask was bought in Naples in 1879 by CDE Fortnum, whose gifts and bequest to the Ashmolean are still the basis of the museum's collections of Renaissance sculpture and decorative arts. Fortnum gave a major part of his collection to the museum in his lifetime, but he kept this favourite ewer until the end of his life.

Shino-style serving dish for the Japanese tea ceremony

Japan
*c.*1590 CE
Stoneware, underglaze blue, 5.8 x 24.6 x 21.9 cm

This is a decorated Shino stoneware dish, a type of ware that was made at kilns in Mino province in central Japan from the late sixteenth century onwards. Shino ware was the first type of Japanese ceramic ware to be emblazoned with painted designs. They were mostly made for use in the Japanese tea ceremony.

The Japanese had drunk powdered green tea as part of a formal ceremony since the thirteenth century. By the fourteenth century, the ruling samurai warrior class had taken up the tea ceremony. They drank tea as part of great feasts held in large meeting halls, accompanied by displays of imported Chinese treasures that showed off the elite status of their owners.

In the sixteenth century, there was a shift away from this flamboyant type of tea ceremony and a new style of tea drinking, known as *wabicha*, emerged. *Wabi* was a term adopted from poetry, meaning quiet or sober refinement tinged with melancholy – the discovery of beauty in the humble, simple and imperfect. Tea gatherings were transformed from a lavish form of entertainment and display into an inward practice emphasising aesthetic and spiritual awareness. Where they had previously used expensive Chinese tea ware, now practitioners incorporated local, rustic-looking utensils into the tea ceremonies.

With a growing demand for Japanese-style tea ware, domestic ceramic production increased at kiln sites such as Mino. Shino ware pieces were admired by tea masters because of their simple, freely applied decorative motifs. The tea masters appreciated the pleasingly uneven texture and the milky-white colouring of the thick, semi-translucent feldspathic glaze, as seen in this rectangular dish. It has foliated corners. It stands on four short feet and was made by pressing a clay slab over a mould. Its painter chose a design of flowering grasses in underglaze cobalt blue, which was probably inspired by blue and white porcelains made in southern China for the Japanese market. The dish would have been used to serve food at the *kaiseki* meal (a series of small, intricate dishes) that took place as part of a tea gathering.

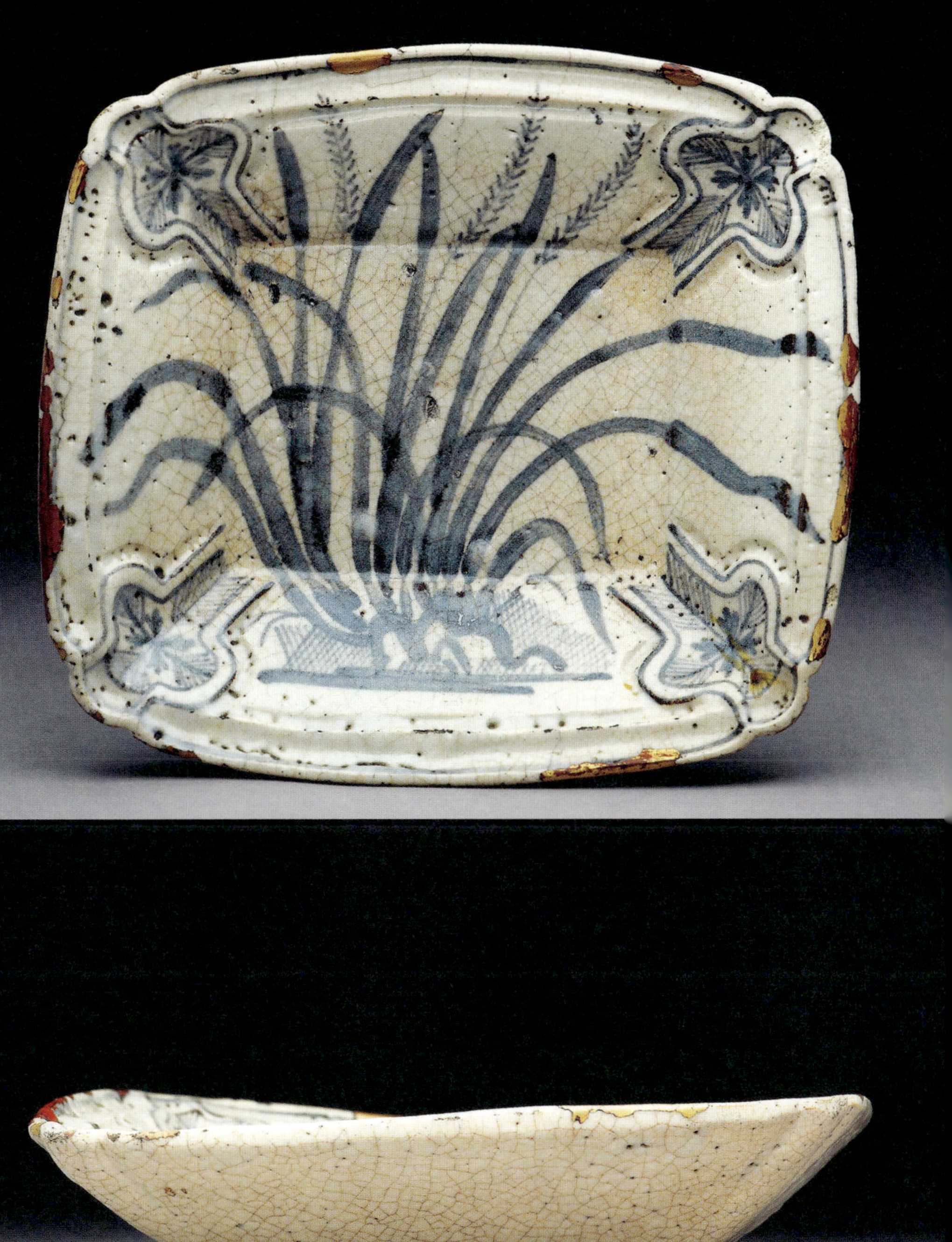

53

Palissy ware dish

Europe
1601–1650 CE
Glazed ceramic, 19.3 x 24.6 x 12 cm

The Ashmolean Museum was established in 1683. These vessels are described in Latin in its catalogue two years later as: "oblong china dishes … [containing] naked women covering their private parts with their hands". They are among only a few ceramic pieces to survive from the Ashmolean's founding collections, and may even have been part of an earlier collection of curiosities held by the Tradescant family in Lambeth, south London, who – unusually for the time – charged visitors a fee to view their display of natural and artificial wonders.

In 1656, Elias Ashmole, the founder of the Ashmolean, helped John Tradescant the Younger to produce a catalogue of the Tradescant collection. This catalogue lists a "variety of china dishes" which may have included these. The dish in the main image is clearly intended as a novelty. It takes the form of a woman lying cross-legged in a bath; she holds two cornucopias and might represent fertility.

The dish is marbled on the underside and has been broken and restored. Behind the woman's head, the outer lip is decorated with cockle shells that seem to have been cast from natural specimens. It's probably an object intended to amuse the viewer, rather than having any specific function at the table. However, we can guess that it might have been used during bawdy drinking games, the naked figure slowly revealed as the contents were drunk. Similar-shaped vessels made of tin-glazed earthenware by Italian potters in Nevers, France, during the early seventeenth century were described as *gondolles* (gondolas). Some experts think that these vessels might have been used by women to pour water on themselves when bathing. Other tin-glazed versions of this shape were made in London – and possibly the Low Countries.

A French potter named Bernard Palissy developed vessels made of earthenware, elaborately moulded with relief figures and decorated with richly coloured lead glazes. Palissy was born around 1510 and died in the Bastille, imprisoned for his Protestant faith, in 1590. To create his unusual relief-decorated ceramics, he moulded natural forms from actual specimens,

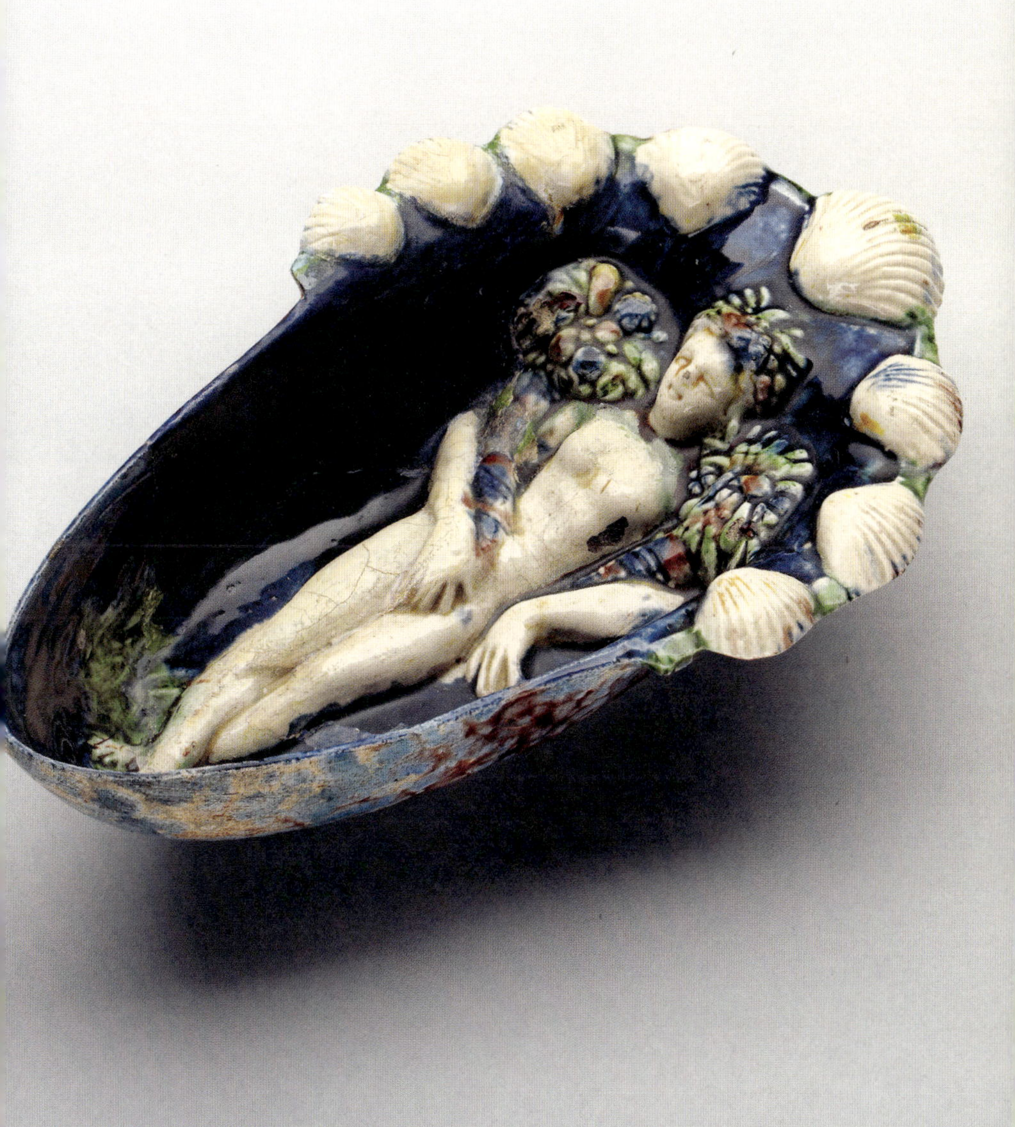

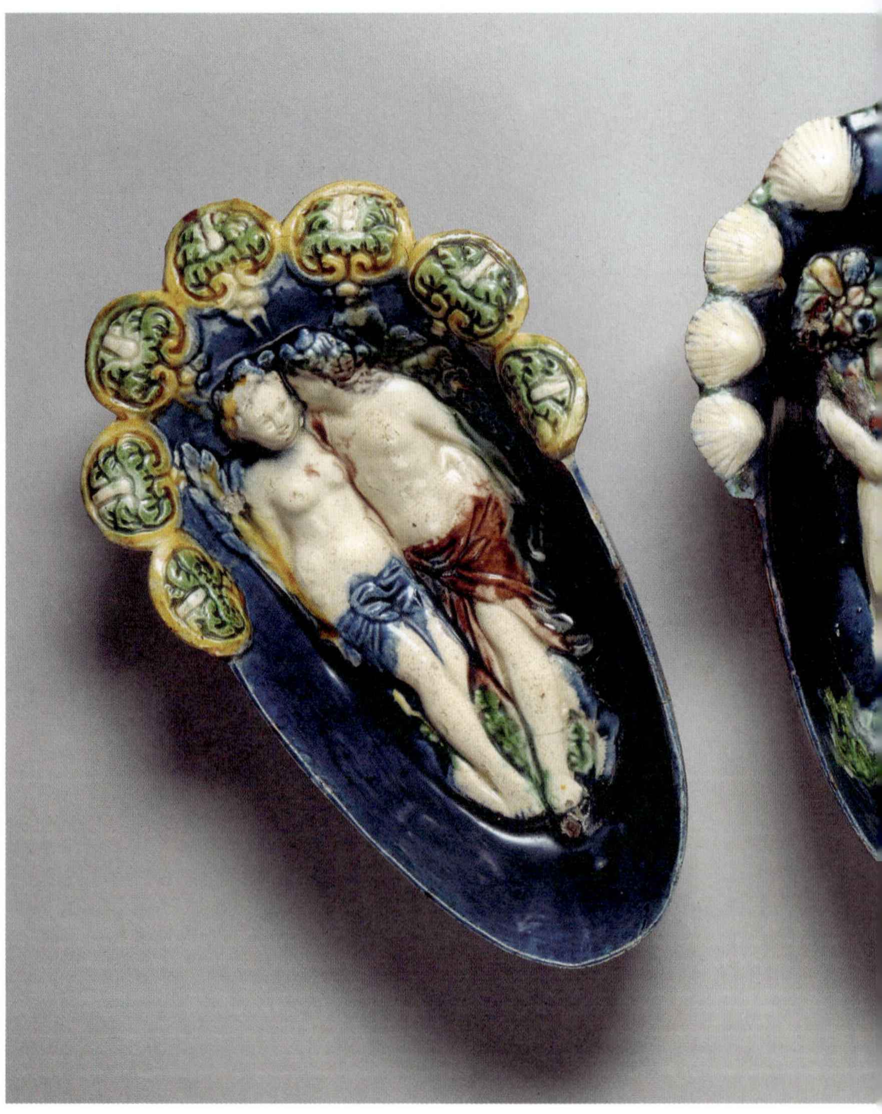

including plants, snakes, frogs, sea creatures and shells. His work was very influential: followers continued to make new work in his manner well into the seventeenth century, in several parts of France. When collectors rediscovered his work, in the nineteenth century, they attributed all ceramics of this type to his workshop. Subsequently, pieces of this sort have been attributed to the workshop of Claude Bertélemy at Fontainebleau and to Pré-d'Auge in Normandy. At the moment, there's no firm documentary or archaeological evidence as to where they were made. They are still described as "Palissy ware".

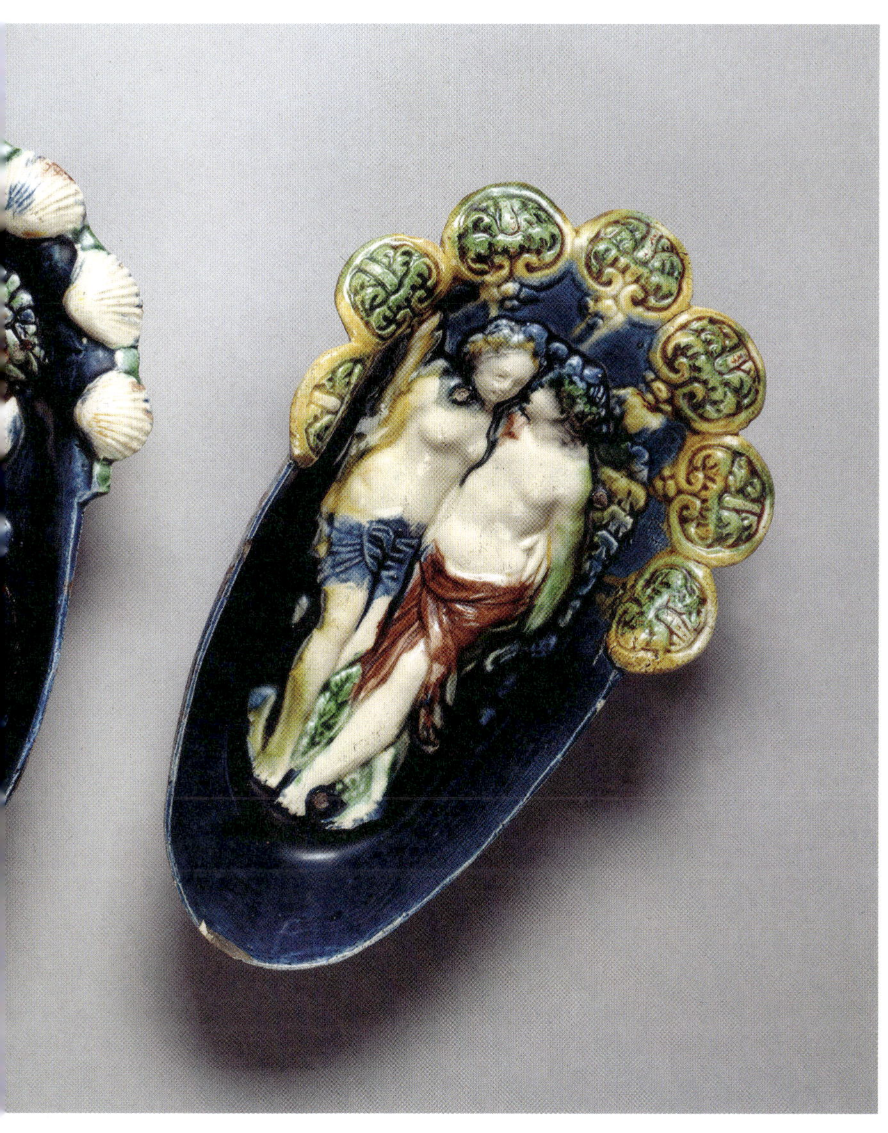

Palissy ware dishes, glazed ceramic. Europe, early 17th century, approx. 20 cm.

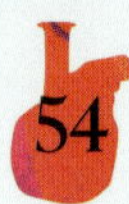

Pouring vessel, or *kendi*, in the form of an elephant

Iran
1601–1700 CE
Fritware, underglaze painting,
22 x 17.5 x 11.5 cm

Traditionally, the use of cobalt in porcelain is celebrated as one of the most critical developments in the history of Chinese ceramics. From the Yuan dynasty (1271–1368) onwards, the production of blue and white wares turned Chinese porcelain into a global success. West Asia soon emerged as one of its most enthusiastic markets.

Islamic rulers such as the Safavids (1501–1736) and the Ottomans (1299–1922) acquired such large amounts of these wares that sites like the Safavid dynastic shrine in Ardabil and the Ottoman imperial palace in Istanbul came to hold the two most extensive collections of Chinese porcelain outside China.

Despite this narrative of Chinese inspiration, cobalt had already been part of Islamic ceramic aesthetics since the ninth century (see, for example, the two tin-glazed bowls on page 118). In the early thirteenth century, it was commonly used in the production of underglaze in Iran. So, in fact, China's breakthrough seems to have come out of the materials and expertise employed in Iran. Recently, early Yuan vessels with Persian inscriptions have been discovered in Jingdezhen: further evidence of a direct technological transfer.

Underglaze decoration in cobalt blue re-emerged in Iran in the late fifteenth century. By the seventeenth and eighteenth centuries, it was the dominant aesthetic in Iran's ceramic production. Centres like Mashhad and Kirman produced wonderful objects that demonstrate both a debt to the East Asian tradition and the originality of local craftsmen. This elephant-shaped vessel is a close replica of its Chinese model, a type of pouring vessel known by the Malaysian name of *kendi*. Unusually, this example keeps the trunk of the elephant – a flourish similar to examples from East Asia. It's rendered as a ribbon-like protuberance gracing the animal's neck. The elaborate trappings hide the rest of the beast's body except for the tail, which is rendered in low relief as it sways to one side.

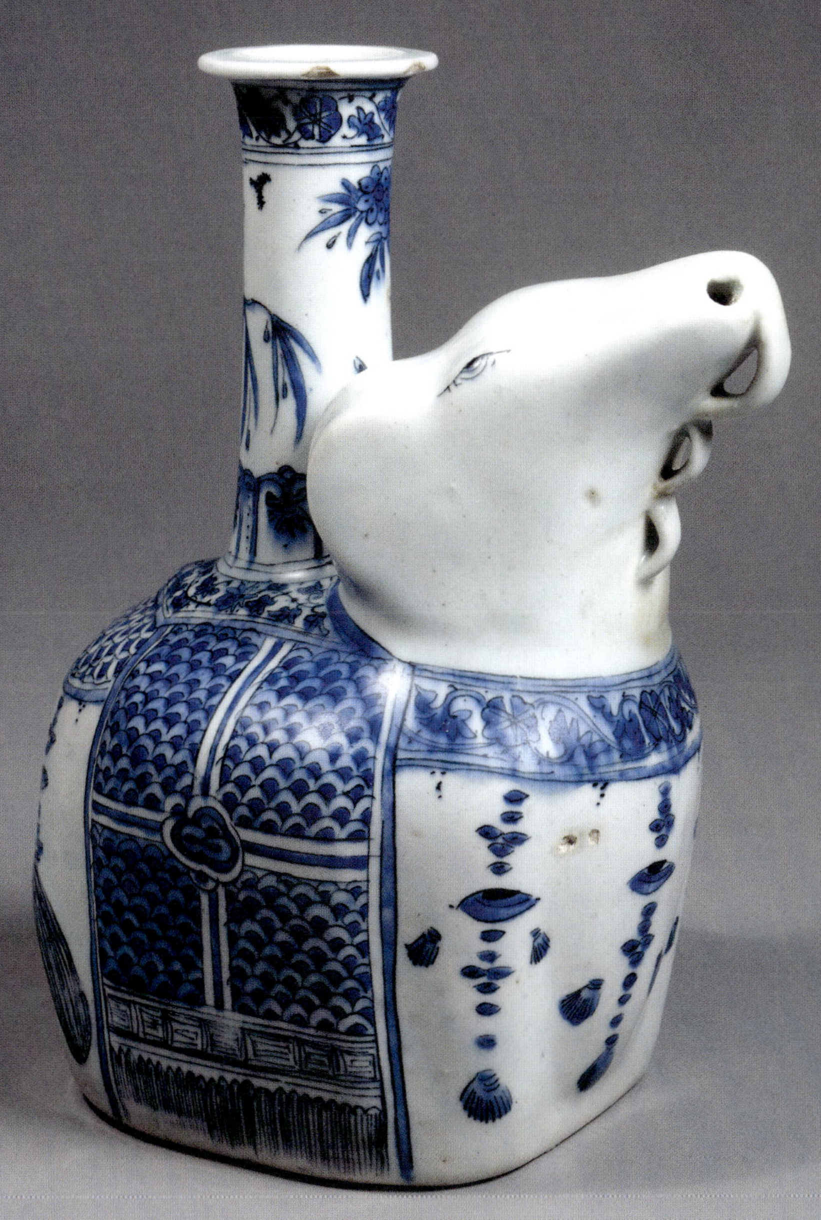

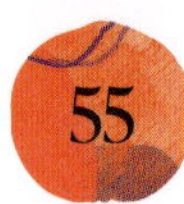

Dish in the form of a peach

Japan
*c.*1630 CE
Porcelain, with underglaze painting in blue,
7 x 27 x 26 cm irregular, max

The export porcelain of Japan is familiar to Europe because vast numbers of pieces were imported in the second half of the seventeenth century and the first half of the eighteenth. It's also familiar because of huge numbers of European imitations and pastiches. The Ashmolean Museum holds one of the finest collections of Japanese export porcelain anywhere in the world.

Much less familiar in Europe are the porcelains made in the 50 years or so of production in Japan that preceded the export trade. These are almost unrepresented in collections. The Japanese started to make porcelain in the stoneware kilns of southern Karatsu, on the island of Kyushu, in an area close to the town now called Arita, in the years around 1600. At first, this porcelain was coarsely made and roughly decorated. The Karatsu potters adapted styles that were originally from Korea. The decoration of their pots was usually in iron brown under a transparent glaze. In Arita porcelain, the decoration was usually in underglaze blue.

In, we think, the first decade of the century, kilns were built especially to make porcelain nearer to the centre of Arita, which was closer to the source of the porcelain clay, Izumiyama. One of these early kilns is called Tengudani, and several pieces found at the kiln site are on display at the Ashmolean. In the first 50 years of production, the styles and shapes of the porcelains evolved as the kilns grew progressively more sophisticated and the potters more skilled. During this time, the use of relatively cheap porcelain became more widespread in Japan. Iron brown and celadon were used almost from the start, but artisans mainly used underglaze blue (underglaze copper-red was used for a few years in the 1620s but was never very successful). The simple, so-called Korean style of this *shoki*-Imari (early Imari) ware soon gave way to a variety of styles.

The Japanese tea ceremony had developed in the sixteenth century as an elite artistic pursuit, incorporating ancient ceramics from China. When Japanese porcelain emerged, its development related to the demands of the

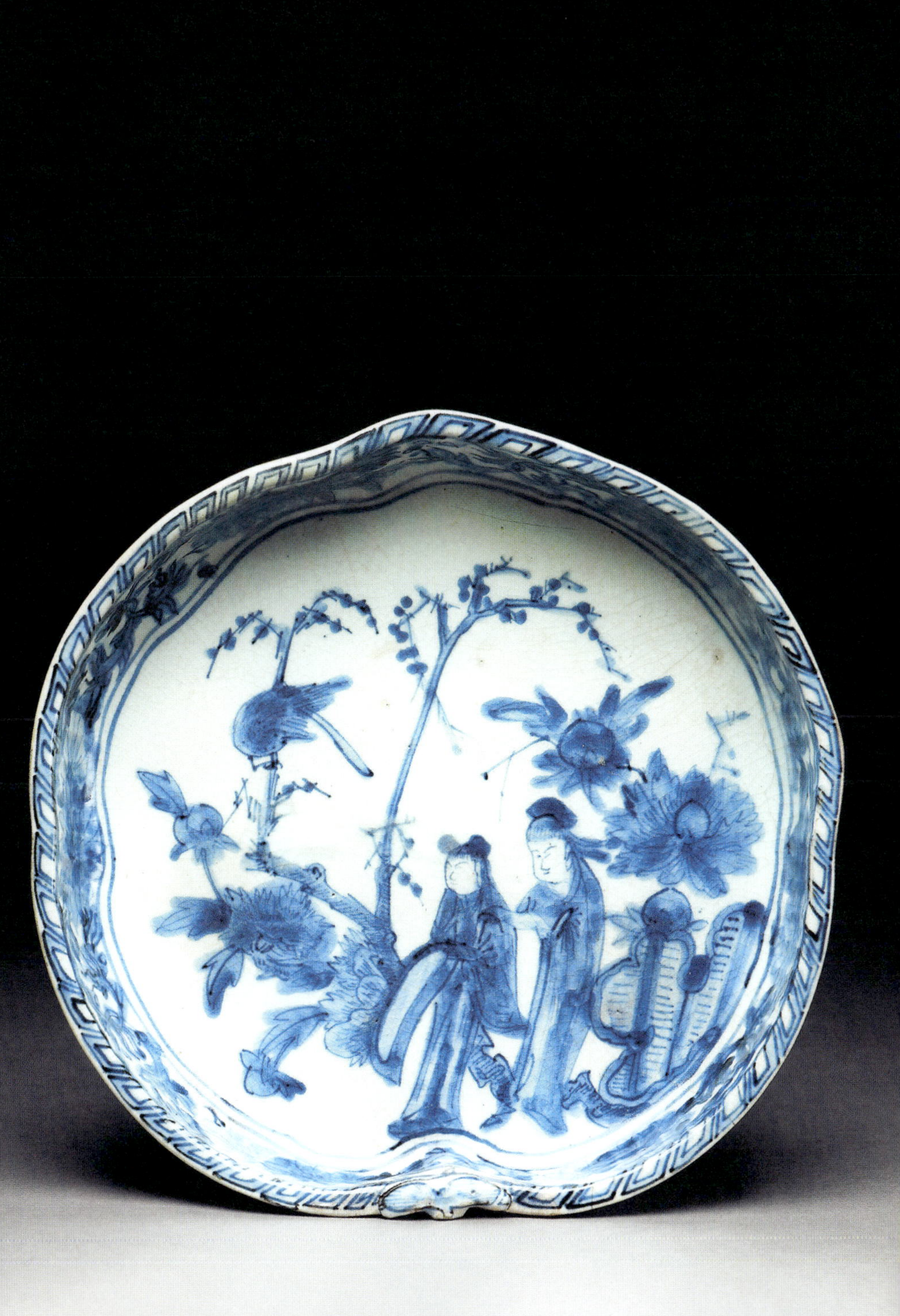

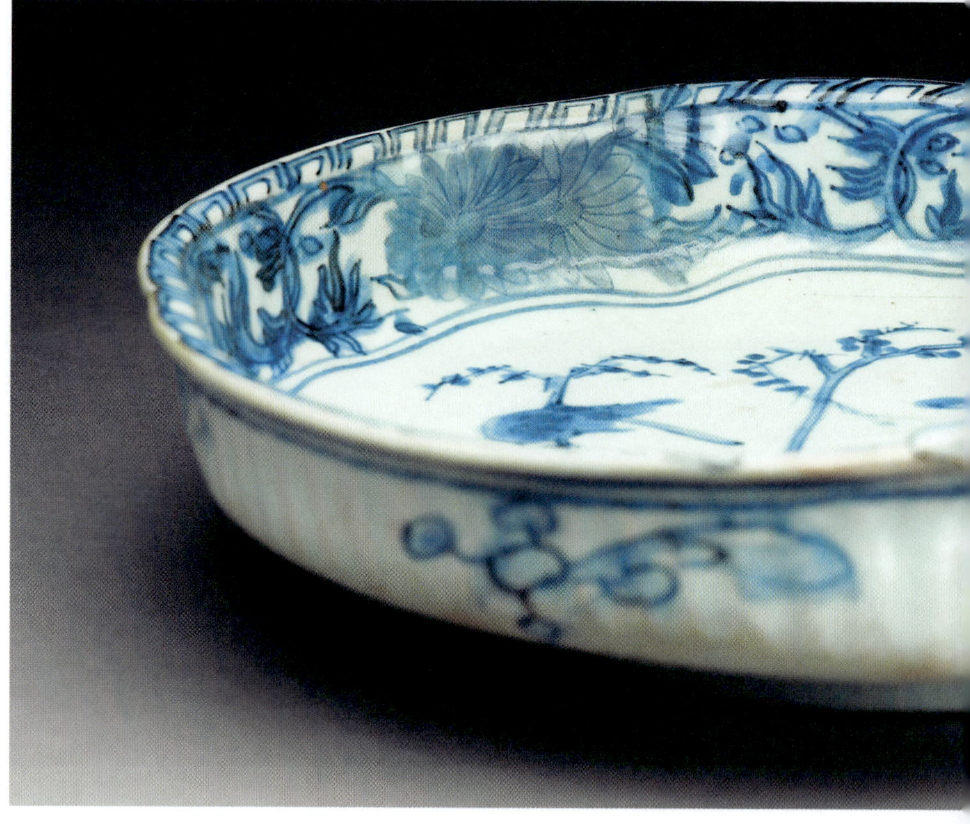

tea masters. The wares might appear basic, but their producers were skilled. As the new Japanese porcelain industry was not yet very efficient, the Chinese entered the market by exporting Tianqi porcelains from Jingdezhen. Created to be suitable for Japanese customers, Tianqi were effectively Chinese imitations of Japanese porcelain.

In the 1630s and 1640s, Japan produced more porcelain as the number and efficiency of the kilns grew and taste diversified. Most of the kilns produced similar wares, but a few were more distinctive. The finest piece in the Ashmolean's collection is this peach-shaped deep dish with ribbed vertical sides, depicting two Chinese sages among a group of trees, in one of which sits an oversized bird. This decoration is typical of seventeenth-century Japanese taste in tea vessels, and this dish is almost certainly one of the very rare pieces actually made for use in the *kaiseki* meal that accompanies a formal tea ceremony. It's unlikely that pieces like this were made much before the 1640s, and yet this dish has many characteristics

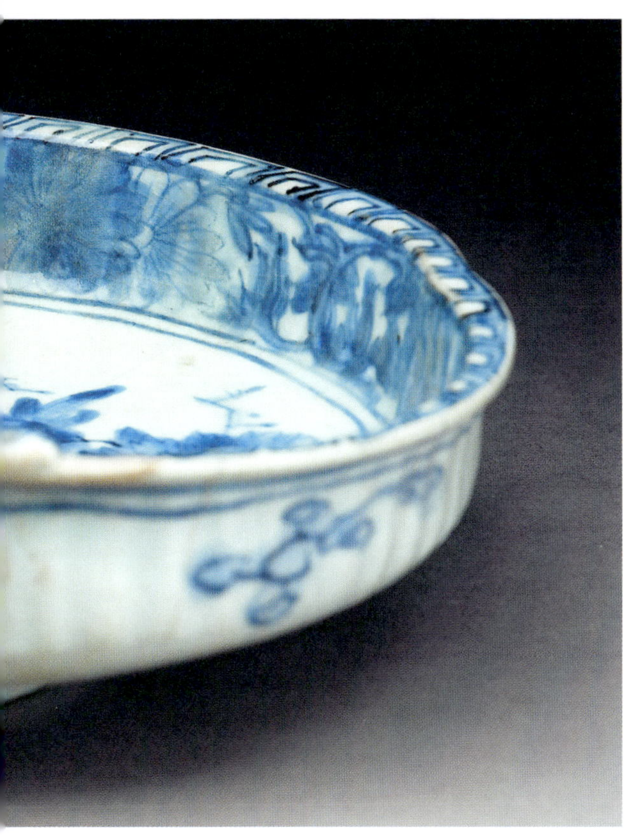

of earlier times. The foot ring is very heavy but fairly small in diameter. In the 1630s, unsophisticated kiln furniture (shelves and other devices used to stack and separate ceramics during firing) was too small to take a large foot ring and forced the potter to make the wall of the pot rather thick. The glaze is uneven, coarse, and bears the fingermarks of a glaze worker. The painting is crude, with its elongated figures, absurd plants and overly large bird.

After the first massive order from the Dutch East India Company in 1659, kilns in Japan changed over to the production of porcelain for export. Ware like this continued to be made by only two or three kilns. We don't know exactly how long artisans continued to make pieces like this, effectively late *shoki*-Imari pieces, in parallel with the wares created for customers abroad. However, it's unlikely to have been for more than 20 years. Most *shoki*-Imari kilns eventually either ended production or vastly expanded, as workers gathered together into the 10 (later 11) kilns that made Japanese export porcelain.

Jug with grotesque decoration

UK
1625–50 CE
Earthenware, tin-glazed (delftware), 28 cm

The technique of tin-glazing pottery, maiolica, was developed to a high level of sophistication in Italy and then spread to northern Europe after 1500. Many emigrating Italian craftsmen headed for Antwerp. Fleeing religious turmoil, two Protestant Italian potters, Joris and Jasper Andries, came to England from Antwerp and set up in Norwich in 1567.

Soon afterwards, the tin-glazing industry was established in London, mainly by immigrants from the Low Countries. It began in east London and branched out south of the Thames. Later on, artisans produced the English version of tin-glazed earthenware, which we call "delftware", in Lambeth, Bristol and Liverpool, among other places. This jug is part of a small group of London-made delftware created in the second quarter of the seventeenth century.

The body of this striking vessel is painted blue and features scrolling leaf fronds and flowers. On the front, among these fronds, is a grotesque winged figure, painted in colour, standing on animal legs and feet. His arms are truncated into inwardly curling scrolls and he wears a pointed hat. Beneath the handle is a figure in a similar style, with crossed arms, his legs and feet formed into fishtails. He wears a round-topped hat. On one side of the jug is a head with a ruff and another pointed hat. A vertical band of interlace runs down the jug's handle and four horizontal lines run round the lower part of its body.

The figures are derived, ultimately, from ancient Roman wall paintings, which were studied closely by curious artists after they were discovered in the 1480s. The Renaissance take on this kind of imagery, in the form of ornament like the decorations on this jug, was known as *grottesche*. The first appearance of grotesques in maiolica included tiles painted in Siena, and this type of decoration became widely popular. The sixteenth-century artist Benvenuto Cellini explains in his autobiography where the term came from originally:

These grotesques (*grottesche*) have been given this name in modern times. They were found in certain underground chambers in Rome by

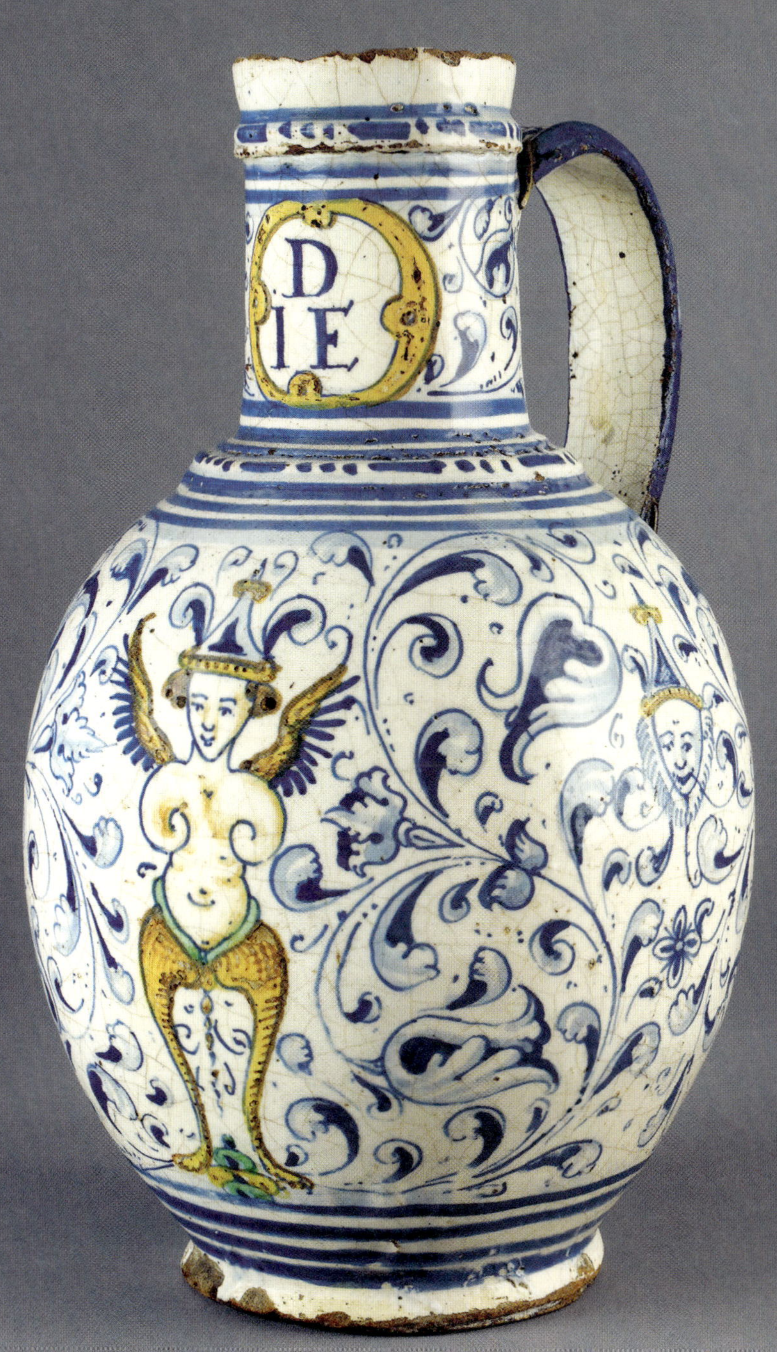

researchers; these underground chambers were bedrooms, bathrooms, studies, living rooms and other such things. These researchers, finding these decorations in these cave-like places, since the ground level has risen since ancient times and these are now underground, and since the word in Rome for these underground places is grottoes (*grotte*), these decorations have acquired the name "grotesques".

The neck of the jug is painted with leaf fronds and on the front there is a frame device, a cartouche bearing a letter "D" above the letters "IE". These are the initials of a married couple, its first owners: "D" (the first letter of their surname) over "I" and "E" (the initials of their first names). This triangular arrangement of initials was an English convention, though not exclusively so. Above and below this ornament are horizontal lines and bands of dashes alternating with groups of three dots.

When they created maiolica, Italian artisans were taking over a technique that was Islamic in origin and transforming it into an artistic effort imbued with all the drive and innovation of Renaissance Italy. The appealing products they created sparked demand in Italy and abroad. Historians of ceramics can track the movements of the potters by the analysis of different styles of painting. But at the time, the potters were chiefly proud of their technical abilities: the knowledge, born of experience, about which clay, which glazes, and which pigments would work together and fire to their satisfaction. Some potters travelled abroad to take their skills to new markets, either as a commercial enterprise or to some extent under royal, civic or aristocratic sponsorship. In the countries where they ended up, they created new forms of artistic expression. During the seventeenth century, these developed in central and northern Europe into creative output that now seems completely characteristic of the respective national cultures where it is found.

The process of opacifying glazes by adding tin was discovered in what is now Iraq, by potters either in or near the port of Basra, some time around 800 CE. Fragments of this early tin-glazed pottery have been found alongside fragments of white-bodied stoneware imported from China. Apparently, the arrival of these technically exquisite high-fired wares from the middle kingdom inspired the Muslim potters of the Abbasid Caliphate (750–1517 CE). But these craftspeople didn't have access to the firing clays or the kiln technology needed to develop a glaze that could look like fine white pottery. From this time almost to the present day, it has been the norm in the Islamic and Western worlds for fine tablewares to be basically white, in contrast to the tradition of red and black pottery inherited from Greece and Rome.

At around the same time, potters in the region developed the ability to add metallic lustre to the surface of pottery. This difficult technique, which seems to have been transferred from the manufacture of glass, gave ceramic products an enchanting sparkle. The techniques behind both tin glaze and lustre spread through the expanding Islamic world and they propelled ceramics to the level of a luxury product. Churches in the great medieval port of Pisa bore witness to the infectious vision of the Islamic potters. From about 1000 CE, for over 300 years, builders set into the buildings' façades brilliantly coloured, often lustred bowls imported from Egypt, Tunisia, Morocco and Spain.

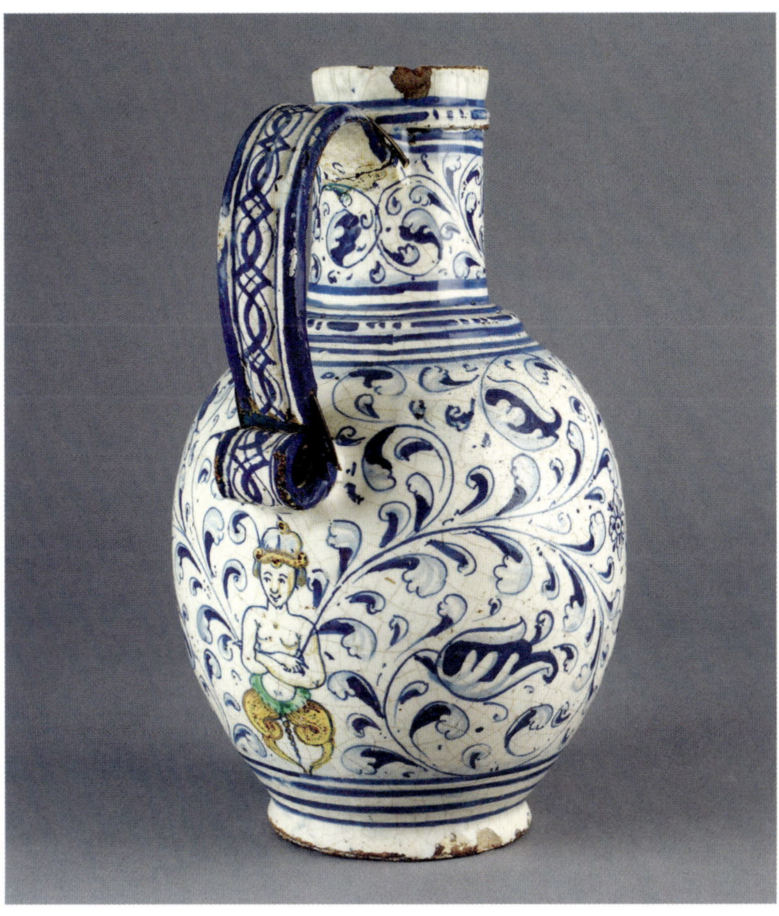

Blue and white vase
with figures and a poem

China
1639 CE
Lei Du (1180–1225)
Porcelain, underglaze painting in cobalt blue, 46 x 22 cm

Ask anyone to picture Chinese ceramics and they will probably call to mind blue and white porcelain like this. This type of porcelain was first made in quantity during the Yuan dynasty (1279–1368).

The famed kilns at Jingdezhen were already well established by then, and during the Ming dynasty imperial kilns were set up there to provide pieces specifically for the emperor and his court. At the time of the third Ming emperor, Yongle (who reigned from 1401–25), reign marks were added, incised beneath the glaze. After Yongle, this practice was replaced with marks written in cobalt blue. In the 1620s, when the dynasty was in decline, the imperial patronage of Jingdezhen ended. The kilns operated privately, producing new styles for new markets, until the Qing dynasty's Emperor Kangxi (who reigned 1661–1722) re-established them in 1683.

The technique remained the same. The potter modelled porcelain clay into shape, allowed it to dry a little, decorated it with designs in cobalt blue, covered it with a transparent glaze, then fired it. For most of the Ming dynasty, the designs were formal and the motifs favoured were dragons, birds and flowers. However, in the mid-seventeenth century, pictorial designs appeared – often scenes depicting figures from well-known plays, novels or popular legend. The porcelain designs imitated paintings in ink on paper or silk. Like the paintings, they included poems and sometimes even seal marks in reference to the classical "three perfections" of poetry, calligraphy and painting. The poem on this vase is by the Song dynasty poet Du Lei (1180–1225) and reads:

On a frozen night a guest arrives, tea is drunk in place of wine, over the rustic fire the water boils as the charcoal begins to redden.

Everything is as normal as the moon shines down outside the window, only when the prunus begins to bloom are things not the same. Written in a month in mid-spring, during the jimao year (1639).

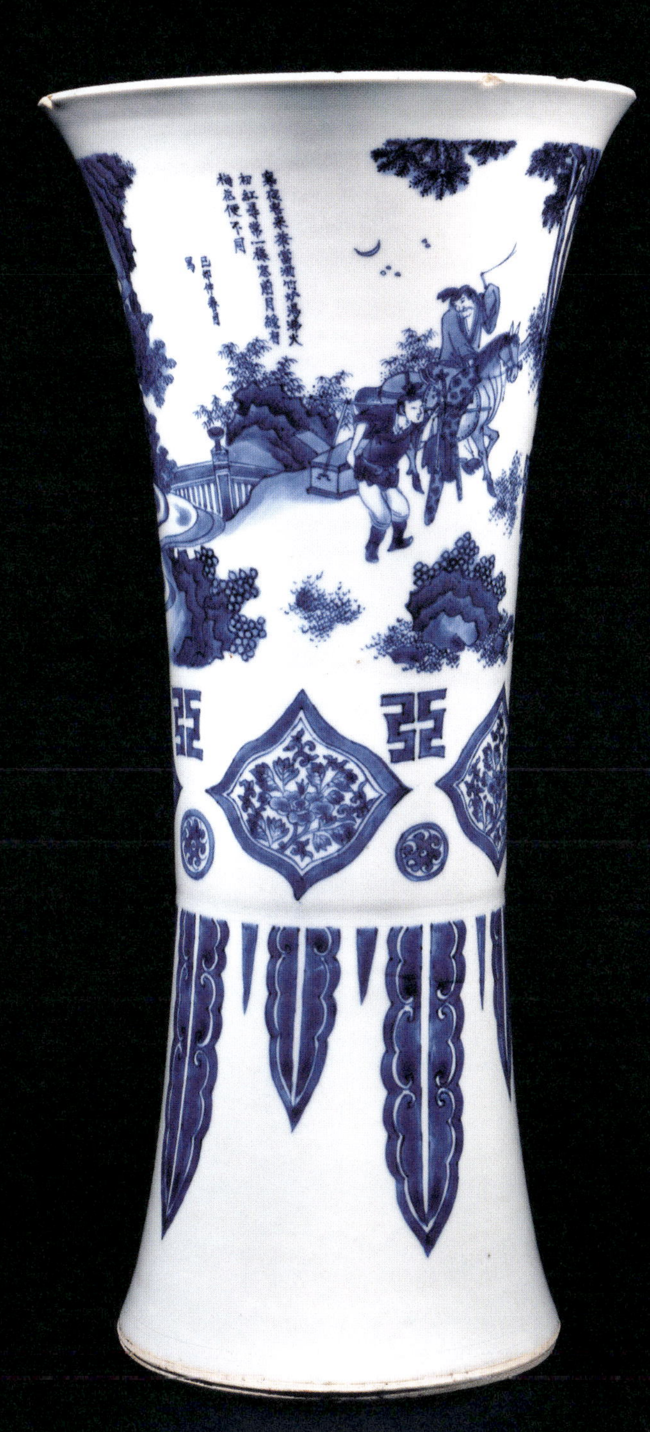

鼻底龍香茶當酒沸河沙鳥乘火
知紅葉第一叢窗前月總有
梅花便不同
己卯仲春月
寫

Kraak-style plate with Dutch East India Company monogram

Japan
1660–90 CE
Porcelain, underglaze blue decoration, 39.7 x 6.8 cm

When this large dish was made in Japan, the secret of how to make porcelain had not yet been discovered by Europeans. The dish was produced in Japan's porcelain centre, Arita, to be exported to Europe by the Dutch East India Company.

At its centre is the Company's monogram, "VOC" (standing for *Vereenigde Oostindische Compagnie*). Around the logo, the dish's painter has depicted two mythical *hō-ō* birds and sprays of camellia and peach. The rim of the dish is decorated with a wide border of alternating panels, showing flowering prunus and bamboo. In between them, narrower panels contain simple floral scrolls on a blue ground.

This style of porcelain – with its panelled borders and floral motifs in underglaze blue – is known as *kraak*. It was originally made at the Jingdezhen kilns in China in the late sixteenth century and exported by Portuguese merchants. In Europe, precious imported porcelain objects were prized as status symbols by wealthy collectors. The term *kraak* is thought to derive from *carrack* or *caracca*, the name given to the large Portuguese ships that carried the Asian trade. By the early 1660s, however, the Dutch East India Company came to dominate the trade between Europe and Asia.

For 200 years after the foundation of the VOC in 1602, the enormously powerful corporation brought the Netherlands great wealth and international power. Not only did they create a vast network of trading posts, but they also waged wars, negotiated treaties, established colonies and traded in enslaved people. Yet it was for its very lack of perceived threat that the VOC was permitted to retain a presence in Japan while other European nationals were not. Portuguese and Spanish merchants and missionaries had arrived in Japan in the 1540s, but were expelled by the Japanese government in 1639 for their Christian interference. The non-Catholic Dutch – concerned with profit rather than people's souls – became the only Europeans allowed to remain in Japan.

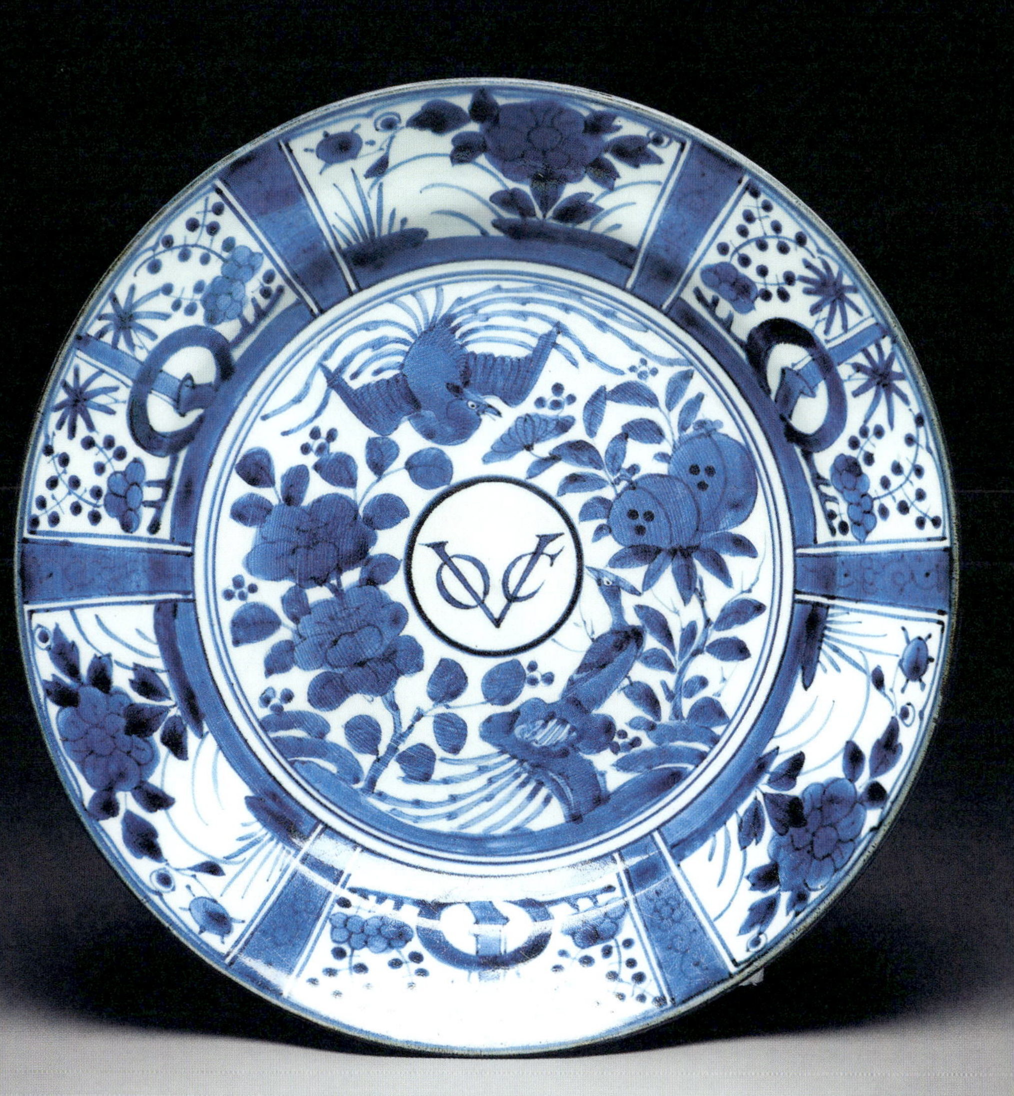

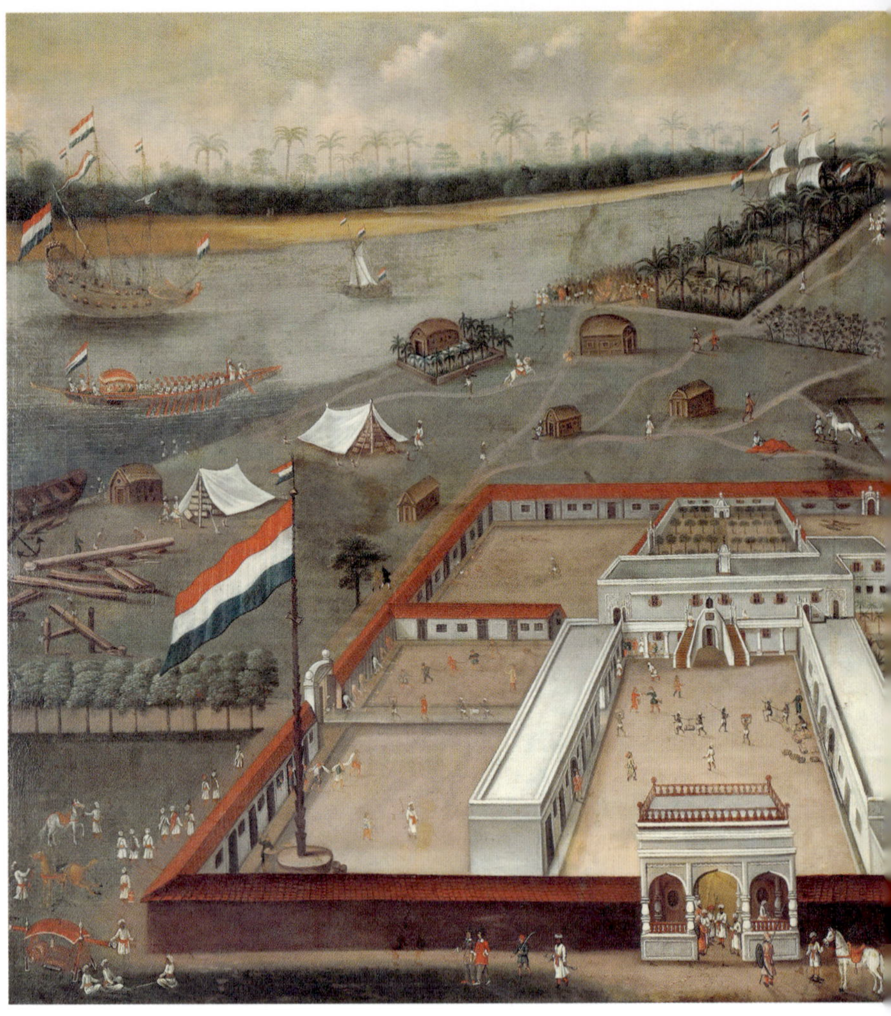

However, they were confined to Dejima, a tiny artificial island in Nagasaki Bay, and were kept under close scrutiny.

The VOC's board of directors always considered porcelain a profitable, if fairly minor, commodity. In the early seventeenth century, Chinese porcelain was so plentiful and cheap that the company initially paid very little attention to Japanese wares. But when the Chinese porcelain trade was disrupted in the late 1640s, because of civil wars in China after the fall of the Ming dynasty, it was a logical decision to turn to Japan to fill the gap. At first, a lot of Japanese export porcelain was made in styles that had been proven to be successful when they were exported from China. These styles had been

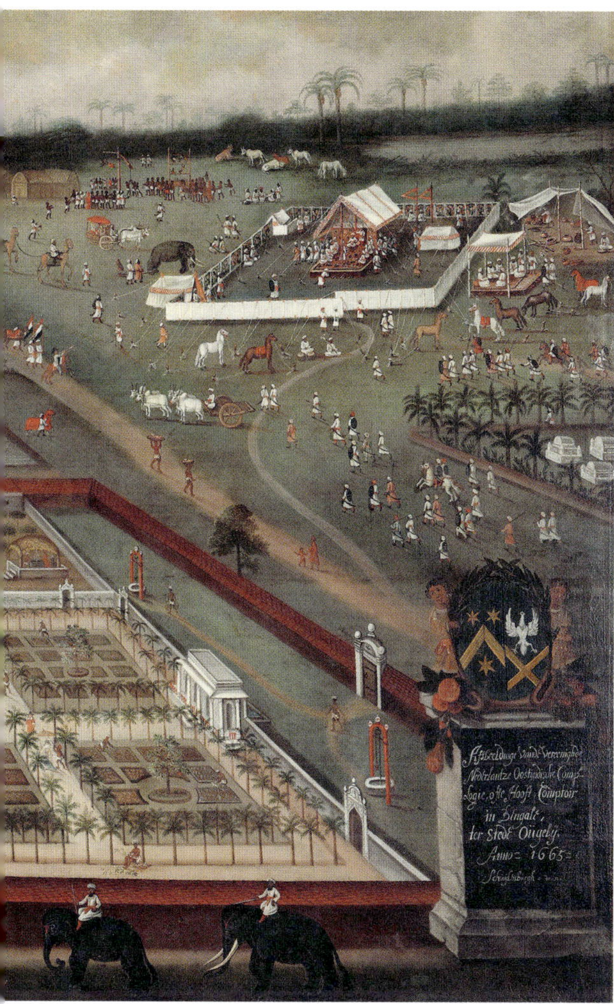

Painting of the Dutch East India Company factory in Hugli-Chuchura, Mughal Bengal. Hendrik van Schuylenburgh, 1665.

developed specifically for export in Jingdezhen, including the *kraak* style. The term *kraak* porcelain (*craeckwerk* and *craeckcommen*) is first found in Dutch house inventories in the mid 1670s. The painting on these *kraak* wares was usually skilful but not especially refined. It was nearly always applied to open shapes such as dishes and plates. Dishes with the VOC logo were probably used as a very early form of in-house corporate promotion. They were likely designed to be used by high-ranking company employees at trading bases in Asia, and probably also in officers' cabins on board ship. Excavators have found the remains of similar dishes in the Dutch residences in Dejima, the company's Japanese base in Nagasaki.

Misshapen baluster jar
with flowers

Japan
1670–80 CE, Kakiemon workshops
Porcelain, polychrome overglaze enamels, 41 x 29 cm

For the British Museum's 1990 exhibition *Porcelain for Palaces: The Fashion for Japan in Europe, 1650–1750*, many pieces of Japanese porcelain were borrowed from the Ashmolean collection. The Ashmolean holds one of the finest and most extensive collections of Japanese export porcelain of this period outside Japan.

This extremely varied ware was in huge demand in Europe at the time, where it was used mostly for display. The Ashmolean also lent many pieces to the Historic Royal Palaces when they recreated Queen Mary II's porcelain display in the Queen's Gallery in Kensington Palace as it had been in 1693.

One of the most beautiful pieces in the Ashmolean's Gallery for Japanese Decorative Art is this magnificent, very large early Kakiemon jar. Kakiemon porcelain is associated with a family, a kiln and an easily recognisable generic style. The jar is decorated in an early palette of the typical Kakiemon enamel colours. It presents the viewer with flowering plants beside a veranda in a bold style, beneath exquisite borders. This piece was made in Arita, in Kyushu, the southern island of Japan, in about 1670. This was a time when the Dutch East India Company was ordering large quantities of enamelled porcelain for Holland and the West, when prices were very high and grand pieces could sell well in Holland.

And this is a very grand piece; it's probably the largest of its type (admittedly a very rare type) that we know of. When it was complete, it would have had a domed cover surmounted by a knob, which might have added 8 or 10 centimetres to its height of 42 cm. This size must have been extremely difficult to make. The Arita potters were not experienced in creating such shapes for the European market and the clay was not easy to work. The firing process was wasteful because of the difficulties of heat distribution in the kiln, and this piece would have taken up a lot of kiln space. The body of this jar is warped. The flames in the kiln were too hot on one side, and it overheated and sagged slightly towards the base. It's astonishing that it still stands upright.

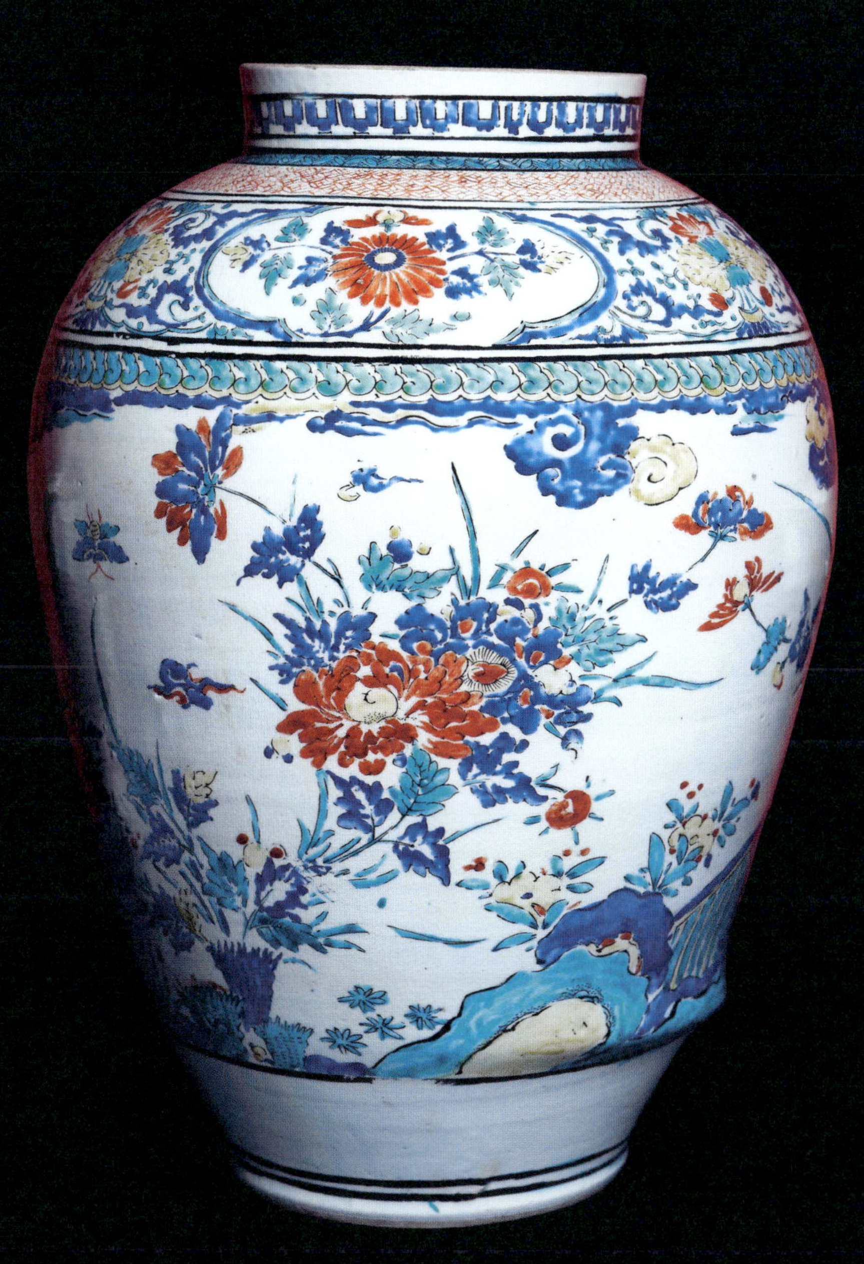

The potters couldn't afford to throw away such a piece. It had simply been too much of an investment. And, after all, not only did the jar stand upright but it was also to be sold to foreigners. Rather than attempting to hide its flaw, the enamellers emphasised the shape in the scheme of decoration, covering the bulge with brilliant green enamel to form the ground for the flowering trees. They lavished their finest skill on the painting. The Kakiemon family appear to have been enamellers before they were potters, and this piece may well have been made before the foundation of the Kakiemon kiln. Even when the Kakiemon kiln began to produce porcelain, it never made closed shapes such as jars or bottles, nor such large pieces as this. The kiln specialised in plates, dishes, cups and bowls – open shapes. Enamelled Kakiemon closed shapes, such as this jar, were made at some other kiln, and only enamelled by the Kakiemon potters.

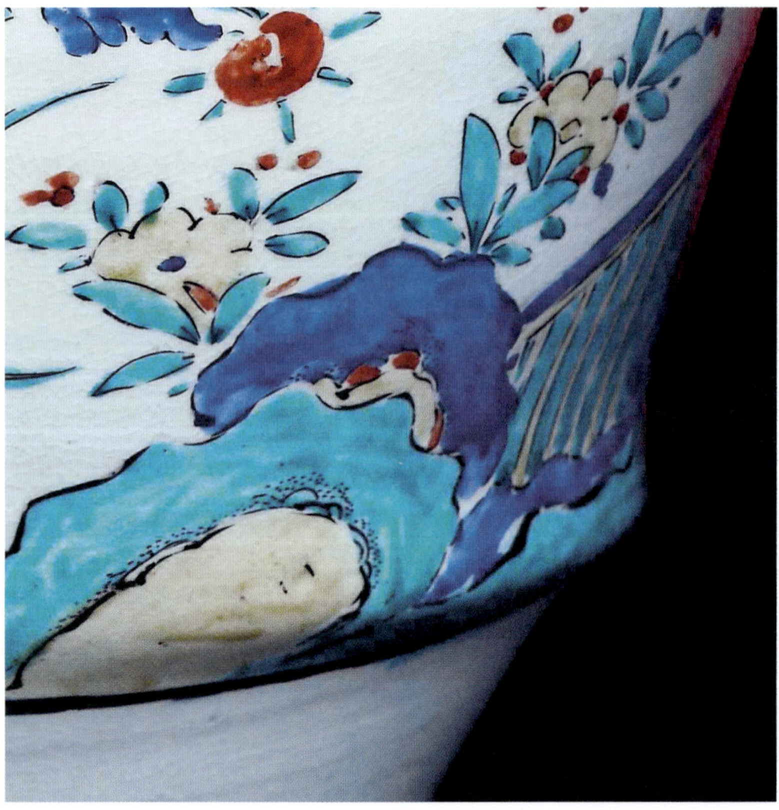

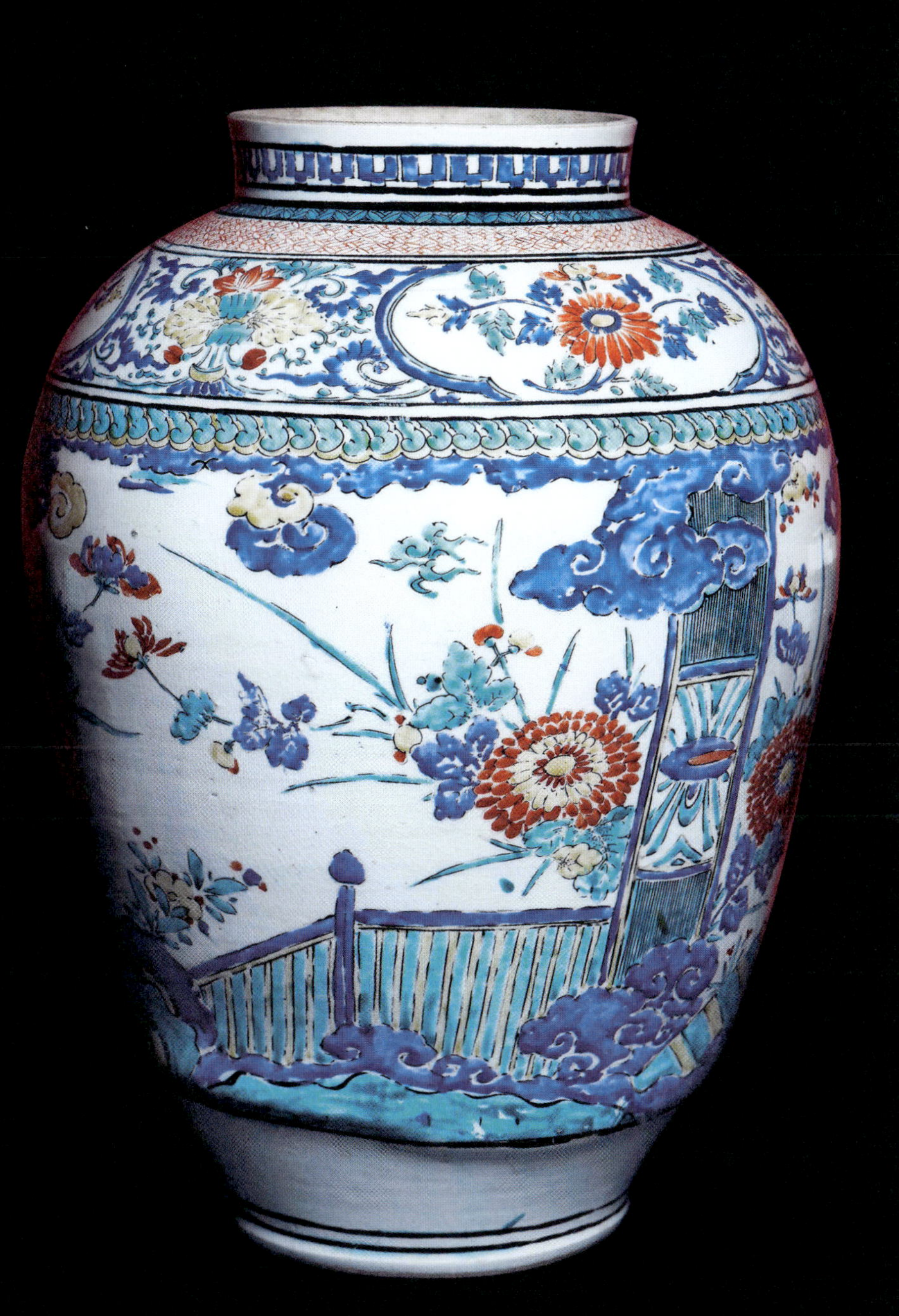

60

Staffordshire slip-decorated earthenware Toft dish

UK
*c.*1600 CE
Ceramic, 43.4 cm

This large, deep platter is an imposing example of the rather rustically decorated dishes made in Staffordshire in the seventeenth century. They were produced in the region around what is now the city of Stoke-on-Trent.

In the nineteenth century, this region – known as The Potteries – became the most productive ceramics centre in the world. But pieces like this one represent the earliest products from this area that demonstrate artistic ambition. They were made in the decades just before factory methods were introduced into the pottery industry.

To create slipware, artisans applied decorative motifs onto coarse earthenware vessels with liquid mixtures of coloured clay, or slip, before firing them in the kiln. The technique flourished in Staffordshire during the second half of the seventeenth century. This dish is decorated in brown, orange and yellow slips, which have been trailed onto the unfired body as if the maker were icing a cake. Next, the dish was covered with a lead glaze and fired once. The galena-lead glaze gives the object a yellow tinge that is especially noticeable on the paler areas. The buff earthenware body has been coated in cream-coloured slip to create a backdrop. On this field, we see the busts of an aristocratic man and woman of the Restoration period, portrayed in a naïve style, above a simplistic lily. The man may be James, Duke of York (later James II), brother of Charles II. The woman may be his wife, Anne Hyde, daughter of the Earl of Clarendon, and mother of Queen Mary II and Queen Anne. If this is the case, the dish is an early example of royal commemorative ware, probably made for the emerging middle-class market. The potter almost certainly intended it to be a display piece, propped up on a dresser or buffet rather than used as tableware.

Two of the principal artisans behind pieces like this one were a father and son from Stoke-on-Trent, both named Thomas Toft. We can see this name traced in large letters around the rim of the dish, under the image of the duke and duchess.

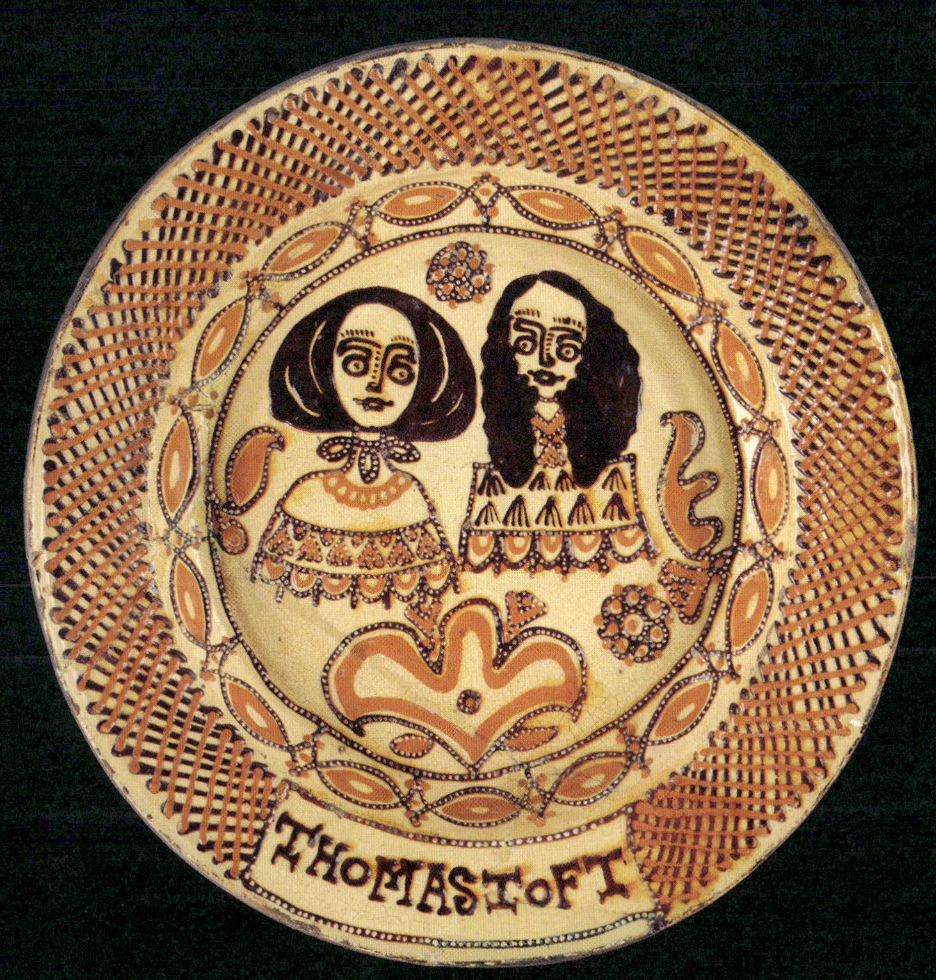

THOMAS TOFT

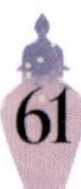

61

Five-piece lacquered Imari porcelain garniture

Japan
1701–20 CE
Porcelain, underglaze painting, overglaze enamels, black
lacquer, *urushi-e* (hand-painted) lacquer painting, mother-
of-pearl, 63 x 45 cm

During the seventeenth and eighteenth centuries, lavishly ornamented Japanese porcelain and lacquer played an important role in the decoration of European stately homes and palaces, where they symbolised the wealth and superior taste of their owners.

This five-piece garniture (set of decorative vases), once owned by the Spencer family at Althorp, is among the largest and most spectacular of all known examples of Imari ware. Imari was porcelain made in Arita for the Western luxury market and shipped to Europe by the Dutch East India Company.

The garniture is decorated in underglaze blue, overglaze enamel and gold, featuring typical Imari-style designs of Japanese temple buildings nestled among spring cherry blossoms and autumn maple leaves. It's a particularly unusual example because large parts of its surfaces are covered in black lacquer, with mother-of-pearl inlay and hand-painted gold designs of peonies and chrysanthemums. The lacquer would have been applied either in Kyoto, the centre of Japanese lacquer production, or Nagasaki.

Porcelain and lacquer were among the most highly sought-after Japanese commodities in Europe at the time, so combining the two materials in this way represents the pinnacle of extravagance. This would have been the ultimate top-of-the-range luxury item. In fact, this garniture seems to be unique. Similar vases, but without the shell inlay, exist in several German palace collections, in the Het Loo Palace in the Netherlands and in the private Usui Collection. An incomplete lacquered garniture of a later style is in the collection of Nostell Priory, West Yorkshire. However, no other complete garniture of this type is known to exist.

At Althorp, successive generations of the Spencer family were enthusiastic collectors of objets d'art. The ceramic collections at Althorp House are among the richest of any of the private collections in the country and include

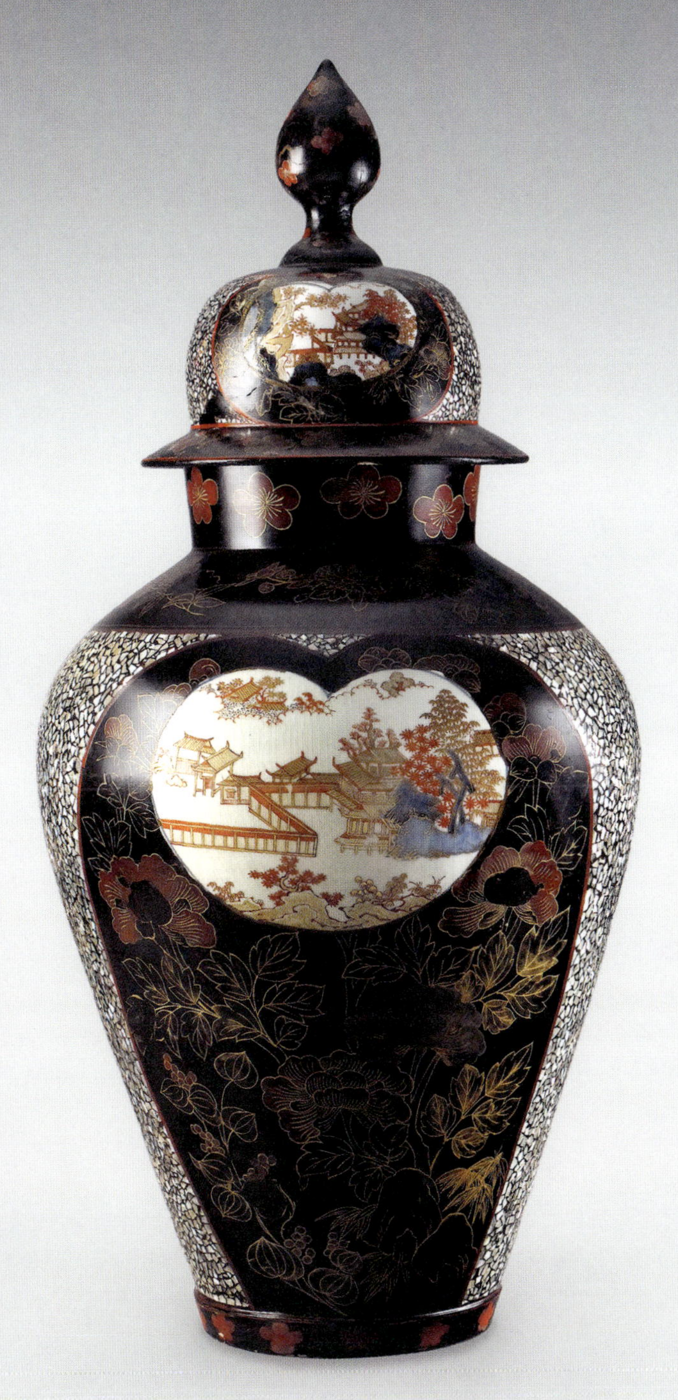

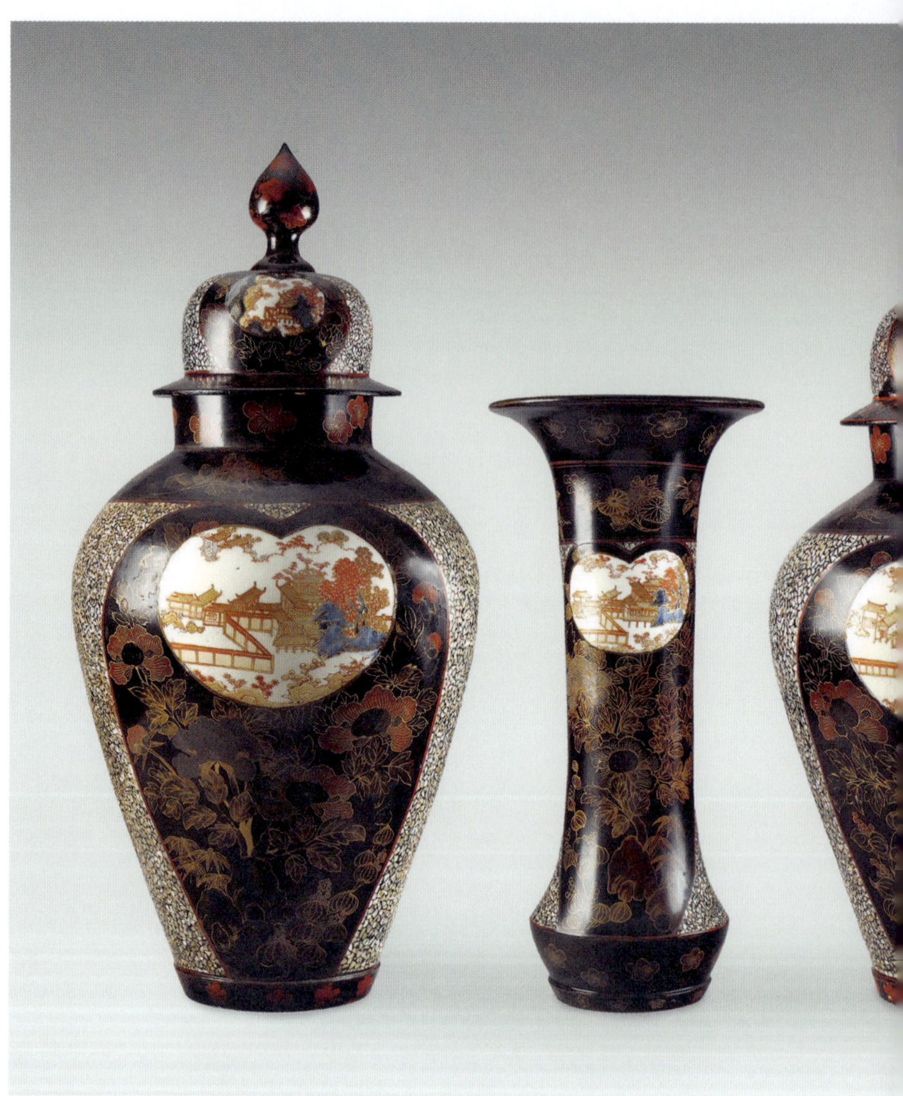

important holdings of Chinese and Japanese export porcelain, much of which was acquired by the redoubtable Sarah, Duchess of Marlborough (1660–1744). She was a wealthy and influential woman, and important eighteenth-century art collector. Her second daughter, Anne, married Charles Spencer, third Earl of Sunderland and owner of Althorp. According to family tradition, the garniture formed part of the Duchess's "Indian porcelain" collection. It's

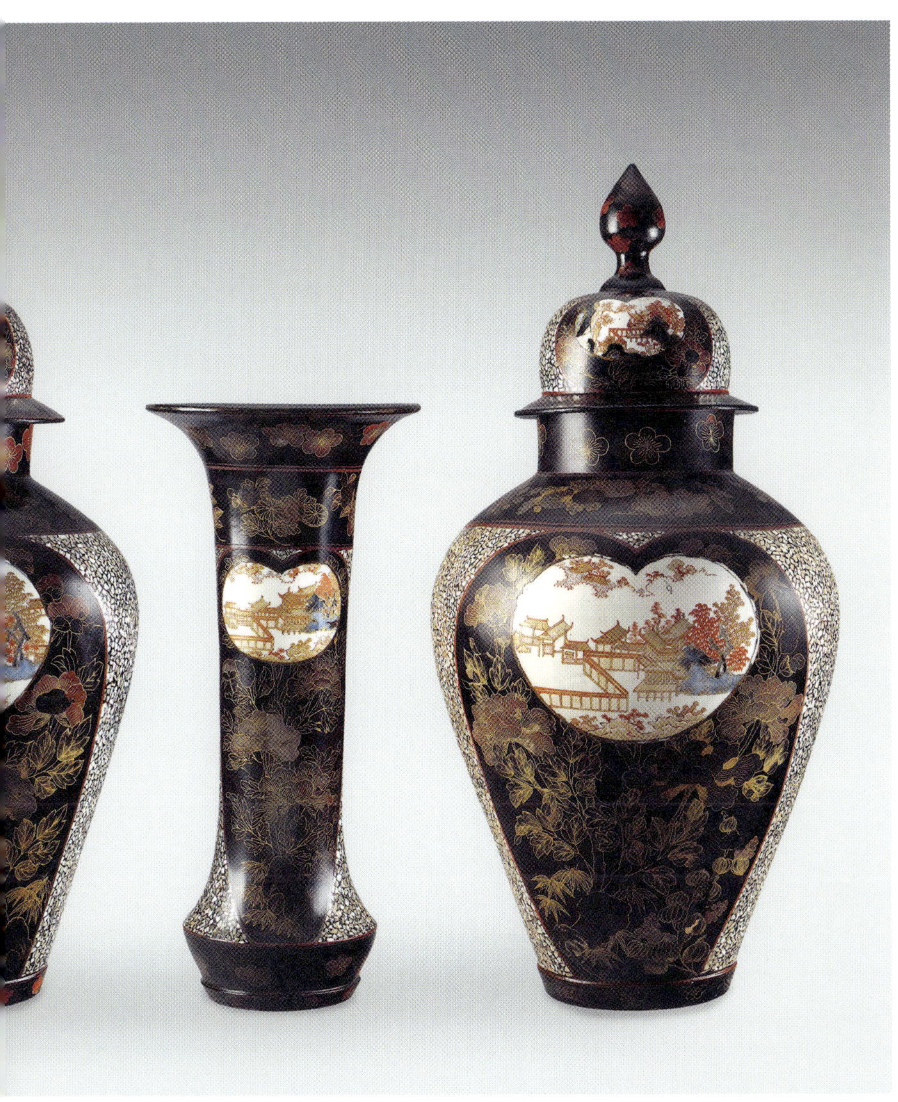

worth noting that Sarah's husband, the Duke of Marlborough, spent time in exile in Germany around 1712, where he probably observed the fashion for such spectacular Japanese objects. A number of "lidded China Jarres" were recorded in Benjamin Goodison's 1746 inventory of Althorp House, compiled two years after the Duchess's death.

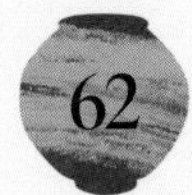

Vase with grape vine

Korea
*c.*1690 CE
Porcelain, underglaze painting, 39.7 x 38 cm

This jar is extremely rare: the closest comparable example is a piece that has been designated a Korean National Treasure. That jar (National Treasure No. 93) is slightly shorter, and its decoration, which is in iron brown only, includes a monkey swinging between two of the vines.

The piece held by the Ashmolean was probably made in the 1690s at the official kilns in Gwangju, Gyeonggi Province, which supplied ceramics to the Joseon dynasty's royal court (1392–1897). At first, the kilns specialised in pure-white porcelain, but eventually the artisans began to decorate the wares, painting them in cobalt blue or iron red and very occasionally, iron brown (like this one). The vine leaves have been subtly delineated, in several layers, and enhanced with grapes in purplish-blue.

By its shape, we can identify this jar as a "moon jar". Most jars like this were made in the early eighteenth century. Usually created from undecorated white porcelain, they are named after their resemblance to the full moon. Most moon jars are the same height as this piece, or slightly larger. They were too big to be shaped as a single piece on the potter's wheel and so were made in two parts, which were joined together along the centre line. The two halves rarely matched exactly and most moon jars are not perfectly spherical. However, admirers felt that this resemblance to a waxing or waning moon made them even more evocative of the skies.

Moon jars are usually pure white, as the Joseon court preferred undecorated white wares to be used in Confucian ritual. The exquisite painting and elegant design of this example make it almost exceptional within Korean ceramics.

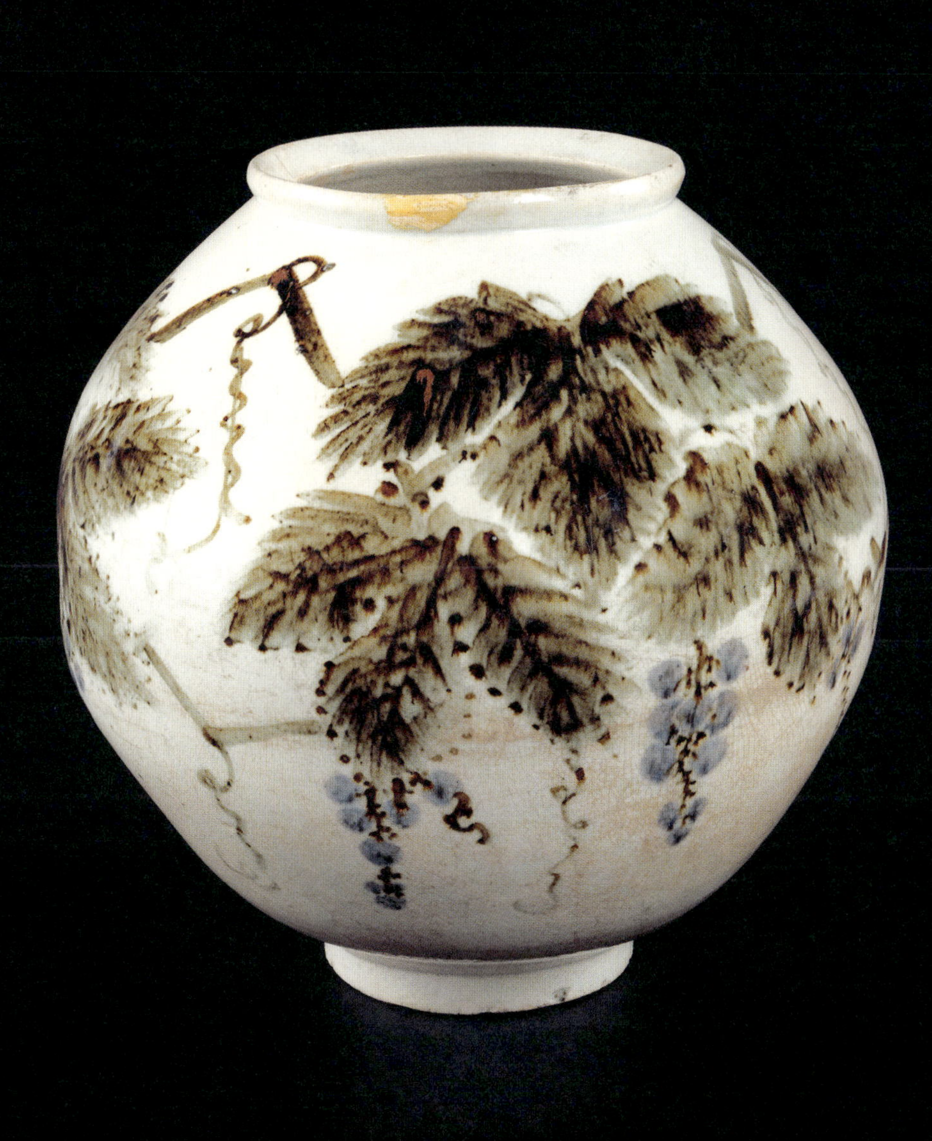

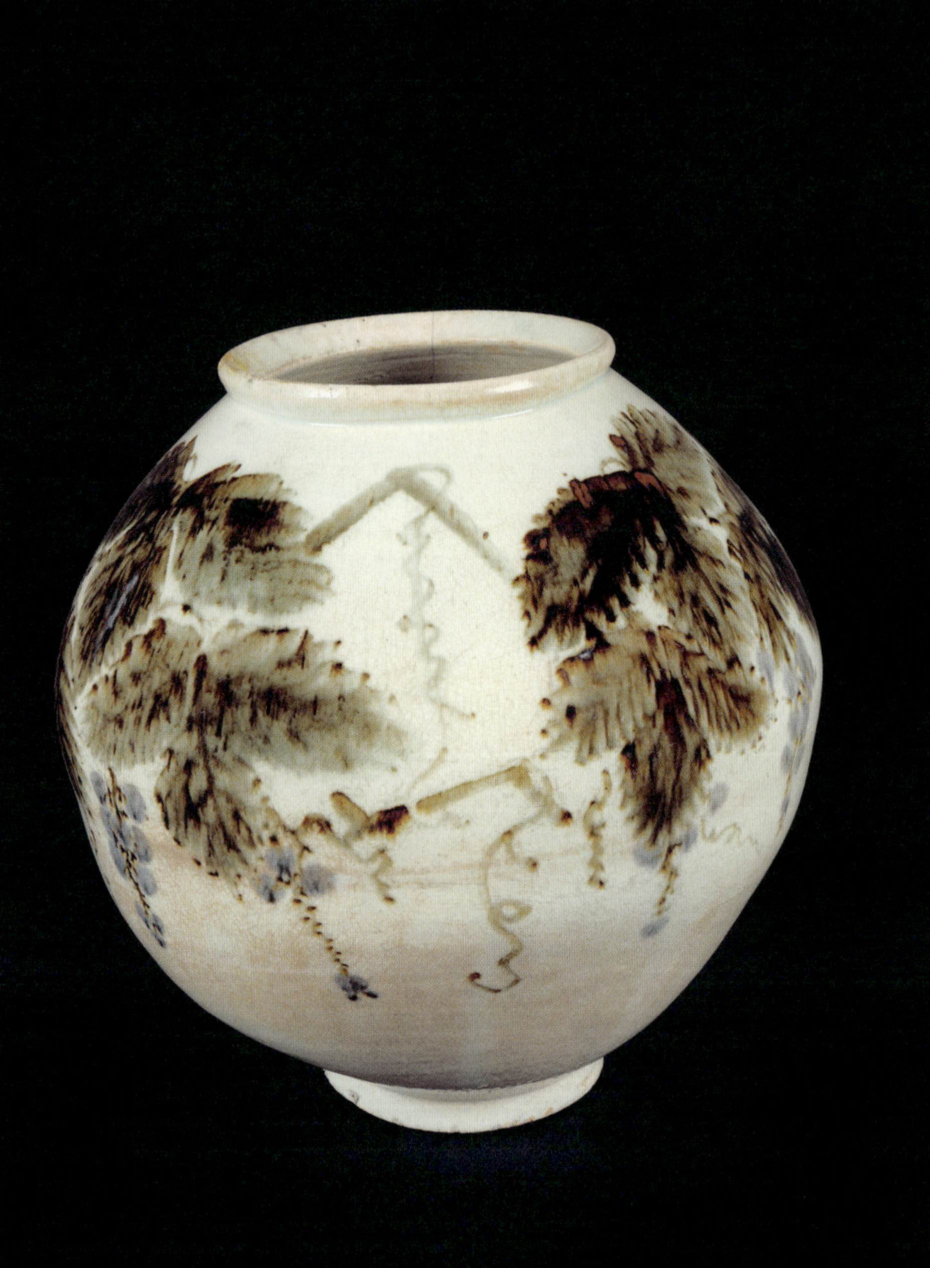

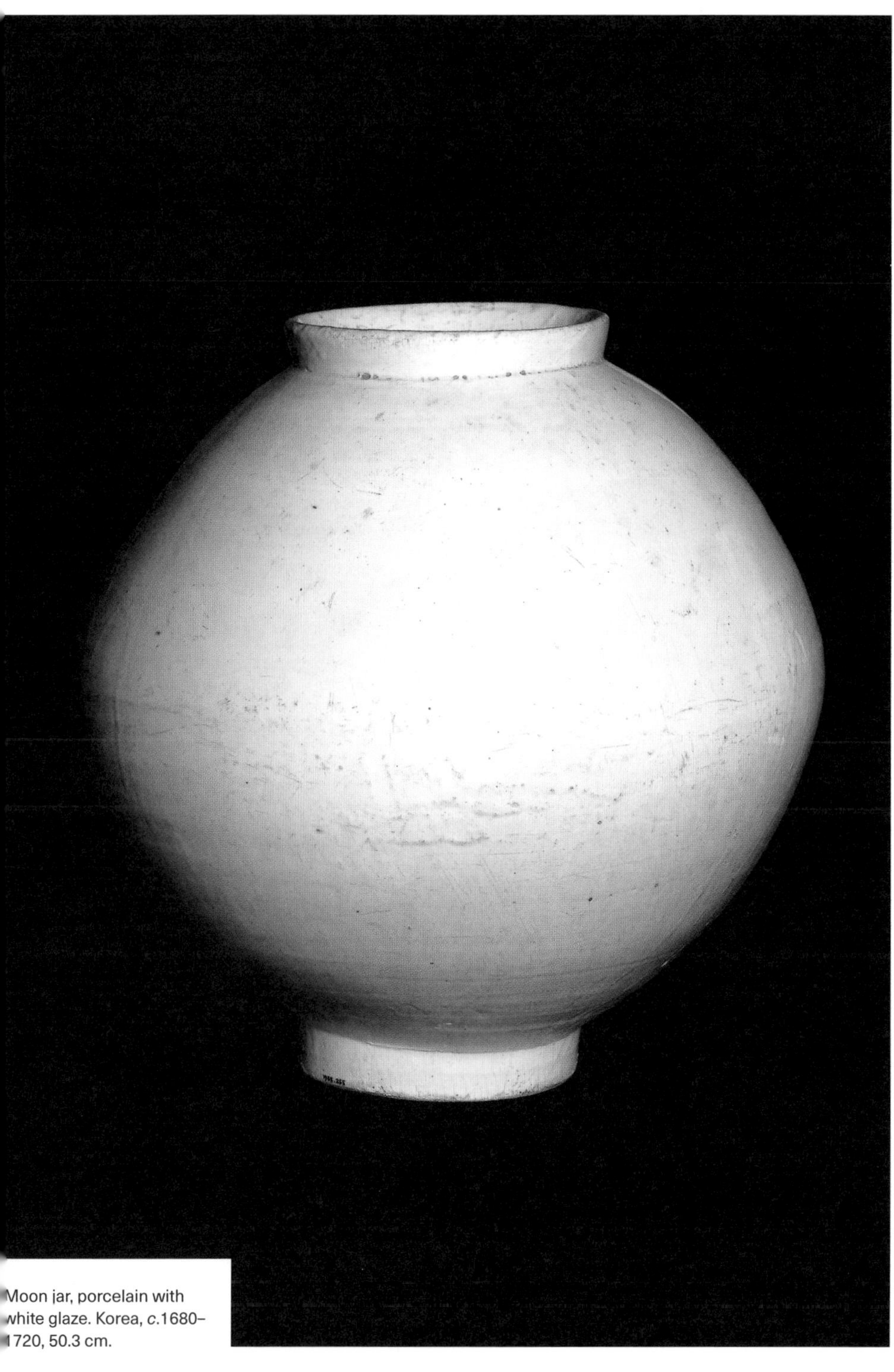

Moon jar, porcelain with white glaze. Korea, *c*.1680–1720, 50.3 cm.

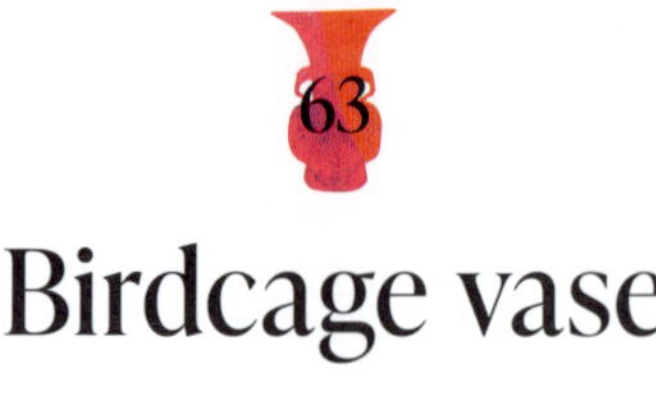

Birdcage vase

Japan
*c.*1700 CE
Porcelain, 53 x 38 cm

This "birdcage" vase is one of the more extraordinary porcelain objects exported from Japan in the period of European export trade (from *c.*1600–1740). It was made in the town of Arita, in around 1700. The vase is decorated in underglaze blue and bears a design of peonies on the outside and two large dragons inside the rim.

Unusually, its body was left unglazed in sections, so that its creator could add elaborate customised decoration: porcelain handles in the shape of elephant heads, gilded papier mâché panels and a wire cage. The cage encloses two porcelain pheasants with metal legs, perched on rocks made of painted wood. From the rocks sprout flowering branches of prunus, most of which are now missing. The unglazed background inside the cage is gilded and painted with trees and flowers.

This lavishly decorated piece of chinoiserie fantasy must have been a special order conveyed by the Dutch East India Company merchants in Japan. We have no concrete evidence to back up the hypothesis, but it's very possible that the vase is connected to one particular eighteenth-century porcelain enthusiast, Augustus the Strong, King of Poland and Elector of Saxony (1670–1733). We know of about 20 "birdcage" vases in existence, nine of which are in the Zwinger Museum in Dresden, part of a collection which Augustus the Strong formed in the early 1700s. The king was a keen collector of Asian ceramics. He was so interested in Chinese and Japanese porcelain that he encouraged the alchemist Johann Friedrich Böttger to search for the formula for white porcelain – at that time, a secret known only in China and Japan. After Böttger perfected his recipe for porcelain in 1709, Augustus founded the Meissen Porcelain Manufactory, the first porcelain factory in Europe.

Augustus amassed over 20,000 pieces of Chinese and Japanese porcelain, which he used to decorate his "porcelain palace" in Dresden. We know that they included a number of birdcage vases, since 20 were listed in the 1721–27 inventory of his collection. But over the years, Augustus's porcelain collection was gradually dispersed. Many "duplicate" pieces were sold or exchanged, and there were further losses during the Second World War, when the collections

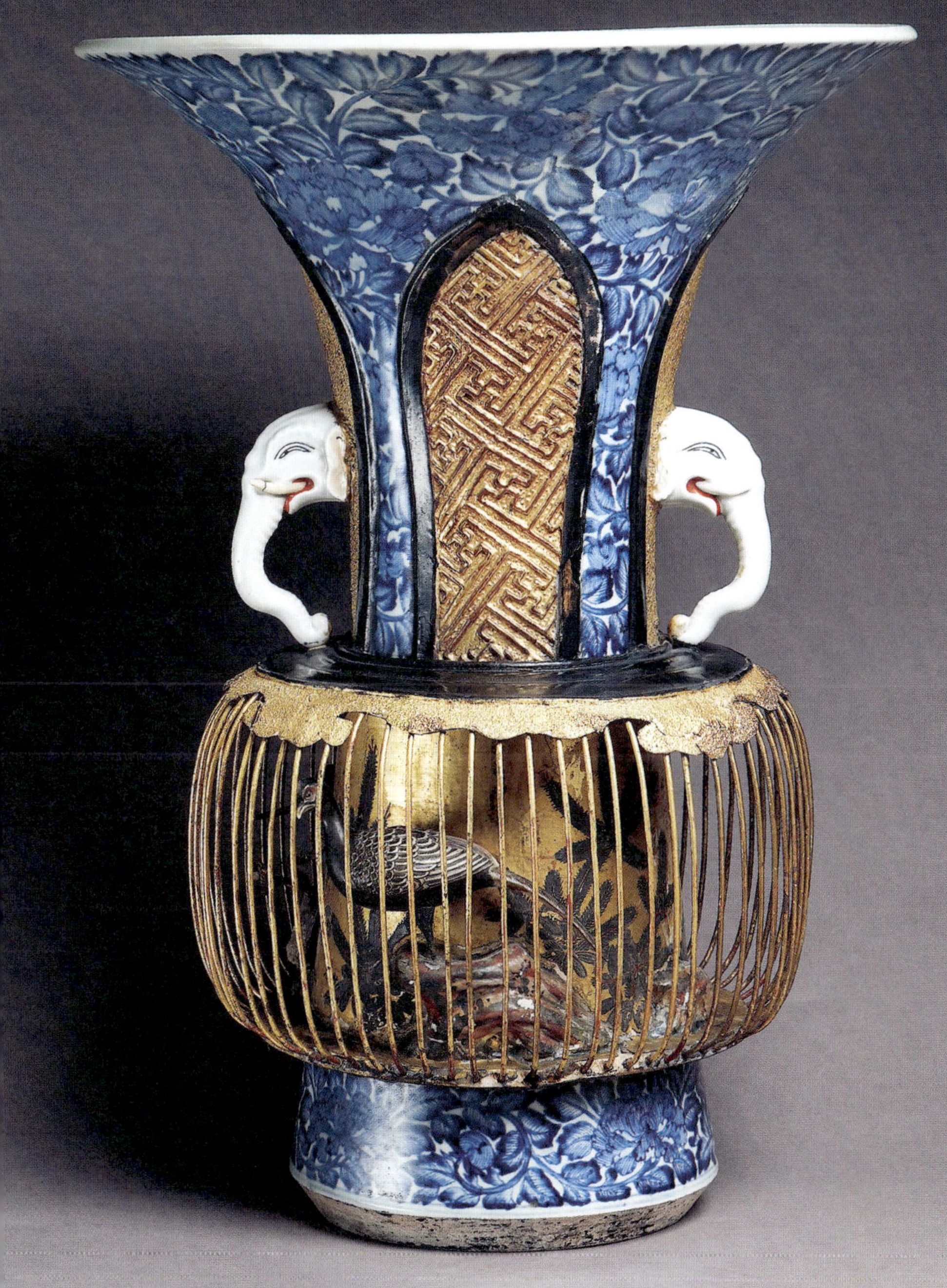

were temporarily moved to stores outside Dresden. By 1779, the collection inventory lists only 14 birdcage vases and eventually, only nine were left. According to surviving records held in Dresden, two of the vases were sold at auction in 1920 and four were returned to the Saxonian royal family in 1999.

This vase was sold at Sotheby's in Amsterdam on 7 May 1992 and was subsequently bought by the Ashmolean from a London dealer. A letter from a previous owner states that the vase was one of a pair acquired by the Gallery Werner in Leipzig before the Second World War, bought directly from a descendant of the Saxonian royal family. One of the vases was destroyed in the Second World War. The surviving piece was restored in Dresden around 1962 and later sold to a collector in Leipzig, thereafter passing by descent to the owner who sold it at Sotheby's in 1992. Could it be one of the Dresden vases? Many of the pieces in the collection of Augustus the Strong were marked with their historic inventory numbers, either painted in black ink or incised through the glaze, the numbers underscored with a wavy line. On the base of the Ashmolean vase there is a very faint ink mark that could, with an imaginative eye, be read as the remains of "N. 18" over the traces of a wavy line. This would correspond to the appropriate number in the 1720s inventory.

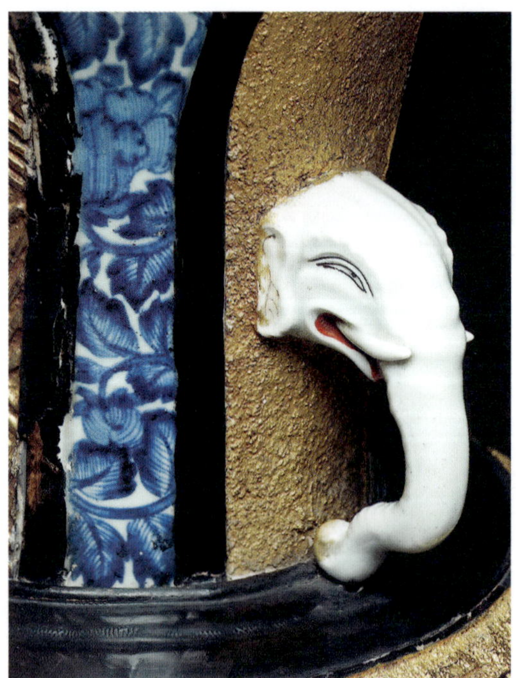

Porcelain vase, papier mâché, underglaze painting in blue and overglaze lacquer decoration in gold; modelled porcelain pheasants, enamelled, and with wire legs; modelled porcelain flowers, with lacquered textile and wire stems; models mounted on lacquered wood, inside a gilded wire cage.

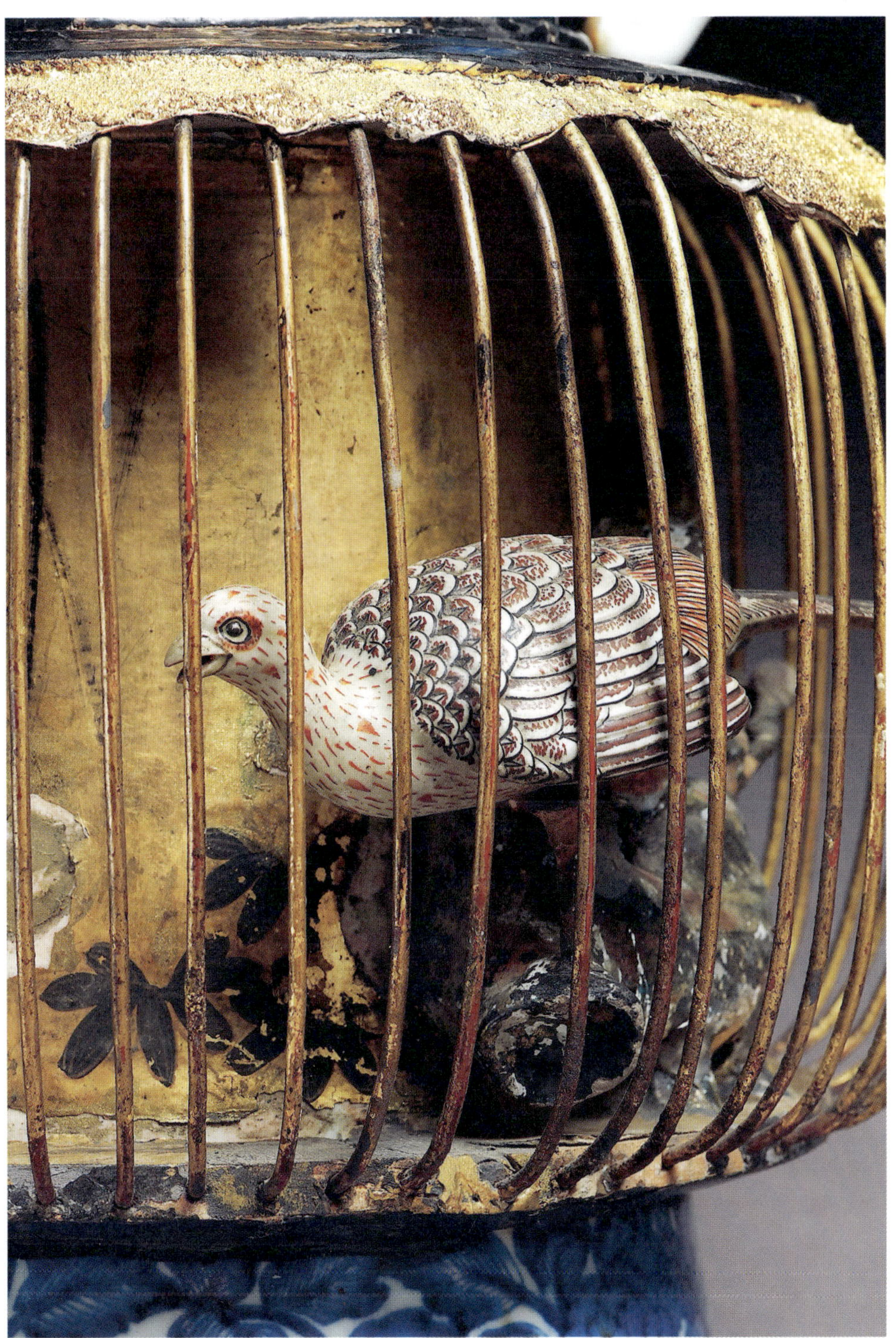

64

Osprey, seated

Germany
1731–32 CE
Porcelain (hard paste), moulded, glazed, 54.5 cm

This sculpture in white porcelain, half a metre tall, was modelled for the Meissen porcelain factory in Saxony in around 1731. It was part of a menagerie of hundreds of porcelain animals and birds, both domestic and exotic, and mostly life-size.

They were commissioned by Augustus the Strong, and displayed alongside his famous collection of Asian porcelain in the Japanese Palace at Dresden.

Most of the creatures have remained in Dresden, but some, including this one, have been sold at various times. Brilliantly modelled by the sculptors Johann Gottlieb Kirchner (1706–68) and Johann Joachim Kändler (1706–75), they are the most ambitious and virtuosic series of porcelain sculptures in the history of ceramics. Kändler, the sculptor of this sea eagle, modelled it after real specimens in Augustus the Strong's personal zoo.

Known as "white gold", Chinese and Japanese "hard-paste" porcelain was highly prized throughout Europe. Augustus the Strong's extreme passion for Asian porcelain was legendary. The term used was *maladie de porcelaine*, or "porcelain sickness": he spent a fortune collecting thousands of pieces. He once exchanged 600 of his own cavalrymen for 151 pieces of Chinese porcelain belonging to Frederick William I of Prussia.

Only 20 years after Augustus founded the Meissen factory, this superb osprey was manufactured there. We can see evidence of the technical difficulties involved in firing such a large and complex piece of porcelain in the many firing cracks around the sculpture's rocky base. The piece remains uncoloured because a second firing would have been needed for the enamels, and this would have risked further cracking. Many of the porcelain animals in the menagerie were cold painted in naturalistic colours, but the paint has now flaked away or been washed off.

This severe and formidable bird of prey was always known in the donor's family as "Birdie". It is still affectionately known by this name at the Ashmolean Museum.

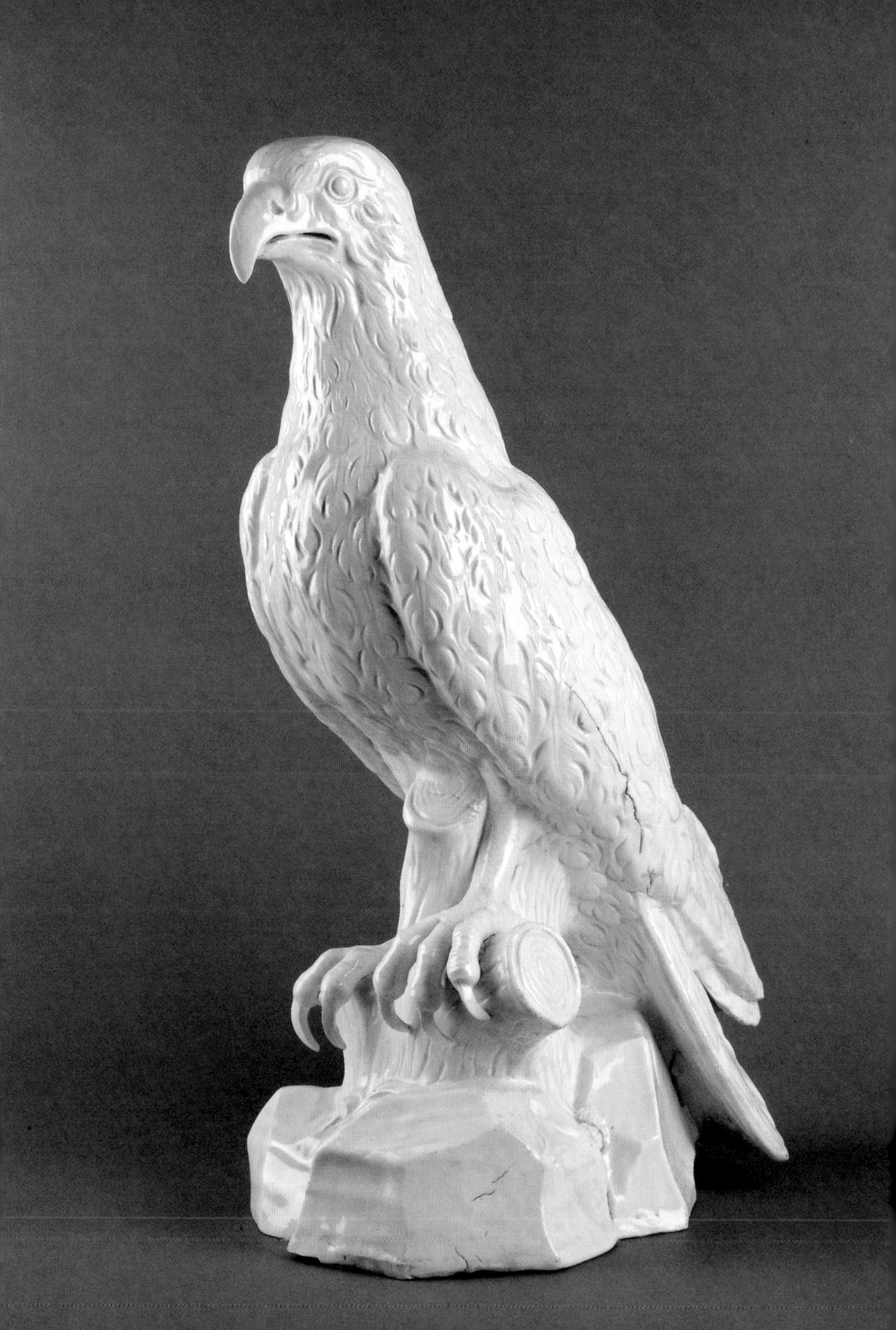

The Oxford Plate

China
*c.*1745 CE
Porcelain, grisaille decoration, famille rose
overglaze enamels, gilding, 23 cm

The Oxford plate tells a remarkable story of cultural exchange between East and West during the eighteenth century. It's an example of Chinese export porcelain which bears an unexpected image on its surface: the Danby Arch, the main entrance to Oxford Botanic Garden. The gateway is a Doric arch, under which visitors enter this site of scientific research, originally named the Physic Garden.

The centre of the plate is black and white, with a gold frame and a spearhead border of gold outlined in red. The flowers on the rim are in famille rose colours; the edge is gilt. In the foreground is a heavily bearded man in a flat cap and short gown, holding a posy of flowers in his right hand. In his left is the snake-twined staff of Asclepius, the ancient Greek god of medicine. This man is Jacob Bobart the Elder (1596–1680), a noted botanist who was the first Keeper of the Physic Garden. He is accompanied by a dog and a goat, both sacred to Aesculapius. An unexplained bird, apparently a stork, is flying above the arch.

It's been suggested that Humphrey Sibthorp, Sherardian Professor and Keeper of the Garden, was responsible for commissioning the plate in 1753. The image originated from an engraving that was published in 1713. It was almost certainly drawn and engraved by Michael Burghers, a refugee from the Netherlands. The processes involved in producing the plate were complex. The engraving would have been transported from England to China and meticulously copied onto the plate before it was fired. The finished object would have then been carried all the way back to England. The plate is the result of a commercial and cultural interchange between West and East that took place over 250 years ago, yet rivals economic interdependence in the modern world.

Several examples of the plate are known. Perhaps they were intended to grace dessert at the professorial dinner table or designed as souvenirs for visitors to the Botanic Garden. In any case, it's gratifying that the memento of Britain's oldest botanic garden (founded in 1621) should be preserved in its oldest public museum (founded in 1683).

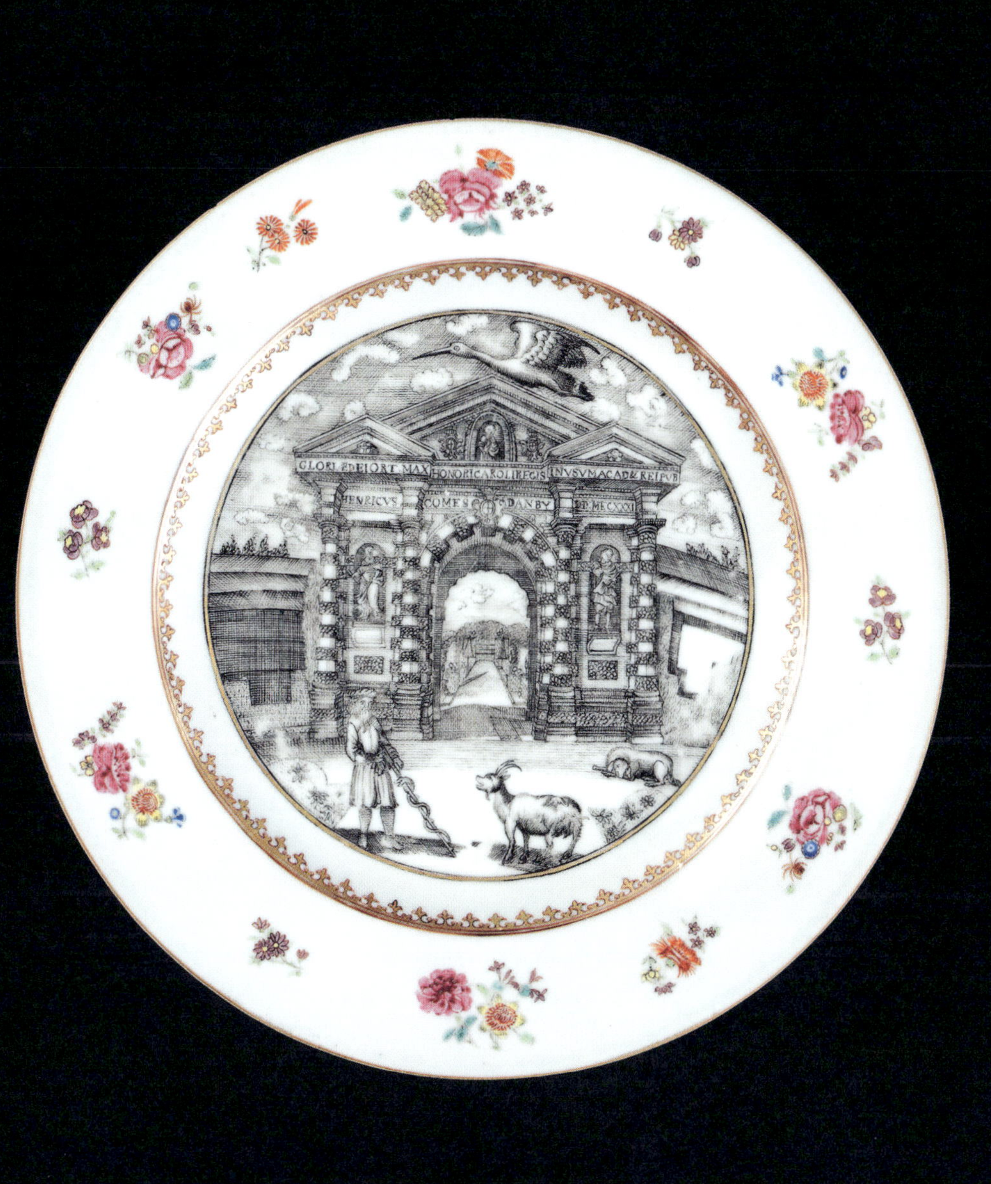

GLORI ÆDEI ORT MAX HONORI CAROLI REGIS IN VSVM ACAD & REIPVB
HENRICVS COMES DANBY DD MDCXXXI

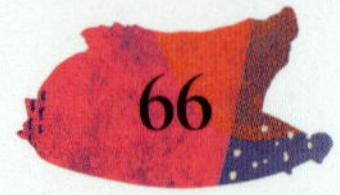

Boar's head tureen

UK
1755–60 CE
Frit paste porcelain, moulded and enamelled,
27 x 53.3 x 39 cm

This wonderfully grisly tureen was the most spectacular piece in a lavish collection of Chelsea porcelain, acquired sometime around 1764 to furnish a grand new house in Yorkshire.

The tureen is in the form of a wild boar's head, while the stand is moulded with a hunting knife, an oak branch and a quiver of arrows. These lie on something that might be intended to be a quilted hunting waistcoat. We think this may be the most expressive example of functional English porcelain tableware that exists today.

Chelsea was one of the first English porcelain factories to be founded. It was established in about 1745 by Nicholas Sprimont, a Huguenot silversmith who was originally from Liège. There were also factories in Bow and Worcester that made porcelain that was marketed at the emerging middle classes. In contrast, Chelsea porcelain was expensive and intended for royalty and the aristocratic elite. In March 1745, the *Daily Advertiser* declared: "We hear that the China made at Chelsea is arriv'd to such Perfection, so as to equal if not surpass the finest old Japan."

This tureen shows Chelsea porcelain at its best. It may have been modelled by Joseph Willems (1715–66), who was born in Brussels and probably started work at Chelsea in 1748. In 1755, the Chelsea Porcelain Factory sale catalogue listed: "A beautiful tureen in the form of a BOAR'S HEAD, in a most curious dish with proper embellishment". Three more were mentioned in 1756. In 1819, an example was included in the sale of personal items belonging to Queen Charlotte.

Like all English porcelain at this time, the tureen is made of artificial or "soft-paste" porcelain. Chelsea's porcelain was particularly glassy and fragile because of the ground glass, or frit, that was added to the clay to imitate the translucency of true hard-paste porcelain. This made the clay very difficult to work with, or fire. When we consider the extent of the challenge, this tureen is an even more remarkable achievement.

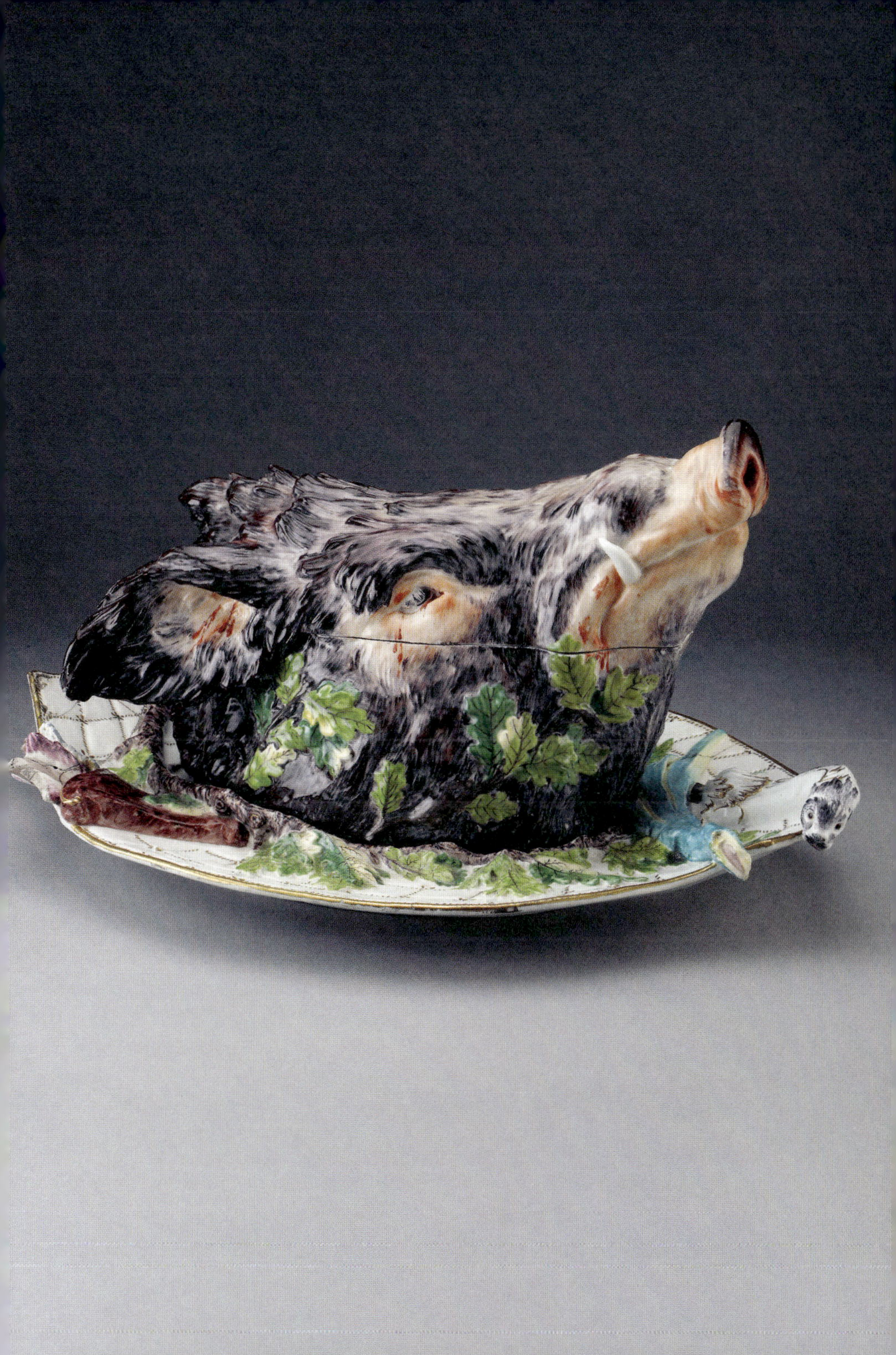

Six tiles with Chinese figures

Mexico
1775–1825 CE
Tin-glazed earthenware, 13 x 13 cm

The creation of these tiles was influenced by potters from Spain, traders from China, artisans in Italy and customers in Europe. The first tin-glazed pottery to arrive in Mexico was brought by the early settlers from Spain in the early sixteenth century, in particular those setting out from the port of Seville.

Potters from Spain were at their heels, establishing potteries in Mexico City in about 1550. They soon created a distinctively Mexican tradition of tin-glazed pottery vessels and decorative tiles. By about 1600, the industry came to be concentrated in the town of Puebla de los Ángeles, south-east of Mexico City, where sources of usable clays were close at hand. By 1698, the Franciscan writer Agustín de Vetancurt claimed with patriotic fervour that "the glazed pottery [of Puebla] is finer than that of Talavera and can compete with that of China".

Spanish technical tradition was at the heart of the industry in Puebla. The impact, specifically, of ceramics from the city of Talavera de la Reina can be seen in the fact that the word *talavera* is still used in Mexico to mean locally made tin-glaze pottery. But in the sixteenth century, Mexicans were already aware of the excellent quality of Chinese porcelain, which passed through Mexico as part of the great trade route involving the cargoes of the "Manila galleons". These trade goods from China crossed the Pacific to Acapulco, on the west coast of Mexico, and were carried overland, before being shipped across the Atlantic to Europe. Chinese blue and white porcelain had a long-lasting impact on Puebla ceramics, much of which was blue and white, and some of which imitated Chinese designs. A lot of seventeenth-century Puebla pottery can be recognised by the use of a heavy blackish blue, appearing slightly in relief on the surface of the glaze. Dated pieces are very rare and we've been forced to give the pieces shown here wide chronological brackets.

The tinted glaze is similar to the *berettino* glazes of Italian maiolica potters; these glazes were also used by potters at Seville. The chinoiserie style demonstrates a more or less distant and often light-hearted imitation of Far Eastern models and modes.

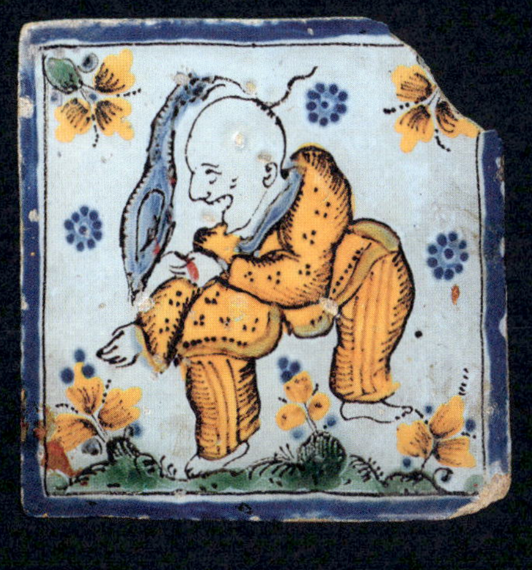
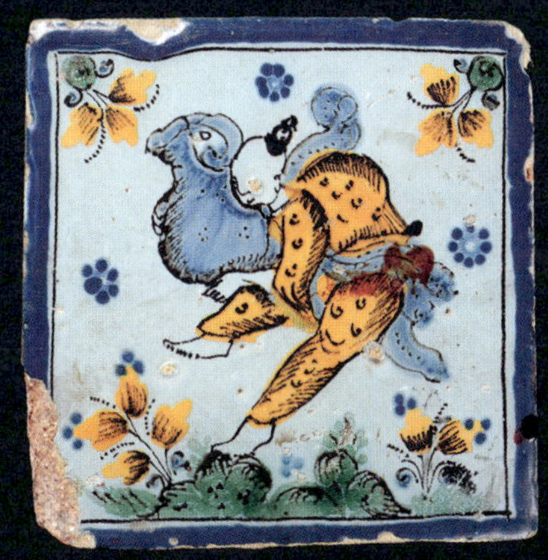
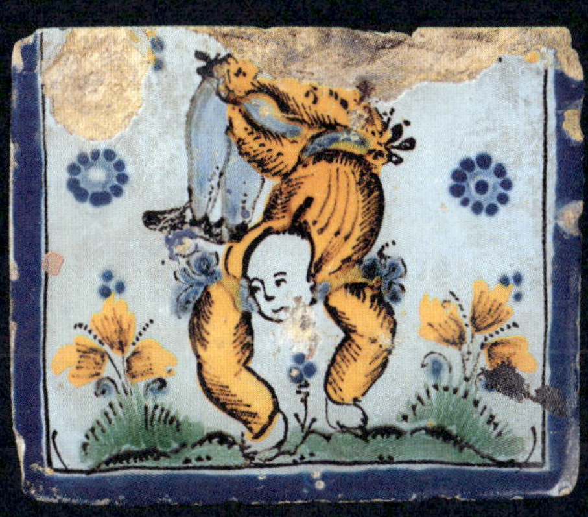
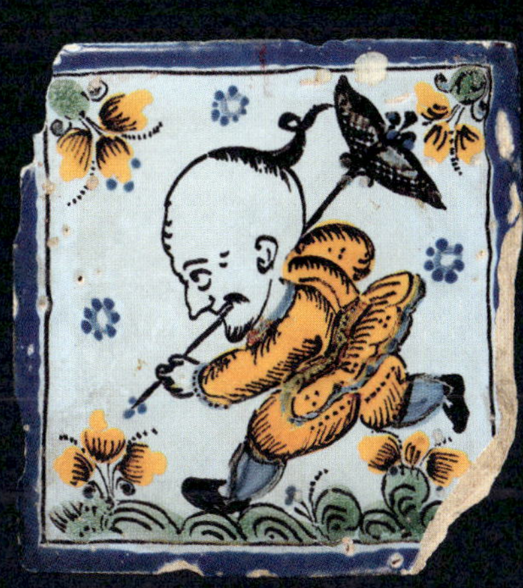
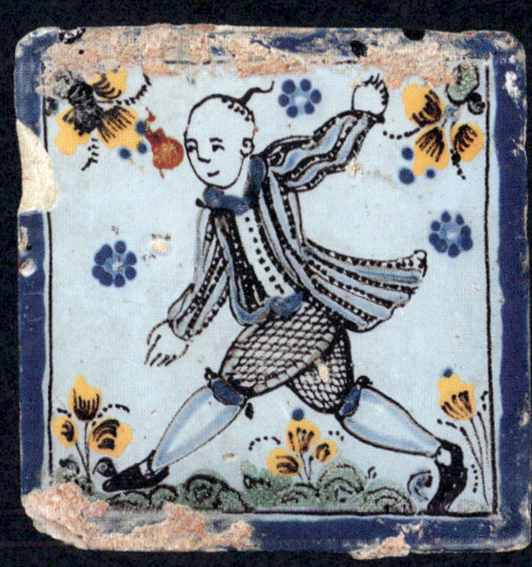
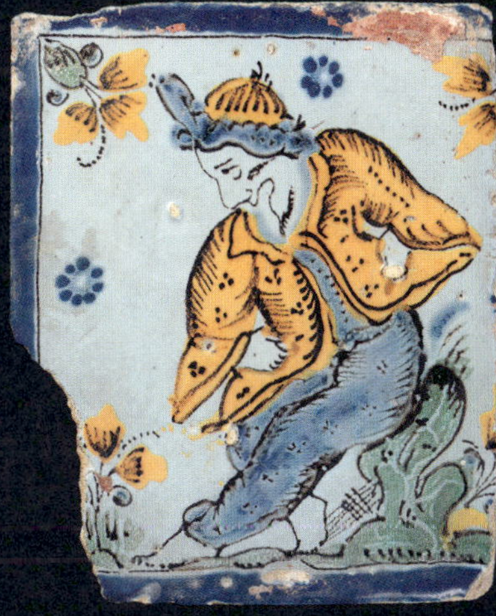

Teapot and lid

UK
*c.*1772 CE
Soft-paste porcelain, painted and gilt, 14.4 cm

The Worcester Porcelain Factory, founded in 1751, was one of the earliest English porcelain enterprises and outlasted many competitors. From the beginning, the factory focused on making functional pieces, which were usually designed to tempt the emerging middle classes.

At first, it decorated its porcelain in close imitation of Asian styles, in blue and white, famille verte and famille rose. But it soon developed its own palette and style. By the mid 1750s, the factory began to draw on Meissen porcelain for inspiration, decorating its products with European figures in landscape settings, harbour scenes and gorgeous depictions of flowers.

The recipe for Worcester's artificial, or "soft-paste", porcelain included soaprock, a naturally occurring mixture of china clay and magnesium silicate from Cornwall. This was the basis for a malleable clay, which produced a body that was thin, white, translucent and could be crisply moulded. Most importantly, this substance could withstand hot liquids without cracking, unlike the soft-paste porcelain produced by other factories such as Chelsea. This made it especially suitable for serving tea. In 1763, the Annual Register gave a survey of English porcelain: "except the Worcester, they all wear brown, and are subject to crack, especially the glazing, by boiling water: the Worcester has a good body, scarce inferior to that of Eastern China, it is equally tough, and its glazing never cracks or scales off". These benefits gave Worcester a commercial advantage over other porcelain manufacturers. The penalty for disclosing the secret formula (the property of the subscribers) was £4,000, although it was never invoked.

This teapot is painted on each side with lovers in a pastoral setting that imitates ancient Greco-Roman style. It's signed "Duvivie[r]" and dated 1772. This signature makes it a key work documenting the career of the ceramic painter Fidelle Duvivier. Born in 1740 in Tournai, in what is now Belgium, and doubtless trained in the porcelain works there, he came to England in 1763 or 1764 and worked at Derby and several other English porcelain factories. This teapot was made at Worcester but Duvivier may have painted it in the influential London workshop of the porcelain decorator James Giles.

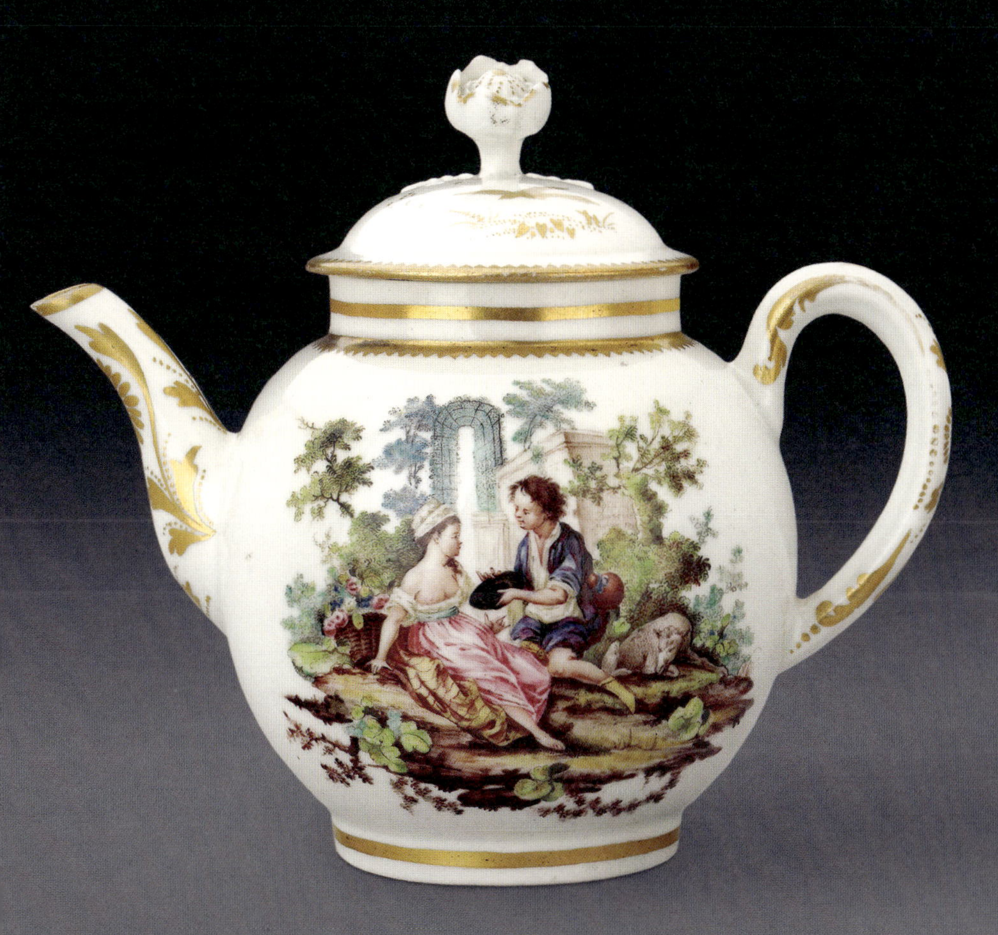

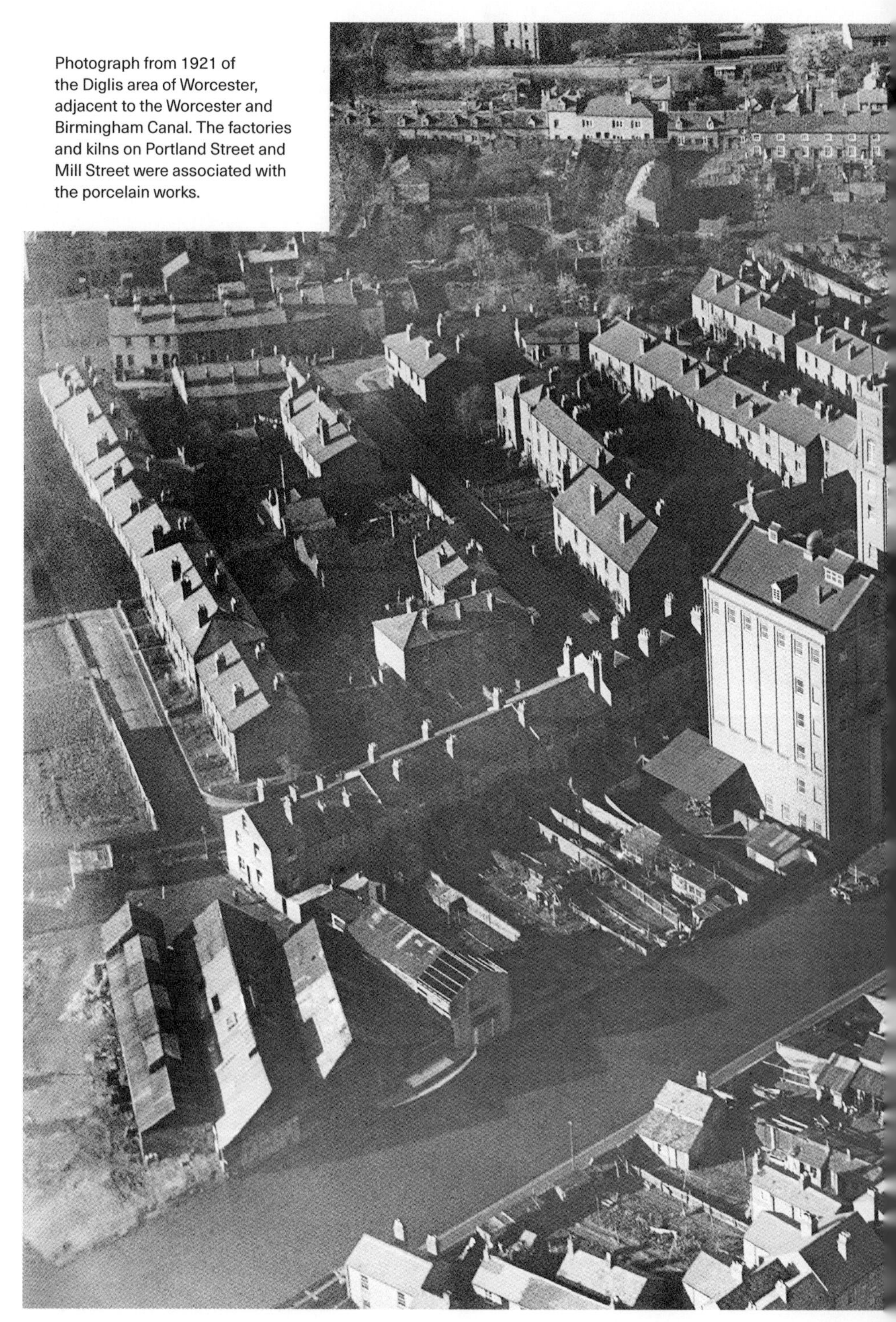

Photograph from 1921 of the Diglis area of Worcester, adjacent to the Worcester and Birmingham Canal. The factories and kilns on Portland Street and Mill Street were associated with the porcelain works.

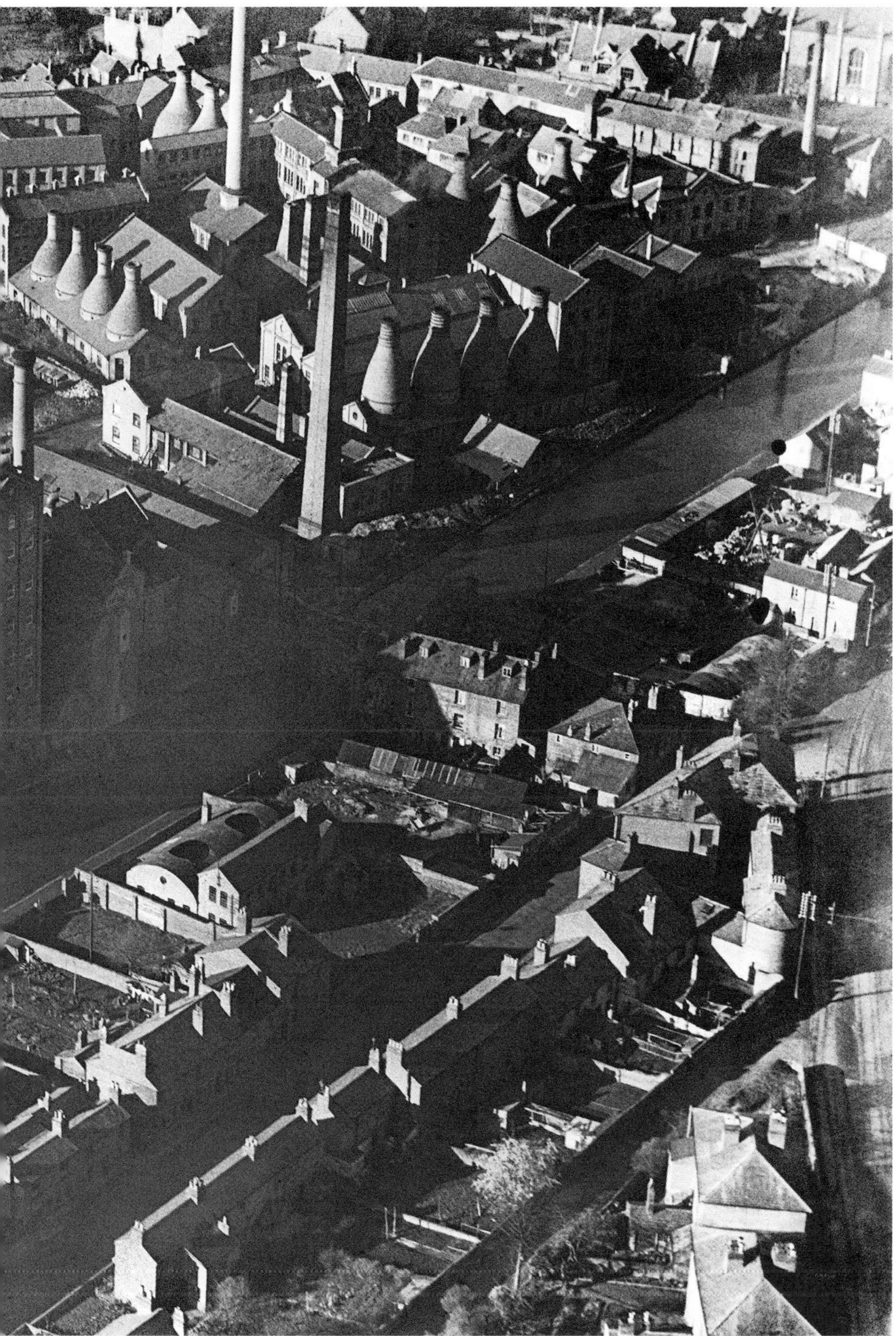

Vase with a design of a bird

Japan
*c.*1910 CE
Yabu Meizan (1853–1934 CE)
Earthenware, overglaze enamels, gold impressed seal
mark on base, 25 x 9 cm

The striking decoration you can see on this vase is typical of the work of the ceramic entrepreneur Yabu Meizan. Meizan was working at a time of rapid change in the Japanese ceramic industry.

In the late nineteenth century, after Japan was forced by aggressive Western nations to open to international trade, the country raced to modernise every aspect of society in order to establish itself as a major player on the world stage. After studying the art of ceramic painting in Tokyo, Meizan established a workshop in Osaka in 1880. This produced sophisticated overglaze enamelled earthenwares in the "Satsuma" style, which featured a heavy use of gold, floral decoration and ornamental borders. These creations were aimed mostly at the overseas market. They earned Meizan a great reputation both in Japan and abroad.

A ceramic decorator rather than a potter, Meizan bought in undecorated blanks from Kagoshima, in the Satsuma region of Kyushu. His early designs featured minute depictions of masses of flowers, butterflies or figures – samurai, farmers, townspeople or children, taking part in festival processions or engaged in everyday activities. The painters who created this work achieved consistency and precision, even with the smallest motifs, by using transfers made from copperplate engraved designs. This gave them precise outlines that they could then colour in intricate detail.

In the early 1900s, probably in response to the popularity of art nouveau, Meizan introduced a new, more open style in which a single motif was arranged dramatically across the surface of a vessel in a way that complemented its shape. This vase, with its painterly design of a bird eyeing up a spider, is a lovely example of this new approach. While the overall decoration is restrained, the level of detail in the motif itself is impressive. The feathers of the bird, the bark of the cherry tree, the surface of the leaves and the body of the spider are depicted in a highly naturalistic style. This style was an up-to-date version of the traditional Japanese Shijō school of painting.

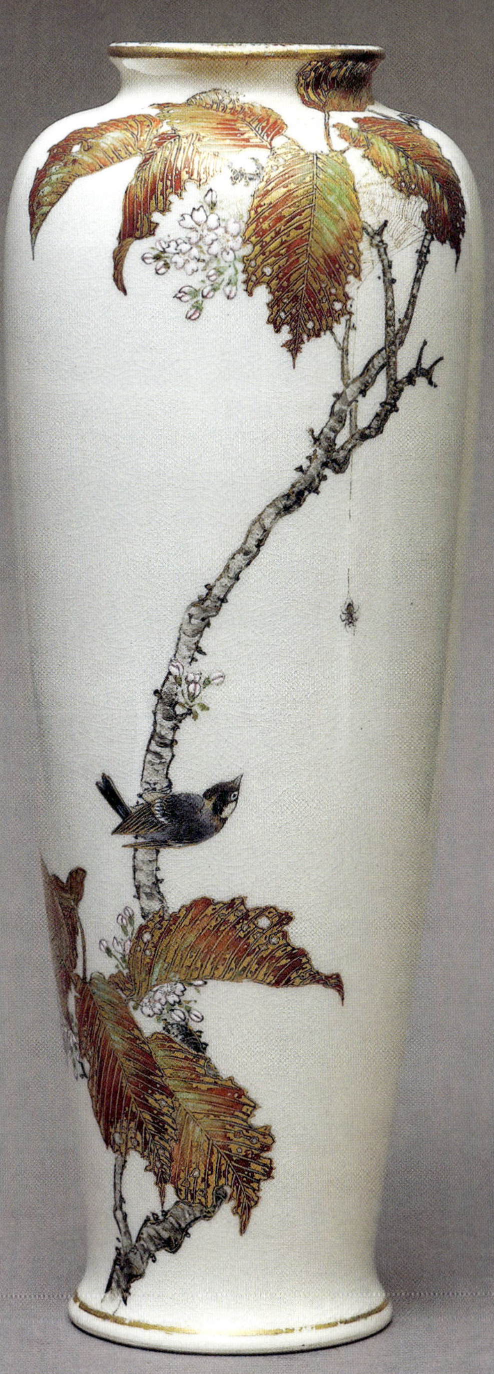

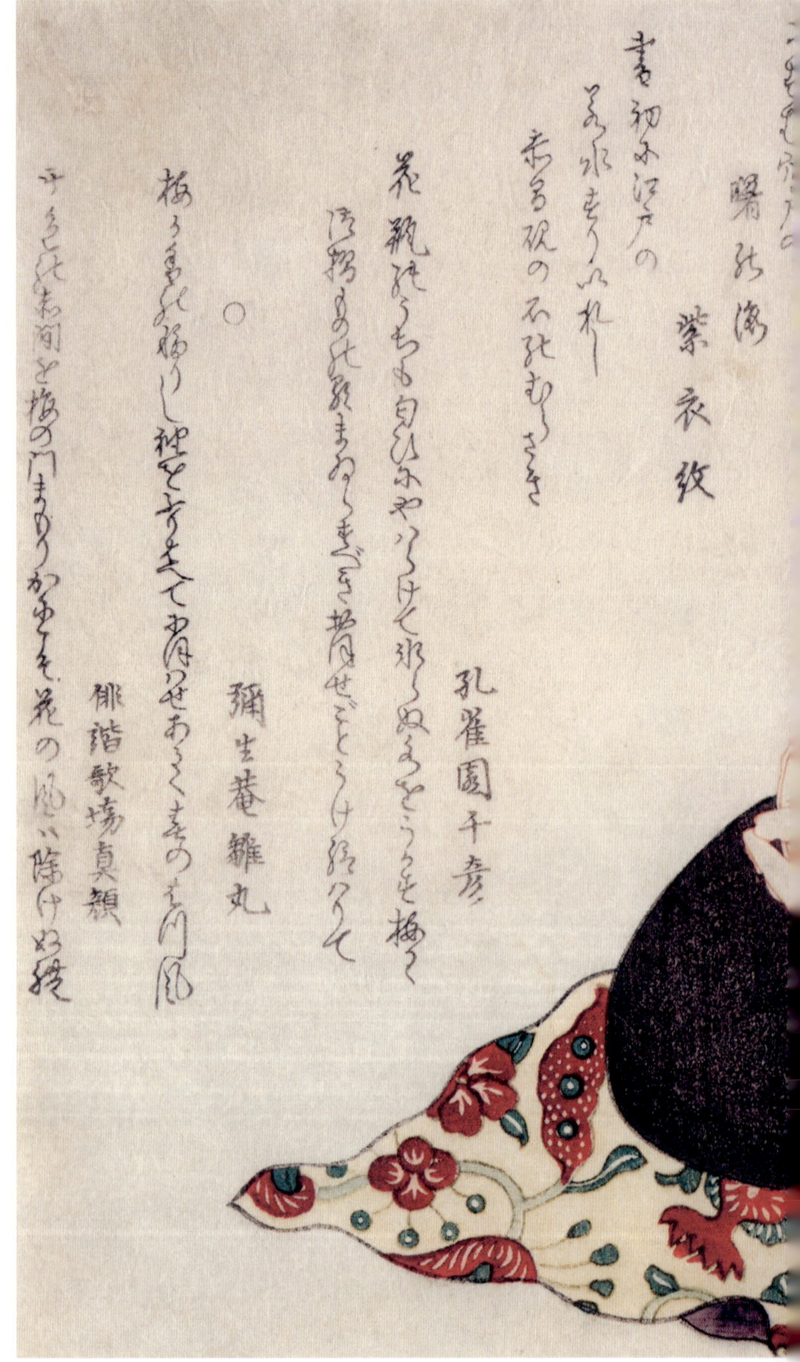

Plum twigs in a vase and a crab on a black court hat, Yūshin Rintei (1780s–1820s), c.1815. Colour woodcut, 16.5 x 20.4 cm.

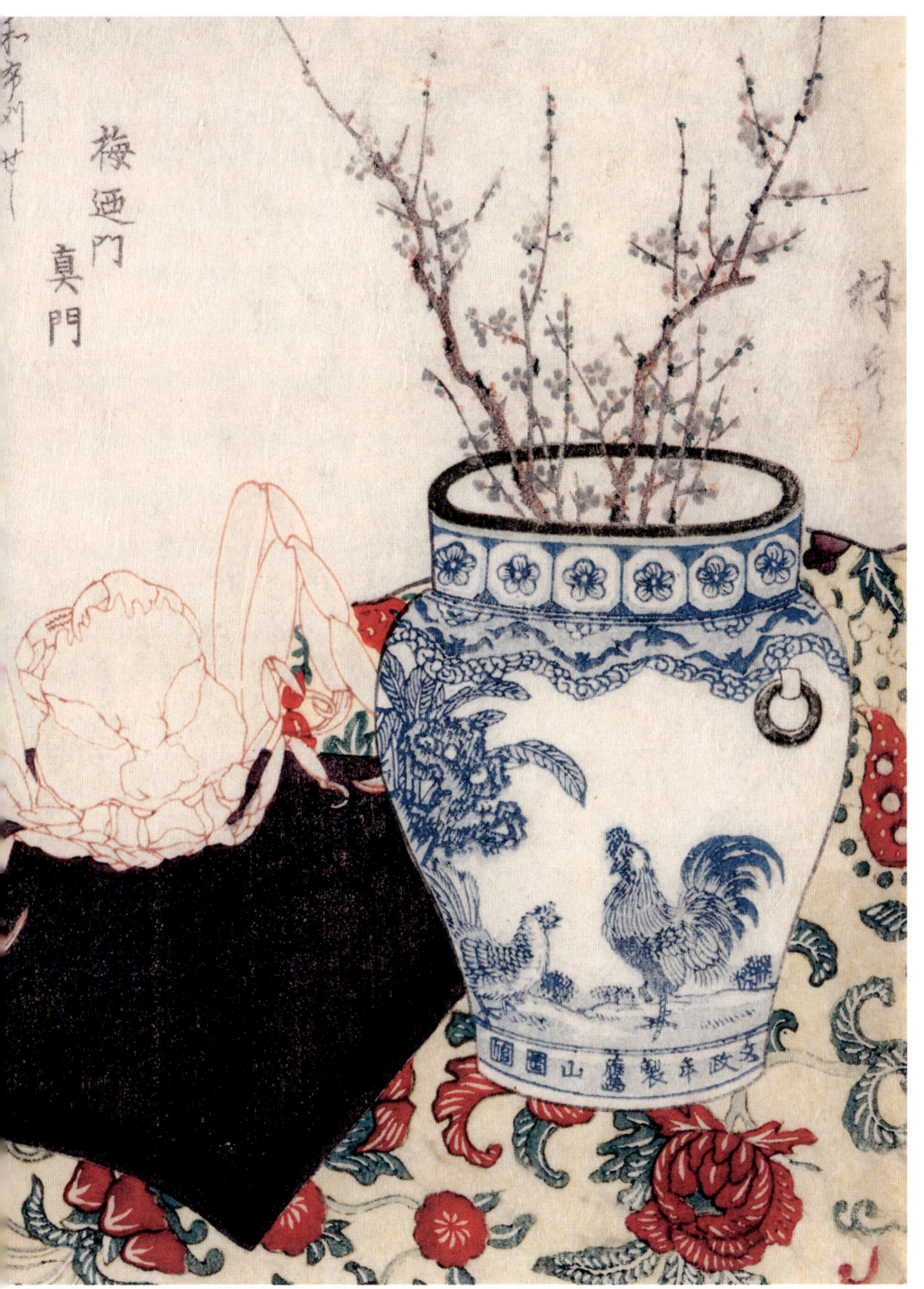

梅逕門
真門
嗊圖山慶製年政玄

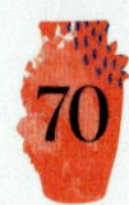

Satsuma-style vase with lotus plants and ducks in high relief

Japan
1870s
Miyagawa Kōzan (1842–1916 CE)
Earthenware, modelled decoration in high relief, overglaze
enamels, gold, 36 x 26 cm

Miyagawa Kōzan, also known by his art name Makuzu, was one of the most respected potters working in Japan during the Meiji era (1868–1912). He produced a wide range of ceramics for both Japan and the West, combining new Western designs and techniques with traditional Eastern ones.

Born into a family of traditional Kyoto tea-ware makers in 1842, Kōzan moved to the port of Yokohama in 1870. There, he set up a workshop producing elaborate Satsuma-style ceramics for the Western market. This market had emerged after foreign powers forced Japan to open up to international trade in the mid 1850s. At some point in the mid 1870s, Kōzan began combining typical Satsuma-style decoration, with its copious use of gold, decorative borders, floral motifs and clouds of gold dots, with intricate high-relief models of birds, insects, plants and animals. These elaborate modelled works were ideally suited to High Victorian tastes, and precisely the sort of singular curiosities that Westerners expected from Japan, a country often described by writers and travellers as an exotic fairyland.

It is possible that Kōzan decided to move away from standard, Satsuma-style export wares towards intricately sculpted relief wares in an effort to make his work more "artistic". Before the late nineteenth century, the concepts of "art", "fine art" and "decorative art" had not existed in Japan as they did in the West. New terms had to be devised to translate the rubrics for the world fairs in which Japanese makers were enthusiastic participants. These new terms were used inconsistently, and their use altered over the years, influenced by changing attitudes to the arts in Japan. However, one thing that was clear from the start was that, in the West, the fine arts were considered superior to the decorative arts. Paintings and sculpture, in particular, occupied a high position.

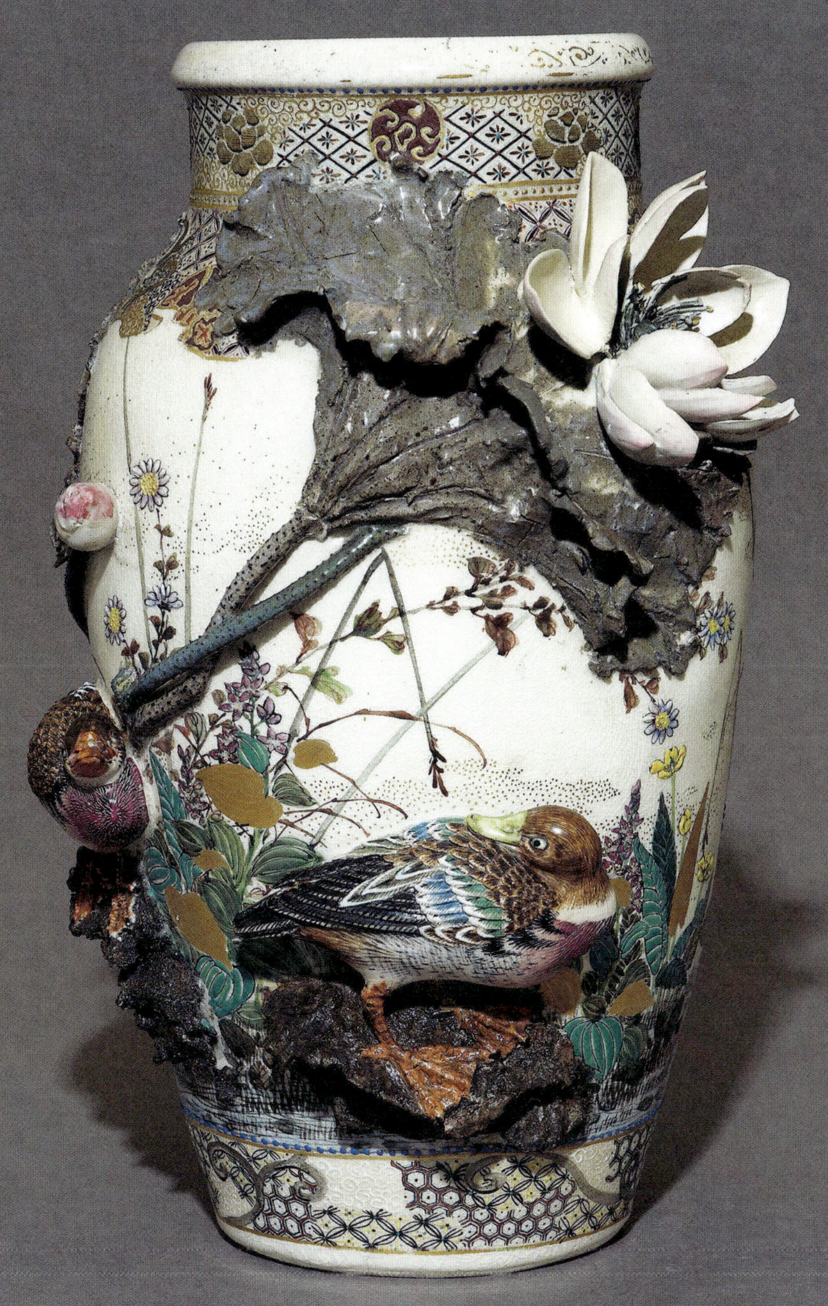

Kōzan's decision to turn to relief work would not have been entirely motivated by artistic aspirations. As the manager of a large-scale workshop, Kōzan had to take a pragmatic approach to ceramic production. According to his personal history, he began carving and modelling the surfaces of his Satsuma-style ceramics as an economy measure – the technique was a substitute for the thick layers of gold pigment that he had previously used. But whatever the motivation, works like the one pictured caused a sensation abroad. At the Paris Exposition Universelle of 1878, for example, Kōzan was awarded a gold medal for a display of vases decorated with modelled crabs, anchors, waterfalls and grazing deer – and praised by critics for his extraordinary imagination and skill in kneading, modelling, tossing, turning, contorting, manipulating, squeezing and even tormenting the potter's clay.

However, Kōzan seems to have stopped making this kind of work in the mid 1880s. At this point, the overseas demand for modelled works was in decline, as Western customers grew more discerning in their taste for Japanese exotica. As Kōzan's heir, Hanzan, later wrote: "Until the early 1880s we produced ceramics decorated with designs of flowers and birds and human figures, in a style closely resembling the Satsuma ware beloved of foreigners, but the foreigners' taste gradually became more sophisticated and they started to become interested in traditional … refined Japanese taste".

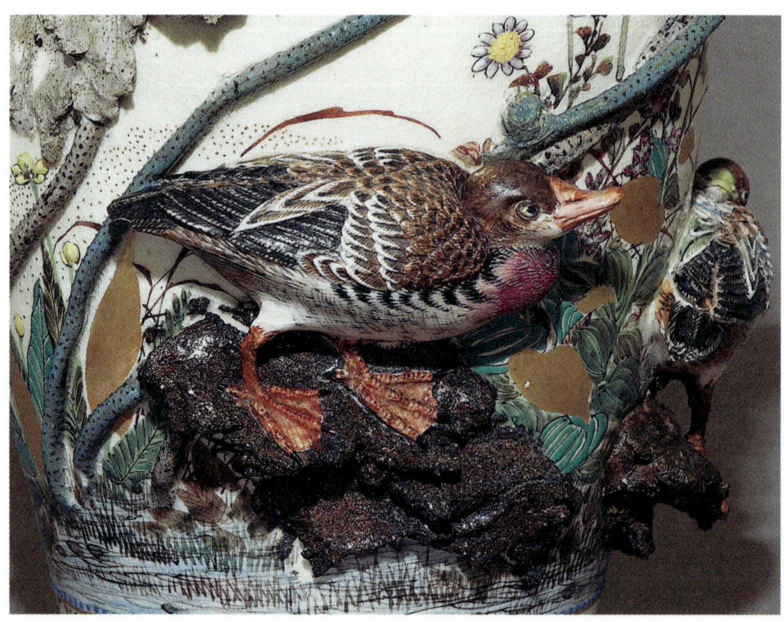

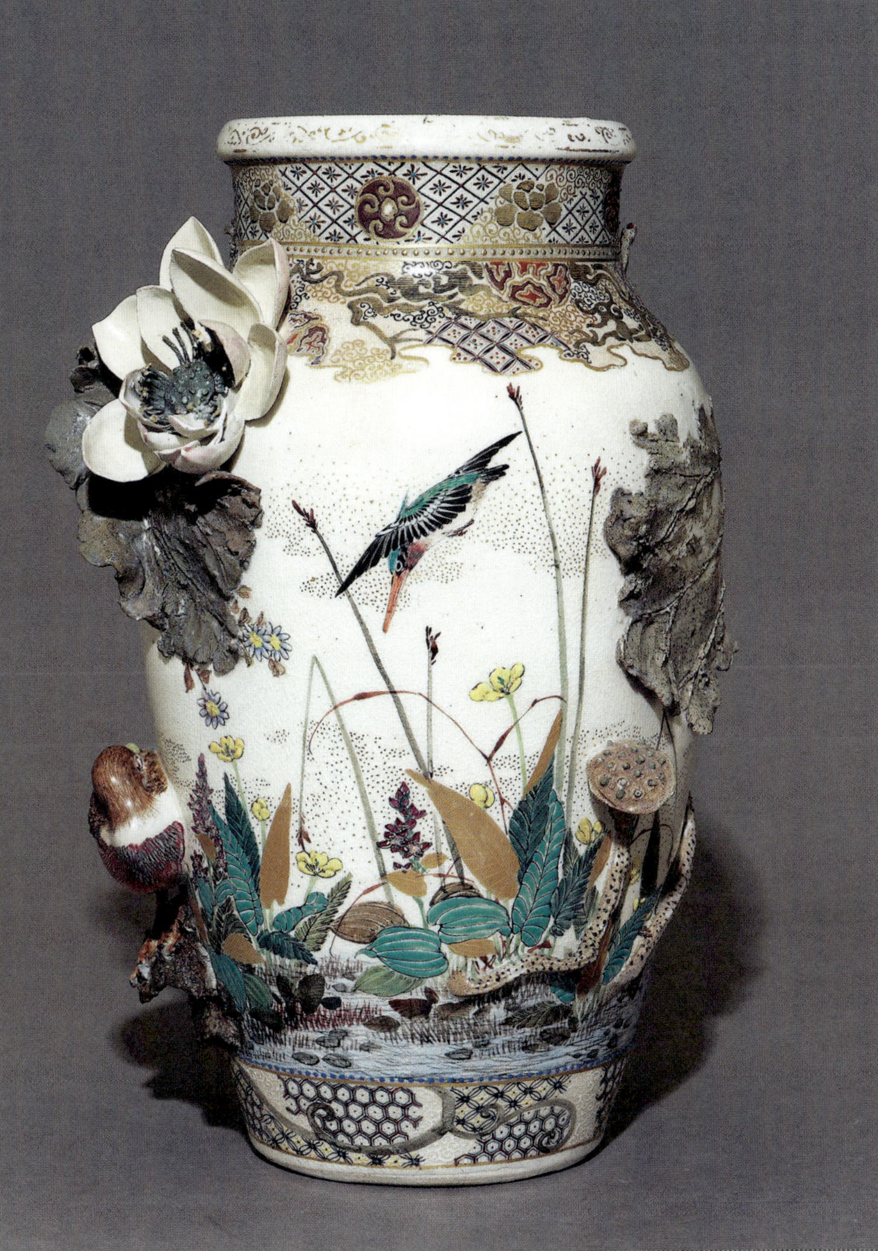

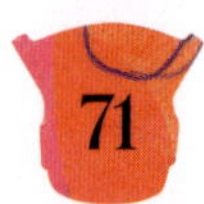

"Peruvian face" bridge-spouted vessel

UK
*c.*1880 CE
Designed by Christopher Dresser (1834–1904 CE)
Manufactured by Linthorpe Art Pottery, Middlesbrough
Earthenware, 18 x 21 x 14 cm

Dr Christopher Dresser is widely regarded as the father of modern design. At age 13, he began his studies at the Government School of Design in London, now the Royal College of Art, set up in 1837.

In the early nineteenth century, Britain led the world in industrial mechanisation and mass production, but Lord Melbourne's government were concerned that the quality of its design lagged behind other countries. They established Schools of Design to train a new generation of industrial designers. Dresser was taught to create new designs by choosing elements from cultures from around the world, both past and present. He was encouraged to seek inspiration in the natural world; every week, fresh flowers were delivered to the Schools from Kew Gardens, for the students to dissect and copy. In fact, Dresser became such an expert botanist that he was awarded a doctorate in the subject from the University of Jena in Germany. He was so proud of the title that he would go on to use it for the rest of his career.

Dresser intended his designs to be mass-produced, so that he could bring "artistic" design to a wider public. He embraced modern industrial manufacturing processes. His refreshingly different work drew on a prodigious range of sources, including China, Japan, pre-Columbian South America and ancient Anatolia.

With its bridge handle, the shape of this vessel is derived from South American design from the pre-Columbian era, before indigenous American cultures were significantly influenced by Europeans. The vessel's experimental glaze owes more to the ceramics Dresser had seen in Japan. The grotesque face carved into the side might be inspired by ancient Chinese pieces.

Unlike his exact contemporary William Morris, who rejected mechanisation, Dresser embraced modern industrial production methods, hoping that his radical designs might reach the widest possible audience. He was a prolific designer who used a wide variety of materials and techniques, and worked for many well-known manufacturers.

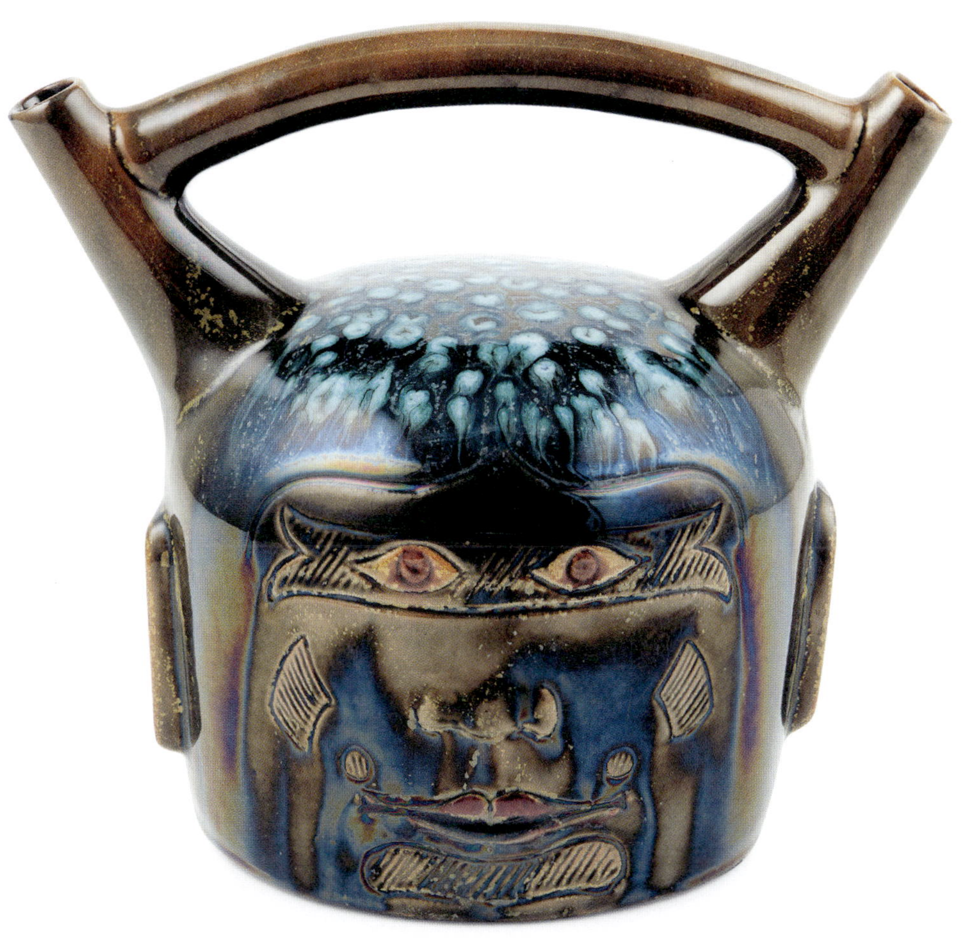

Tiles with red lustre decoration

UK
1882–88 CE
William Frend De Morgan (1839–1917 CE)
Earthenware, red lustre, approx. 15 x 15 cm

William De Morgan was one of the most noteworthy ceramic artists of the Arts and Crafts movement. He trained as a fine artist, but his artistic direction changed when he met the influential poet and textile designer William Morris in 1863.

De Morgan set up his own ceramics company in 1872. His designs were often based on medieval or Islamic patterns, and he experimented with groundbreaking glazes and firing techniques. Around 1873–74, he made a breakthrough. He had rediscovered the technique of lustreware – pottery with a metallic surface such as that made in Muslim Spain – and Italian maiolica.

These cheerful tiles are decorated with De Morgan's signature red "ruby" lustre, which was much admired at the time. The set was made when De Morgan was working at Merton Abbey in Surrey between 1882 and 1888. The creatures perfectly demonstrate William De Morgan's whimsical approach to design. There is a dodo, a peacock and a pelican. A kingfisher and an otter both brandish gigantic fish. De Morgan didn't paint the pottery himself: he made drawings, in various media, which were then transferred onto vessels or tiles by the artists in the factory. For lustre tiles, they used a stencilling technique. At the height of his popularity, De Morgan employed 10 or more decorators in his workshop. By the beginning of the twentieth century, the popularity of his wares had begun to wane and he started to wind up business in 1904.

On 4 March 1887, a group of similar red lustre tiles were ordered from De Morgan by the mathematician Charles Lutwidge Dodgson (1832–98) for his rooms in Christ Church college at Oxford University. Dodgson is better known to us as Lewis Carroll, the author of *Alice's Adventures in Wonderland*. A guest of Dodgson's reported: "He used to make the creatures have long and very amusing conversations between themselves ... Mr Dodgson's allegorical explanation of several of the pictures – for instance the bird which is running its beak through a fish, and the dragon which is hissing defiance over its left shoulder – was that they were representations of the various ways in which he was accustomed to receive his guests."

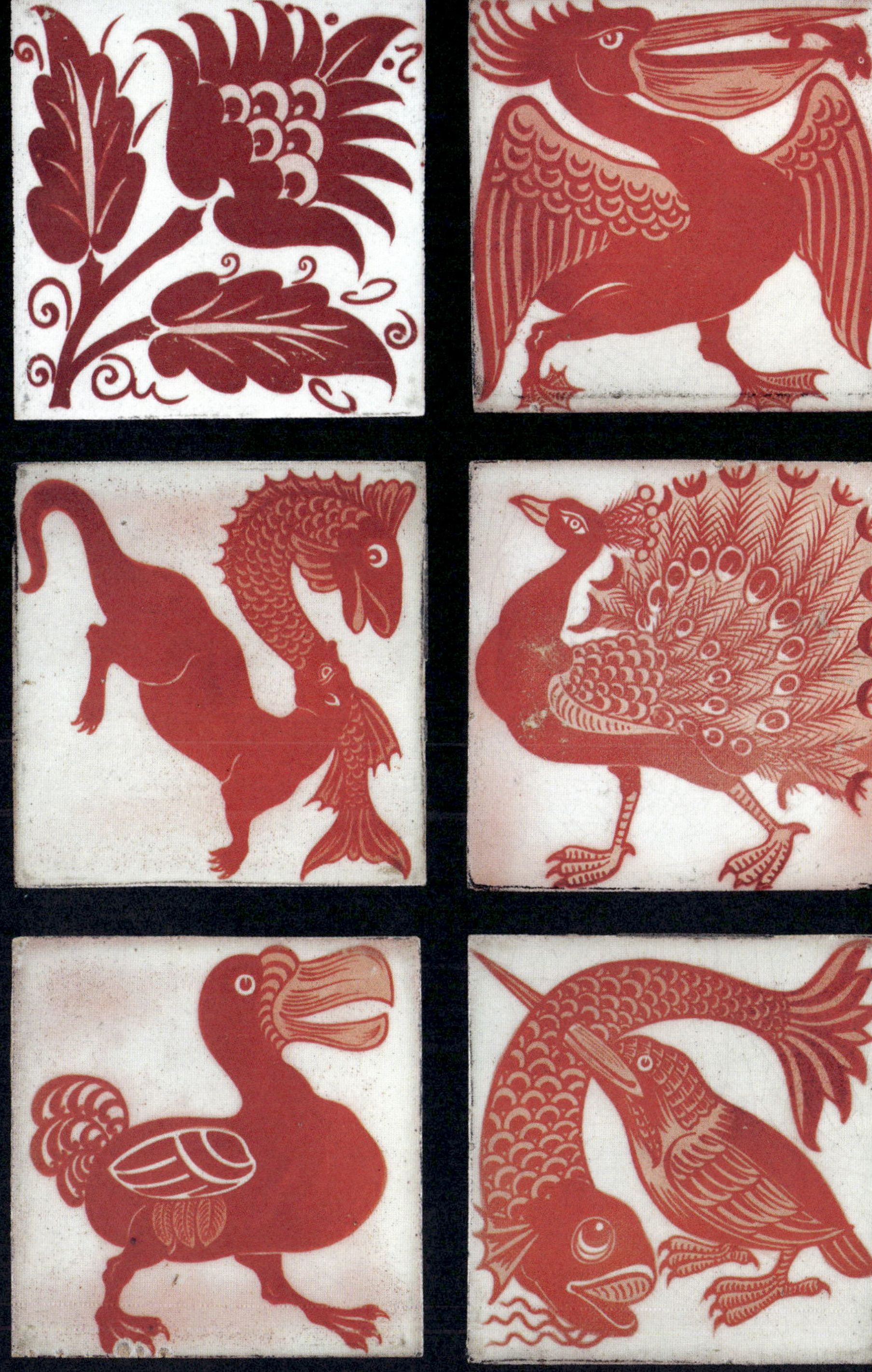

Baluster vase with flattened shoulders

Japan
*c.*1885 CE
Makuzu workshop (1871–1959 CE)
Porcelain, thrown, with "peach-bloom" glaze, 6.4 cm

This little porcelain vase has a "peach bloom" glaze – a gorgeous pinky-red glaze made from copper oxide that was originally developed in China. For many years, it sat in the Ashmolean's China collection stores, blending in among other similar ceramics.

Clare Pollard, a graduate student at Oxford in the 1990s, was writing her doctorate on Japanese potter Miyagawa Kōzan, an imperial household artist and one of the most influential potters of the Meiji era (1868–1912). She was studying under the supervision of the curator Dr Oliver Impey, then Keeper of Japanese art at the museum. One day, they were discussing Kōzan's extraordinary skill at producing Chinese-style ceramics, which looked so authentic he was even accused of forgery. Impey joked that there were probably Kōzan pieces hiding in Chinese porcelain collections around the world.

Pollard decided to check the Ashmolean's China store and her eye was eventually caught by this vase, only 6.5 centimetres tall. After taking a closer look, she established that it was indeed made by Kōzan in Yokohama, Japan, in the mid 1890s. The vase now sits on display in the Shikanai Galleries of Japanese Art. As an aspiring scholar of Japanese art, Pollard gained a confidence in her abilities as a researcher and expert that eventually led her to her current role as a curator at the Ashmolean.

The vase had been presented to the Ashmolean in 1956 by Sir Herbert and Lady Ingram, who bought it in Japan on their honeymoon in 1908. The couple had spent three months in Japan, sightseeing and hunting for curios. We know from Sir Herbert's diaries that the Ingrams acquired this vase in Tokyo for one yen, but not as a piece by Miyagawa Kōzan. It was sold as a Chinese work from the reign of the Emperor Qianlong (1735–96), despite the fact that it's clearly stamped on the base with the mark of Kōzan's workshop.

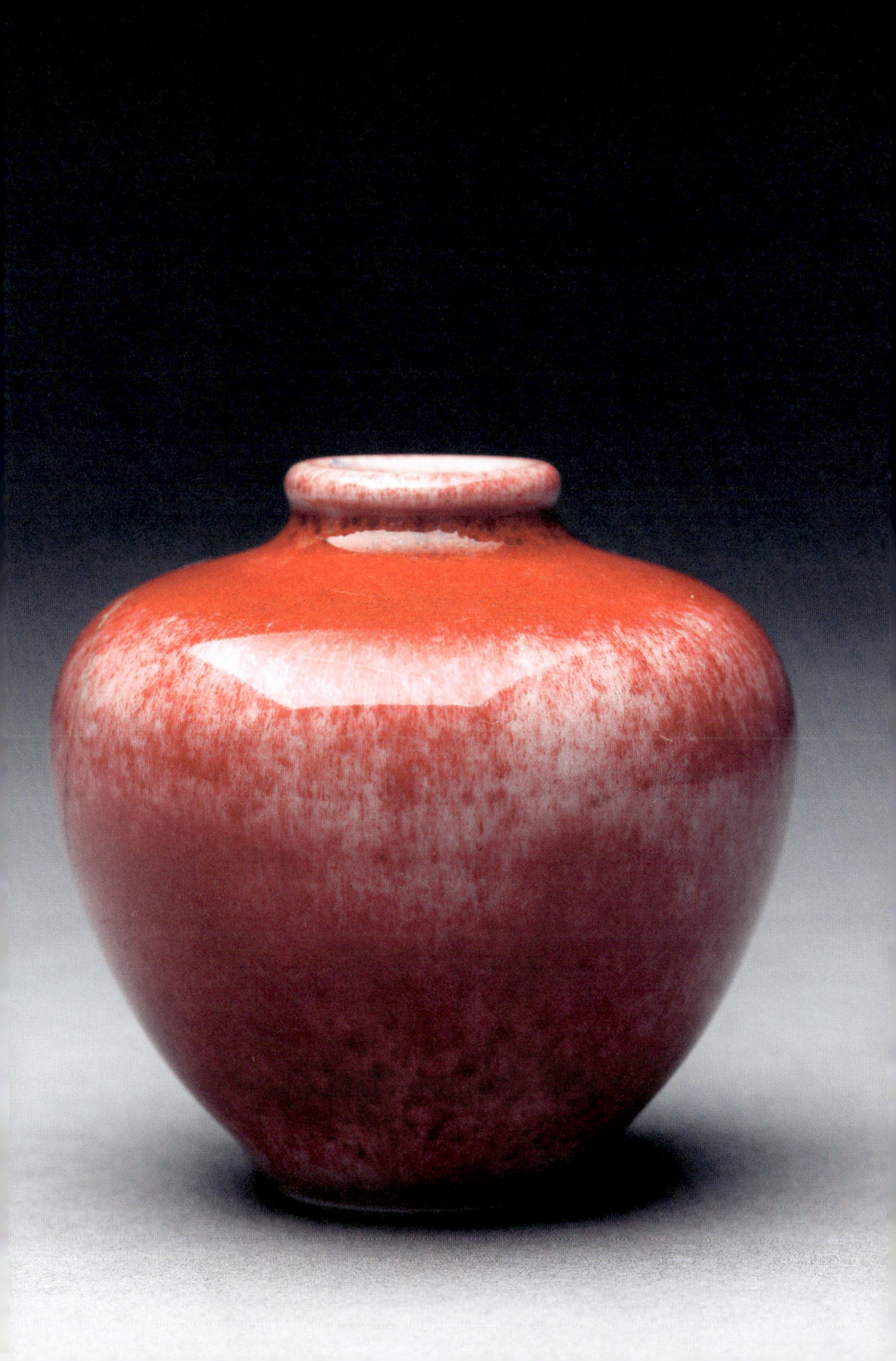

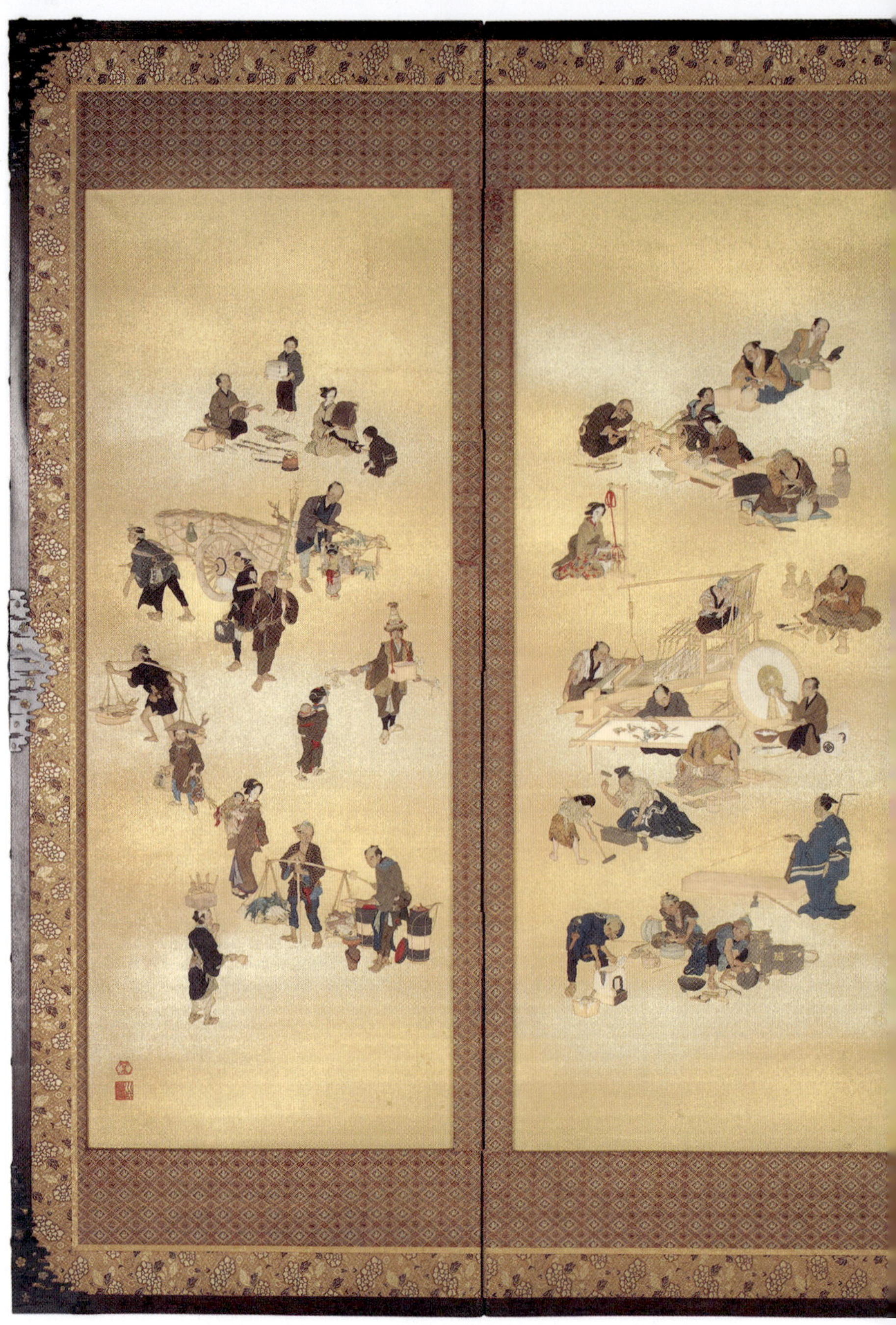

Screen depicting the four classes of Edo Japan, including a group of potters. Silk, mounted on paper, padded, and decorated with dyed and painted silk appliqué (*oshi-e*) and probably cotton; hard-wood frame covered in transparent lacquer; shakudō (gold and copper alloy) mounts, ink and colour on silk. Japan, *c.*1883, 182.4 x 276 x 3.2 cm.

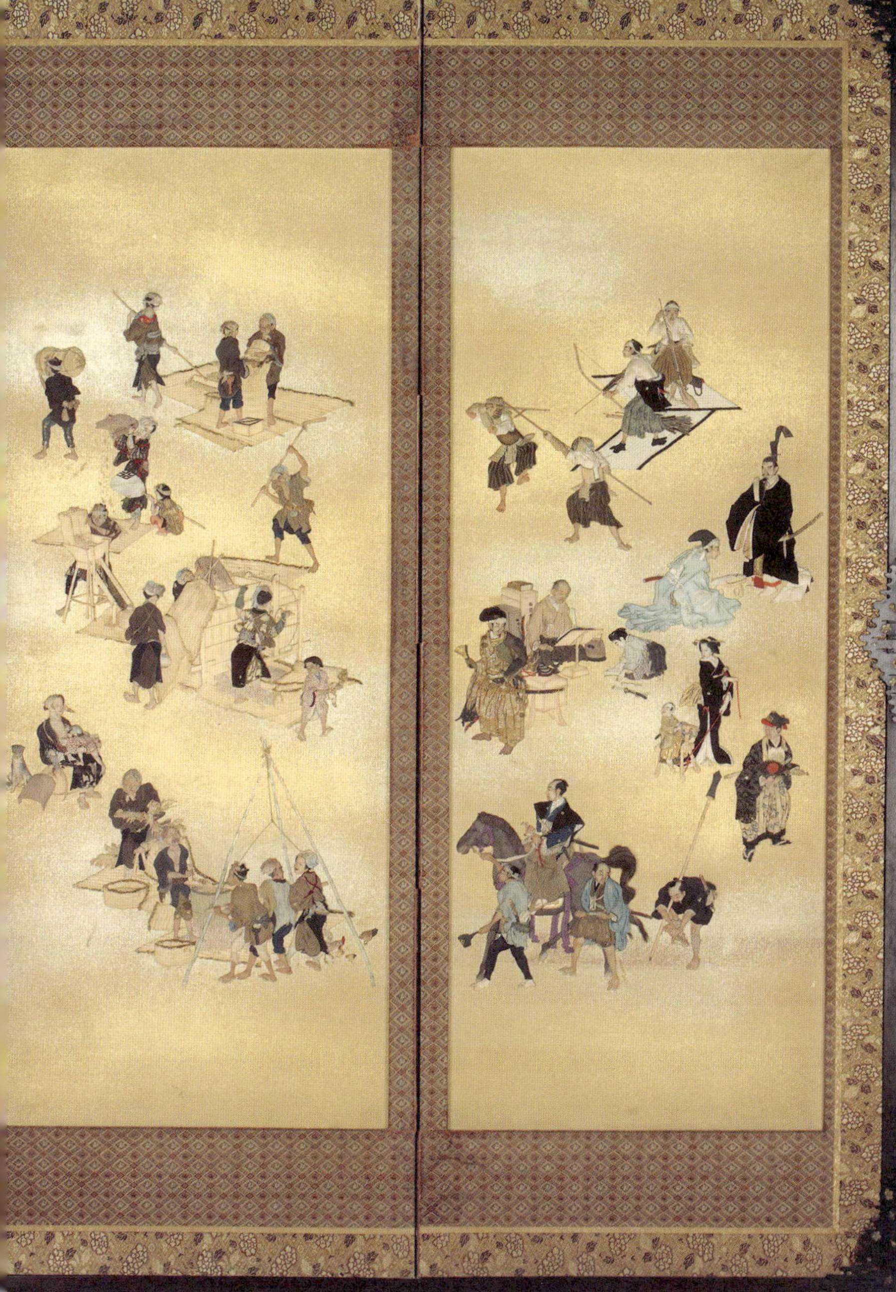

Satsuma cup with chrysanthemums and key-pattern border

Japan
*c.*1900 CE
Tōkōzan Kiln
Earthenware, overglaze enamels and gold, 7.3 x 6.9 cm

This small cup was created at the Tōkōzan workshop in the Kumamoto district of Kagoshima, in Satsuma domain. It arrived in the Ashmolean museum in 1856 as part of a major gift of over 3,000 Chinese and Japanese objects from Sir Herbert and Lady Ingram.

Herbert Ingram (1875–1958) was the grandson of the founder of the *Illustrated London News* and the son of William Ingram (whose baronetcy he inherited in 1924 on the death of his father). Most of the Ingrams' Japanese objects were acquired in 1908 when Herbert Ingram and his new bride Hilda spent two and a half months touring Japan during their honeymoon, enjoying a busy schedule of sightseeing and curio hunting.

One of the Ingrams' particular interests was Satsuma earthenware. Satsuma ware was first made at Kagoshima in the late eighteenth century and is known for its eggshell-coloured body and finely crackled transparent glaze, decorated with overglaze enamels and gold. It attracted international attention at the Paris Exposition Universelle of 1867, and then became so popular with overseas customers that it began to be imitated in Kyoto and other areas of Japan. In their enthusiasm for Satsuma ware, Western collectors gradually became convinced that there was a huge difference between early Satsuma ware, which was actually from Satsuma domain, and its modern reproductions. By the early 1870s, a demand arose for pieces that were deemed "early", and their prices went up.

Japanese manufacturers were more than happy to supply the demand. Satsuma-style works were often deliberately stained and distressed to achieve an air of antiquity and were passed off as "old Satsuma ware" to naïve foreign buyers, at great expense. By the late 1870s, Western collectors were becoming more discerning about Japanese ceramics and were more often able

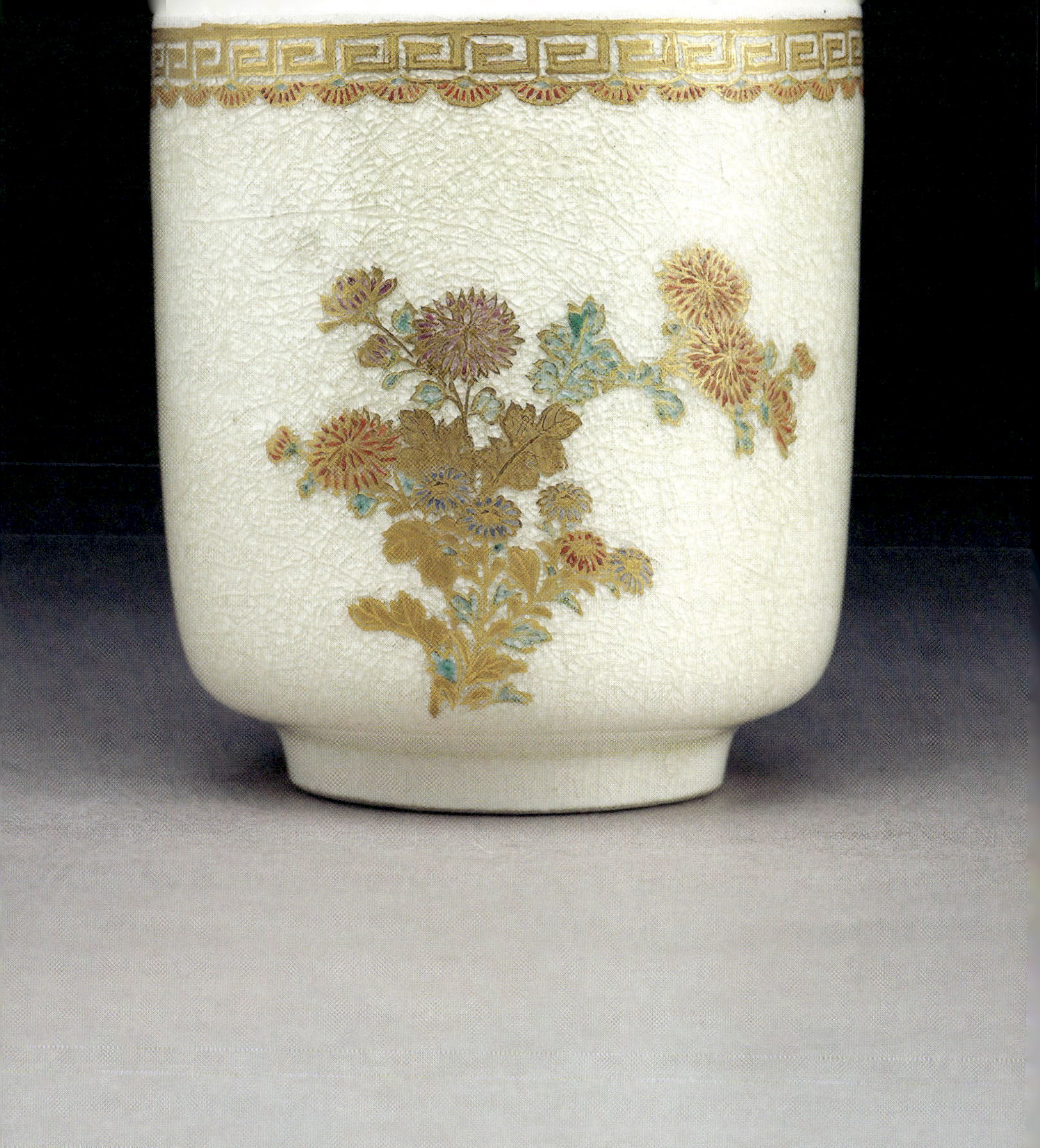

to distinguish between shoddy export ware and better quality objects. But unscrupulous dealers were still keen to convince potential buyers of the rarity of their wares.

Mr Ingram disliked the heavily decorated and gilded Satsuma-style ceramics that dominated the export market. He preferred the more restrained decoration found on some pieces made in Satsuma itself, such as this elegant Tōkōzan cup, with its single spray of chrysanthemums and simple key-fret border. One of the curio dealers who accompanied the Ingrams on their travels through Japan was called Matsuzawa. Mr Matsuzawa introduced Mr Ingram to a retired sumo wrestler by the name of Sakahoko, living in Tokyo, whose collection of Satsuma ware Mr Ingram bought on the couple's penultimate day in Japan. His diary for 23 May 1908 begins, "The great Satsuma day!!" and ends, "after somewhat prolonged discussions, I got the pieces I wanted but did not get to bed until nearly eleven". Mr Ingram recorded that these specimens were "exceedingly scarce ... practically all made for the prince's court". In fact, most of the Sakahoko Satsuma pieces were no more than 30 years old, at the very most, and had no aristocratic heritage. However, they weren't typical export pieces and were very different from some of the poor quality, over-ornate ceramics made for the less discerning end of the tourist market.

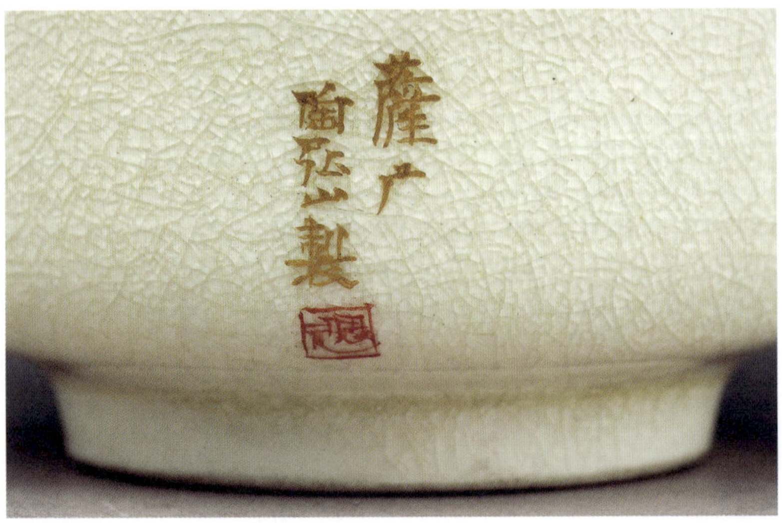

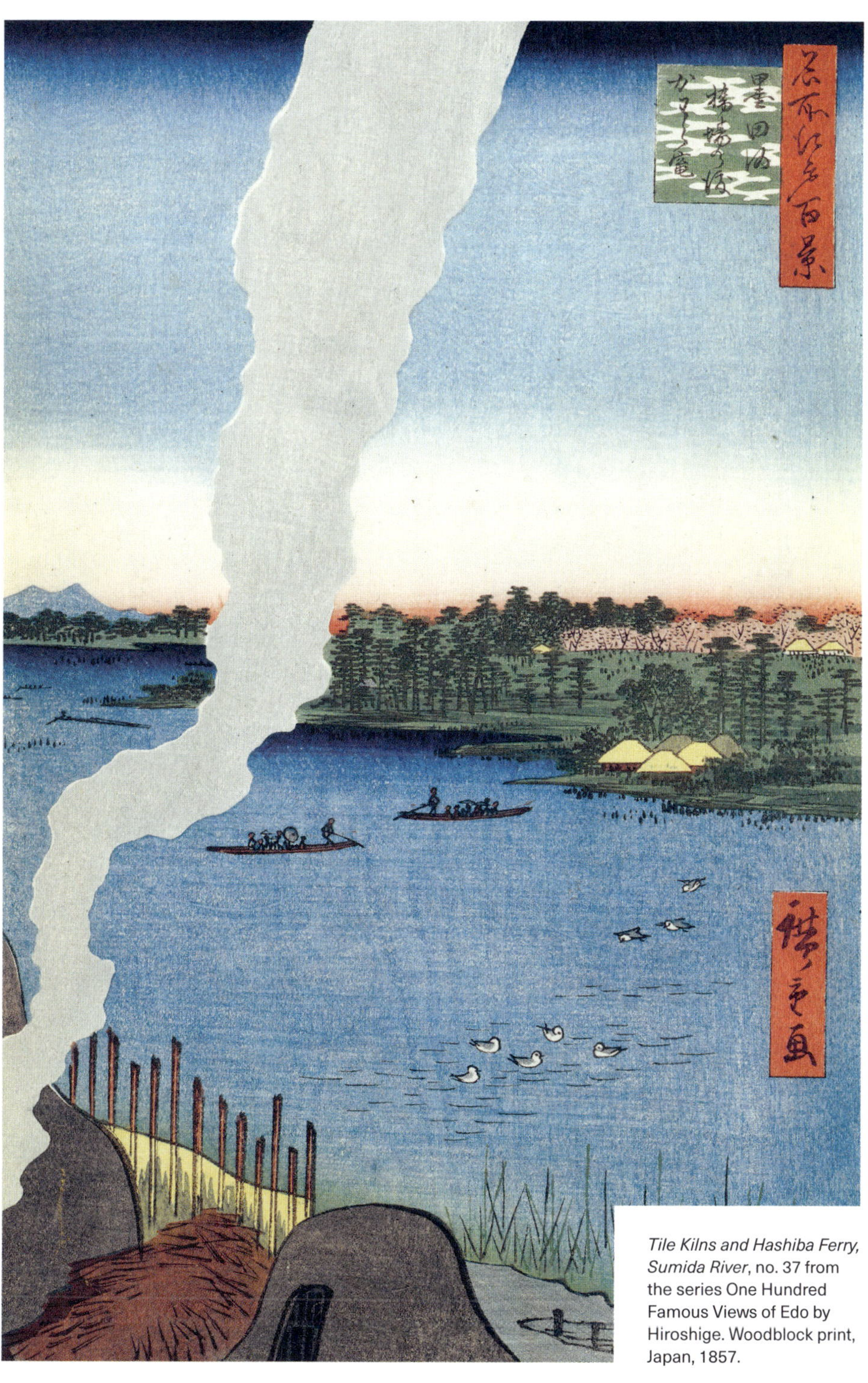

Tile Kilns and Hashiba Ferry, Sumida River, no. 37 from the series One Hundred Famous Views of Edo by Hiroshige. Woodblock print, Japan, 1857.

Art nouveau-style vase with a design of chrysanthemums

Japan
1900–05 CE
Kinkōzan Sōbei VII (1868–1927 CE)
Glazed earthenware, pierced and modelled, 37 x 21 cm

This arresting vase, with its elaborate pierced design of chrysanthemums, was a product of the Kinkōzan workshop in the Awata district of Kyoto.

For generations, the workshop produced exquisite enamelled earthenwares for the Japanese market. Then, in the 1870s, Kinkōzan's father, Kinkōzan Sobei VI (1868–1928), began making Satsuma-style ceramics for the foreign market. Kinkōzan VII succeeded his father in 1884 and continued to produce elaborately decorated earthenwares for export. The workshop became one of Japan's largest and most successful ceramic manufacturers, employing hundreds of workmen and designers.

The chrysanthemum vase was made in a new style developed after Kinkōzan went to the Paris Exposition Universelle of 1900. The exhibition was dominated by the art nouveau style. Despite the fact that art nouveau was significantly inspired by Japanese art, Japanese crafts at the exhibition were criticised as out-of-date and over decorated. They were also critiqued on the grounds that their decoration did not reflect their form. After his return to Kyoto, Kinkōzan decided to modernise his ceramic designs. He hired the Ishikawa potter and researcher Suwa Sozan (1851–1922) to help revitalise the workshop. Sozan worked as artistic manager at the Kinkōzan kilns between 1900 and 1907. He encouraged the creation of more organic vessel shapes and stylised floral motifs that complemented those shapes. The Kinkōzan workshop continued to sell and display Satsuma-style wares but also began to produce innovative vases like this one, with modelled, naturalistic designs of flowers, leaves and branches sprouting all over the vessels. These pieces were displayed at several national and international exhibitions in the years following Paris, and won a grand prix at the Louisiana Purchase Exposition held in St. Louis in 1904.

The curving shapes of the chrysanthemum petals, the semi-stylised patterning of the flowers and leaves and the arrangement of the stalks over the body of the vase epitomise art nouveau taste.

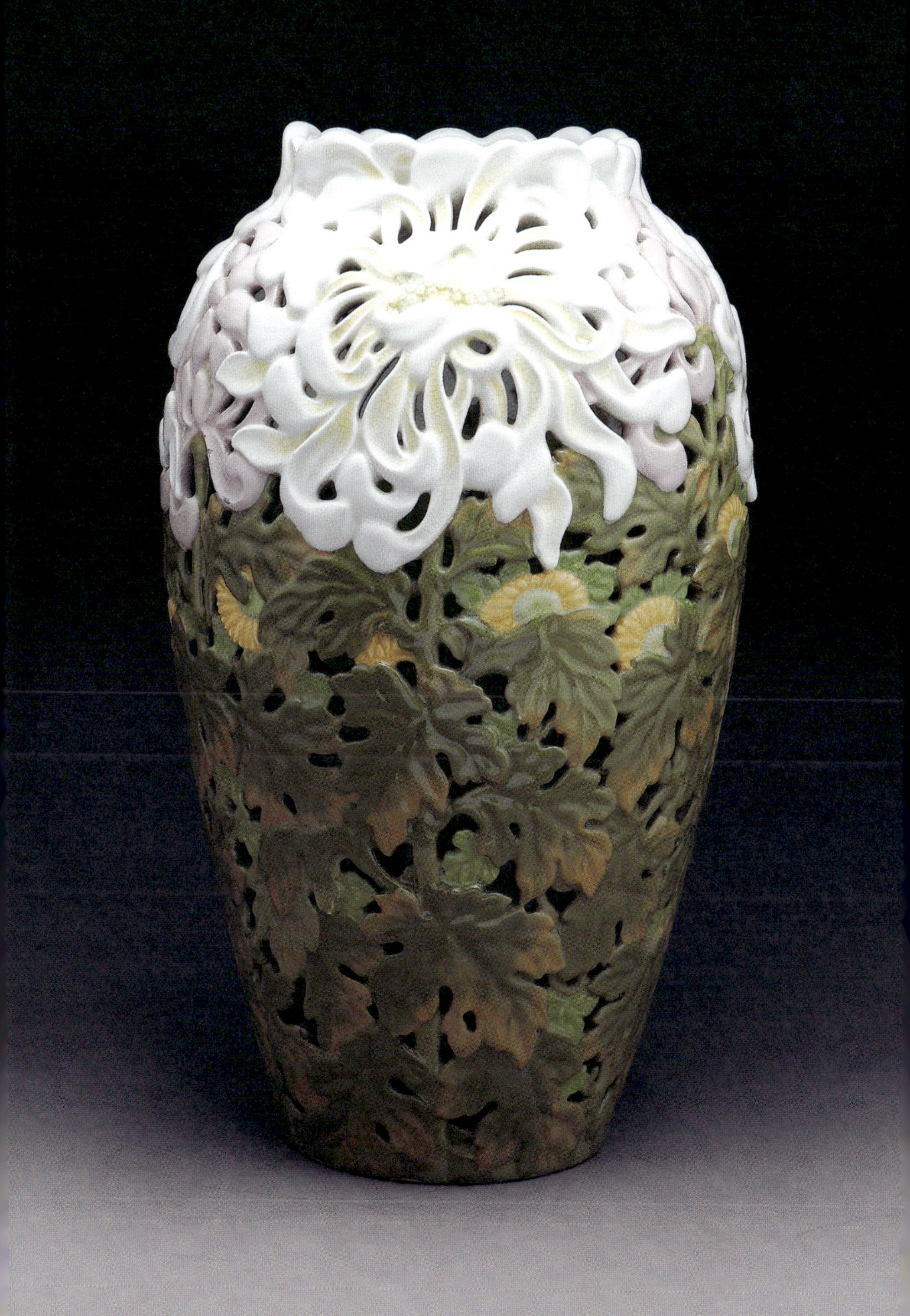

76

Vase with winter landscape

Japan
*c.*1910 CE
Miyagawa Kōzan (1842–1916)
Porcelain, underglaze painting in blue, 86 x 38 cm

In the mid-nineteenth century, Japan's feudal government was abolished and a new government was established. As part of an effort to open up Japan to the West after a long period of national isolation, this new government actively encouraged artists to show their works in competitive national and international exhibitions.

With its impressive size and enchanting continuous design of a winter landscape in various tones of underglaze blue, this grand vase by the eminent potter Miyagawa Kōzan was probably made for just such an exhibition. It depicts a Japanese landscape, but the misty effect of the blue glaze is reminiscent of works produced at the same time by the Danish Royal Copenhagen Porcelain Manufactory. In this way, this vessel illustrates the fact that in this period, there was artistic exchange taking place between different nations.

Kōzan established his international reputation in the 1870s, when he produced enamelled earthenwares for export. From the 1880s, he began developing sophisticated underglaze-decorated porcelains. He made use of newly imported Western glaze technology to achieve a broad spectrum of different glaze colours and effects. One of the most impressive features of these new underglaze porcelains was their use of what was called *bokashi* or "gradation". This was a soft, layered glaze technique that was used to give the impression of the ink washes in Chinese or Japanese painting, or as part of realistic, more Western-style pictorial decoration. This effect would have been achieved using a technique called *fuki-e* (literally "blow-painting"). It was a refinement of a technique that had been used for centuries in Japan. In *fuki-e*, coloured glaze was brushed onto a fine metal mesh and the artisan carefully blew it through the screen to create a fine spray of glaze. By the early twentieth century, artists often used an airbrush instead. A metal stencil or baffle was used to mask the sections where the glaze was not wanted, and in this way artisans could achieve extraordinarily subtle effects. The technique is beautifully illustrated on this vase, its bare-branched trees and bamboo grass silhouetted against a background of misty mountains.

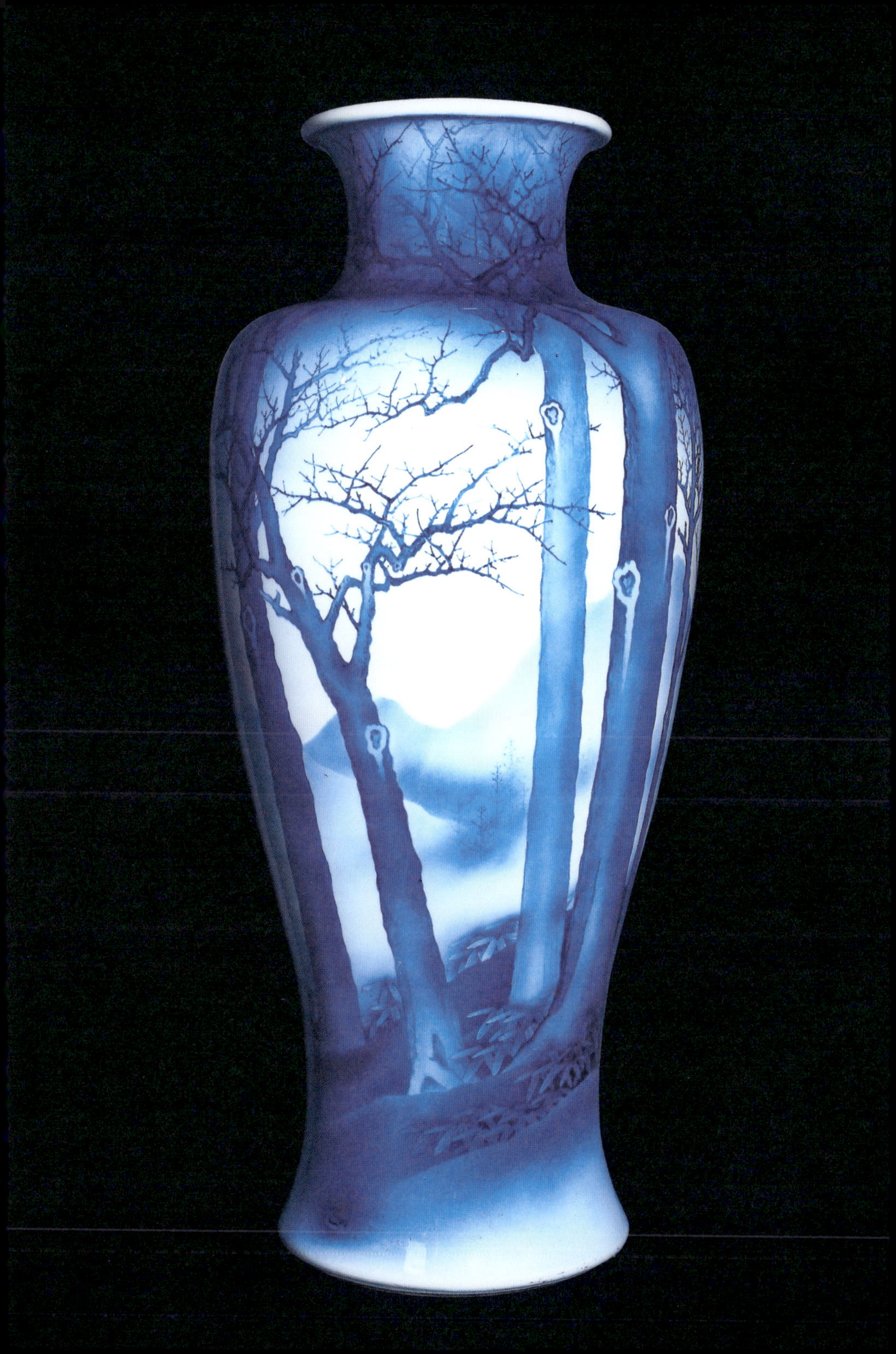

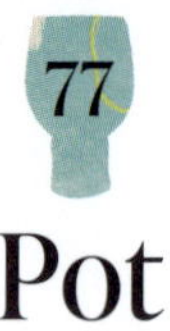

Pot

UK
*c.*1975 CE
Hans Coper (1920–81 CE)
T-material, a high grog clay, thrown and assembled, 18.3 cm

One of the world's best-known twentieth-century ceramic artists, Hans Coper, redefined the concept of studio pottery over a period of just 30 years. Using clay as his medium, he distinguished himself from a previous generation of artisans by fusing together wheel-based pots, creating vessels that resemble sculpture. Each pot has its own presence, created within a limited palette of burnished black or creamy white, often ornamented with scored lines or etched surfaces.

Born in the town of Chemnitz, Saxony, Coper escaped from Nazi Germany in 1939 and arrived in London, only to be sent to an internment camp in Canada. The Austrian-born modernist potter Lucie Rie, who had herself fled Nazi occupation, employed him to assist in her Albion Mews studio in 1946, even though he had no previous experience as a ceramicist. But by 1959, he was well known in his own right. He moved to Digswell, Hertfordshire, and began a phase which he referred to as his "architectural period", during which he was commissioned to make seven-foot-tall pottery candlesticks for Coventry Cathedral.

In 1963, he returned to London, working at the peak of his productivity, and in 1967 moved to Frome in Somerset, where he created a series of "Cycladic" pots, reminiscent of Bronze Age artefacts from the islands of the Cyclades. These are considered his finest work. Coper taught at Camberwell College of Arts and the Royal College of Art, and his work was exhibited widely in his lifetime.

Coper's quietly ethereal modernism emerges from his ability to distil ideas from the history of art – Neolithic, ancient Egyptian, early Mediterranean, Sub-Saharan – and bring them to life in a mid-century context. His signature forms – spade shapes, pedestal and thistle pots, hourglasses and "Cycladic" pots – are masterpieces of ingenuity. They have archetypal qualities, but there's something very intimate about them, too. Their shape, feel and weight are sensuous, reaching beyond self-expression to something deeper. For this reason, perhaps, they suit communal and ceremonial spaces. As objects, they reach across time and across tradition.

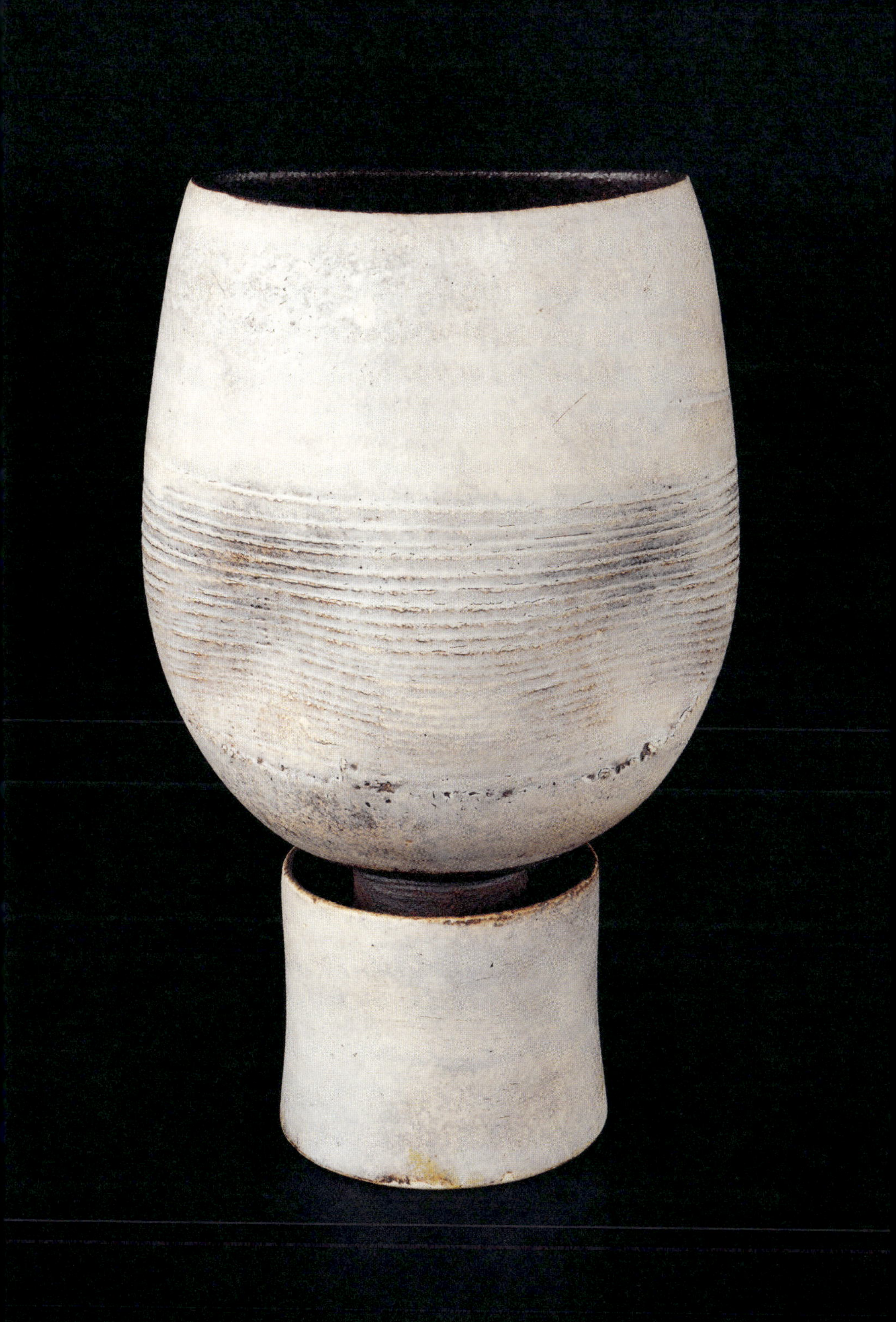

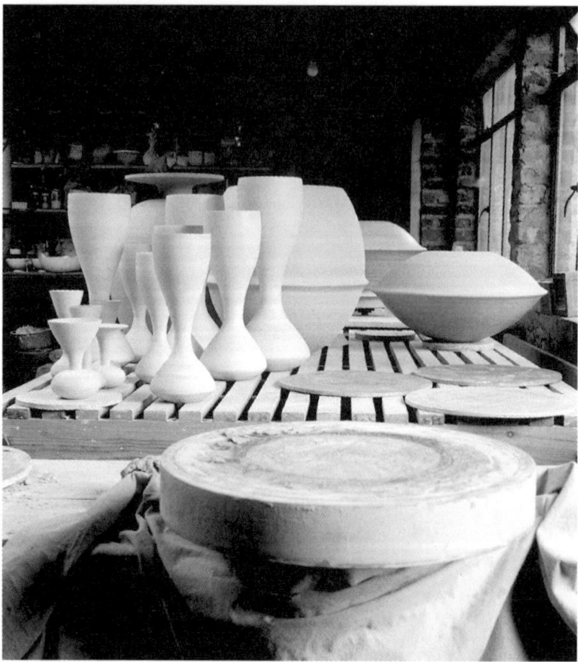

Hans Coper at his wheel,
London, 1963–7.

Ceramics in Hans Coper's
studio, London, 1963–7.

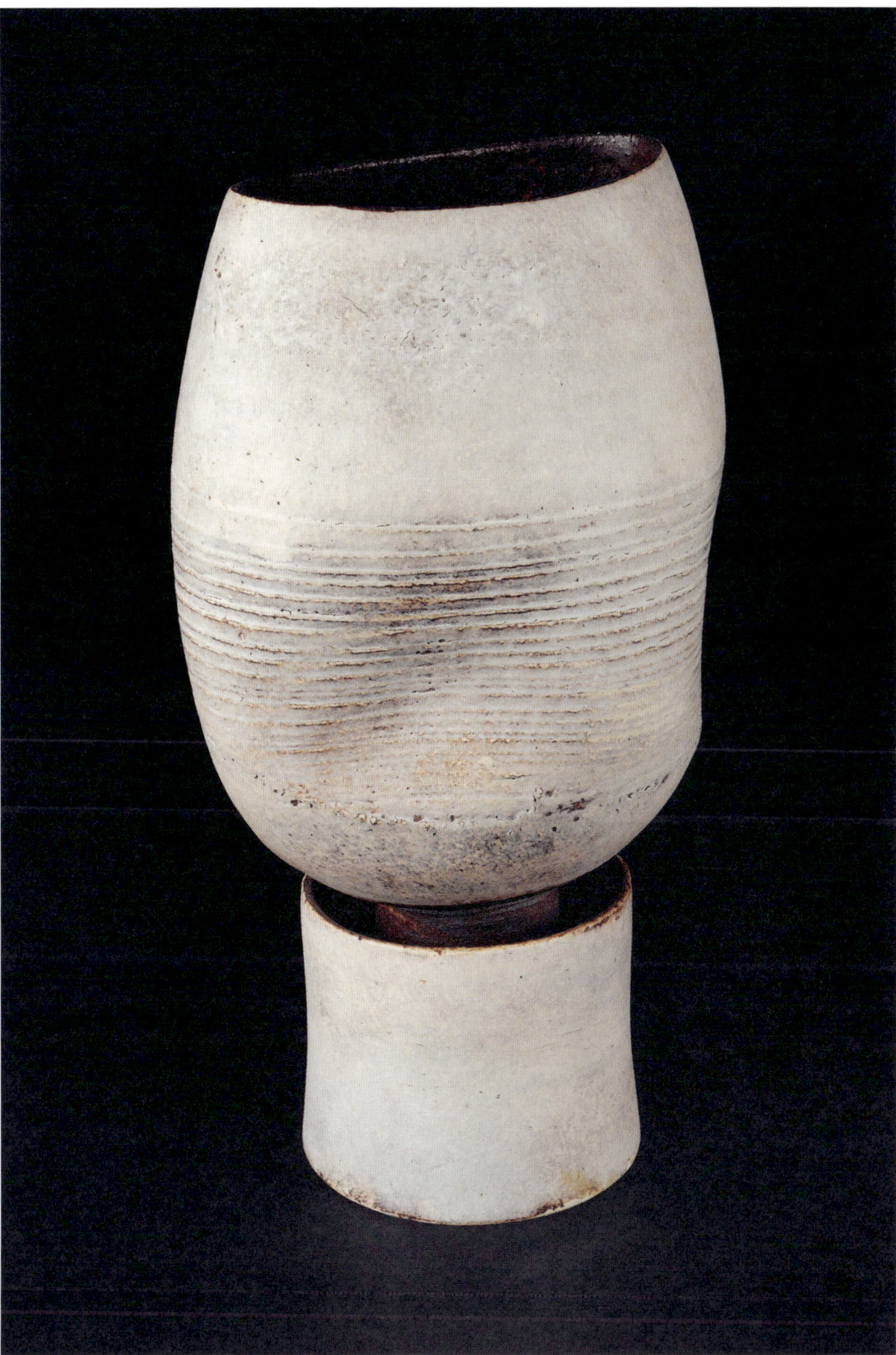

Lustred bowl

UK
2005 CE
Alan Caiger-Smith (1930–2020 CE)
Earthenware, tin-glazed, lustred, 45 x 17.5 cm

The heritage of this bowl stretches back for more than a millennium and takes in much of Christendom and the Islamic world. Tin-glazed pottery is the oldest of the European traditions of art pottery and the best adapted to creative brush-painting.

When it was developed in around 800 CE, in what is now Iraq, it was apparently a response to the arrival from China of high-fired, white-bodied ceramics. These Chinese imports were impossible to reproduce with the kiln technology and clays available to the Islamic potters. The invention of an opaque white glaze that could cover a ceramic body was a passable imitation with its own aesthetic potential.

Spreading west, sometimes exploiting the mysterious technique of reduced-pigment lustre, tin glaze eventually reached southern Spain, the westernmost outpost of the Islamic world. In the fourteenth century, potters from Málaga created the monumental lustred masterpieces for the Alhambra Palace in Granada. For people at the time, lustre could seem quite magical, as if it was the fulfilment of the alchemist's dream of making gold out of base materials.

In sixteenth-century Italy, tin glaze was transformed into a form of Renaissance painting. In Gubbio, Maestro Giorgio produced masterpieces of lustreware in which copper reds provide a perfect counterpoint to silver-based golden lustres. As Italian potters travelled across Europe in search of opportunity, national traditions of tin glaze developed. Potters in France and Germany made "faience"; in central Europe, "Haban" ware; in the Netherlands, "Delft"; and "delftware" in England. The application of lustre became rarer. The artistic use of tin glaze itself was dealt a fatal blow in the late eighteenth century by the success of English industrial white wares, especially those made by Josiah Wedgwood. Subsequent historicist revivals of tin glaze and lustre – which took place, for example, in Italy and in the designs of William De Morgan in England – tended to be sporadic.

In 1955, Alan Caiger-Smith bought the old smithy in Aldermaston and opened the Aldermaston pottery. Caiger-Smith wanted to return to tin glaze

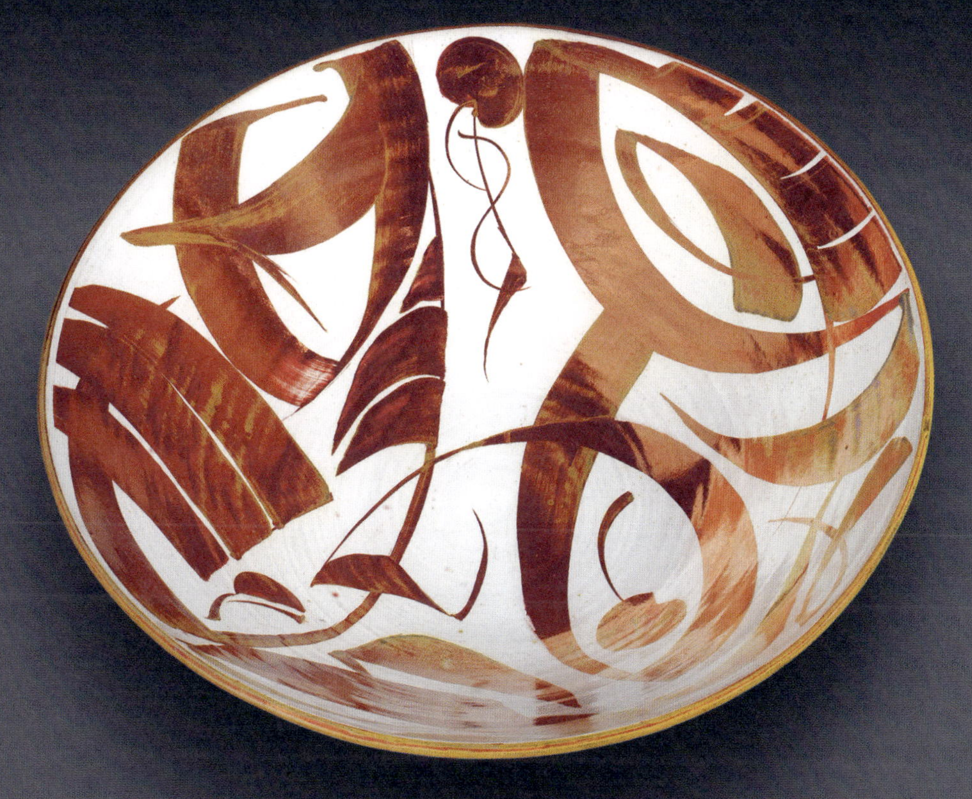

in reaction to the predominant aesthetic of
"brown pots" among British studio potters
in the 1950s. This phrase is shorthand for
a trend in ceramics inspired by Bernard
Leach, whose rustic pots were shaped
by Japanese and European traditions.
Caiger-Smith experimented with
recreating the technological virtuosity of
Islamic and Italian potters in painting on
tin glaze and rediscovering the iridescence of
reduced-pigment lustre. However, in his hands, painting on the raw glaze was
an original medium of artistic creativity.

A community of potters gathered at Aldermaston under Caiger-Smith's
leadership for half a century. Their creativity was guided by an ethical
underpinning. For example, the group avoided any division of labour, so that
each potter would have a sense of ownership of the works they created and
could develop the range of skills they needed to follow an individual path of
artistic expression in the outside world.

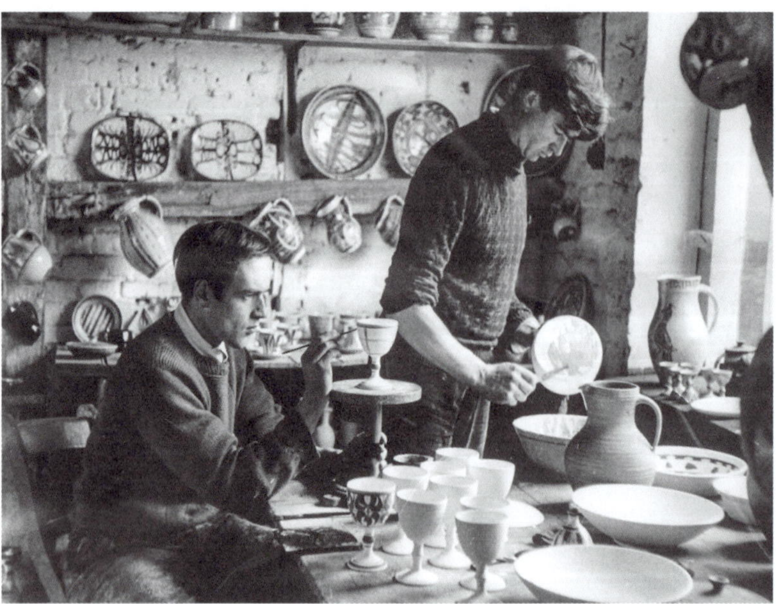

Alan Caiger-Smith (left) and
Geoffrey Eastop at work at
Aldermaston Pottery, *c*.1961.

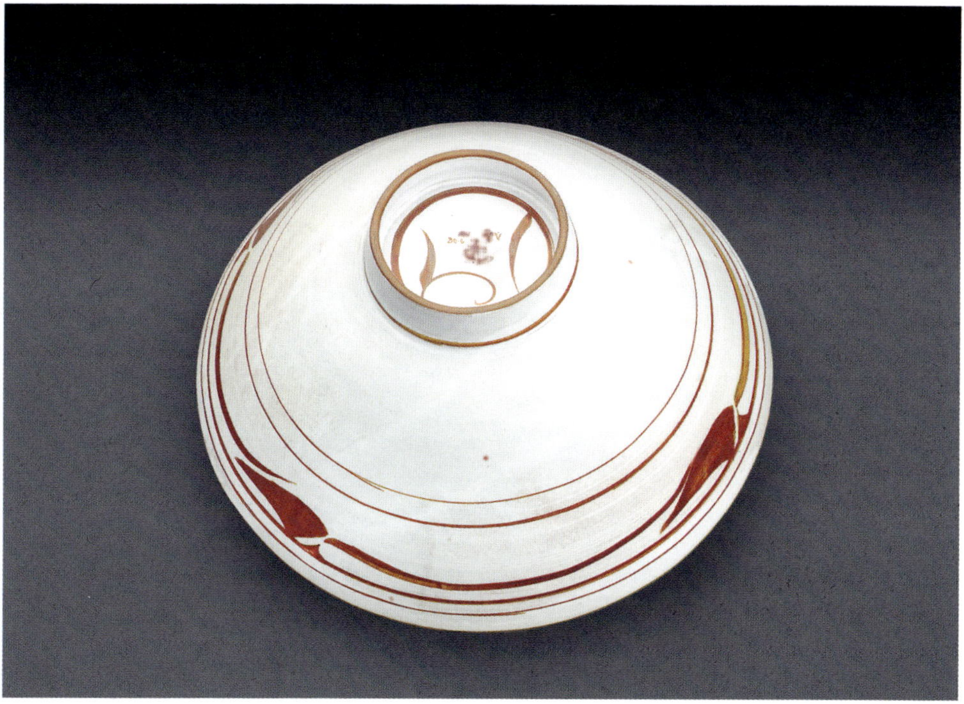

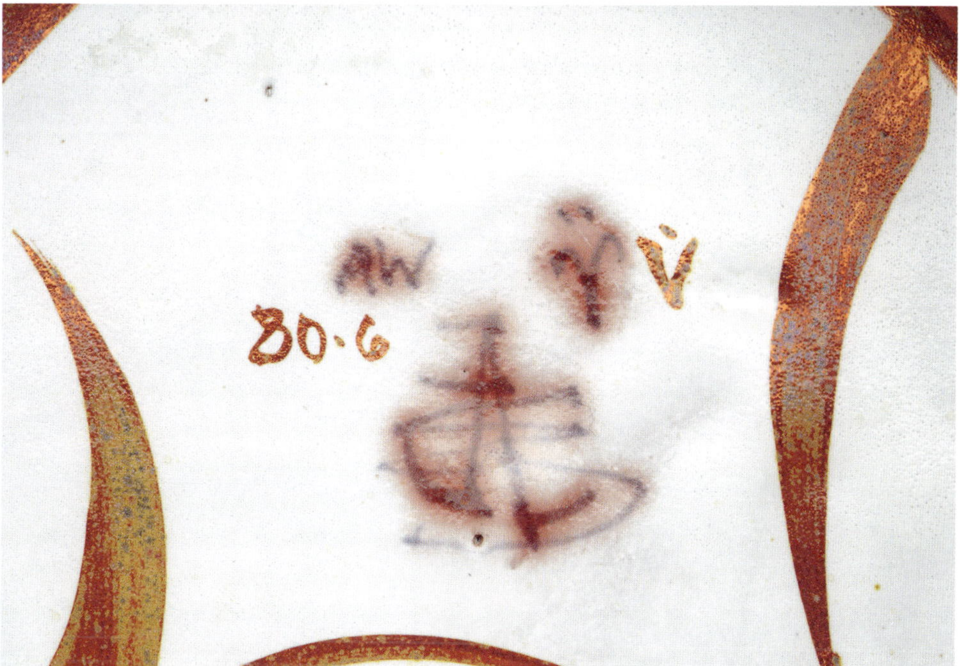

80.6

Asymmetrical "Betu"
Series I

UK
2009 CE
Magdalene Odundo, DBE (b. 1950 CE)
Terracotta, burnished and carbonised, 47.5 cm

Many people believe that Magdalene Odundo has the greatest command of pure sculptural form of any potter working today. Born in Nairobi in 1950, she moved to Cambridge in 1971 to take up a place on an Art Foundation Course at Cambridge School of Art. She originally intended to study graphic design but, inspired by her Zimbabwean-born pottery teacher, Zoë Ellison, she switched her focus to ceramics.

Odundo trained at the West Surrey College of Art and Design and at the Royal College of Art. She's travelled extensively to explore vernacular ceramic traditions, including to Kenya, Nigeria and New Mexico, USA. This pot embodies the formal grace of Odundo's work, and played a starring role in a solo show in Liverpool in 2009. Its title is a reference to the Mangbetu people who live in eastern Congo.

Odundo's work takes inspiration from the human body. It's informed by a long list of influences: traditional ceremonial vessels from Kenya and Nigeria, carving from the Congo, British studio pottery, ancient ceramics from Cyprus, Japan and Peru, modernist sculpture, and even Elizabethan costume. She often takes as her starting point traditional African techniques and forms. Like much of her work, this pot is unglazed. The surface of the clay has been painstakingly burnished and polished, and then fired in a reducing atmosphere to create the blackened finish, reminiscent of ancient ceramics.

Odundo has exhibited her work all over the world. In June 2018, she was appointed Chancellor of the University for the Creative Arts, and in the 2020 New Year Honours she was awarded a DBE for services to art and art education.

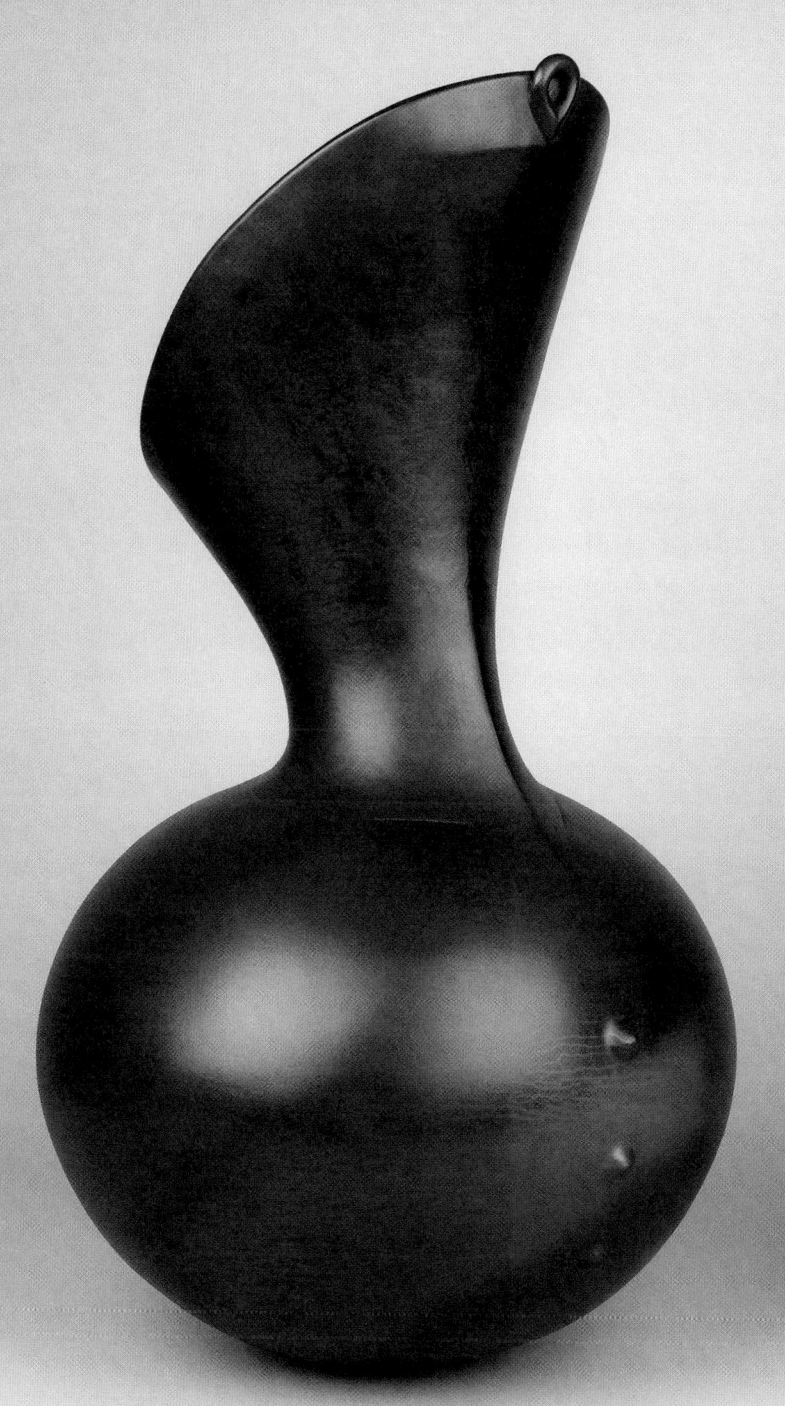

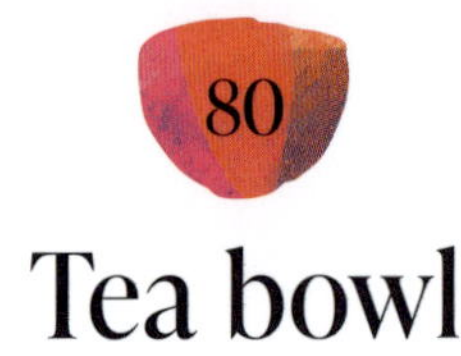

Tea bowl

Japan
2019 CE
Ogawa Machiko (b. 1946 CE)
Stoneware, gold glaze, 9.5 cm x 10.5 cm

This tea bowl by leading Japanese ceramic artist Ogawa Machiko came to the Ashmolean in 2020. Many of Machiko's ceramics resemble strange mineral formations or archaeological artefacts that have just been excavated from the earth.

Her carefully crafted bowl, which belongs to the refined world of the Japanese tea ceremony, nevertheless has an elemental quality, as if scooped out of raw clay. The earthy cracks and deep fissures of the roughly textured exterior contrast with the creamy-smooth, matte-gold glow of the interior.

The tea bowl is one of the key utensils of the Japanese tea ceremony. It's more than just a functional vessel: it represents the essence of *chanoyu*, the way of tea. According to the ceremony, serving tea is an act of hospitality that's ostensibly simple but, in actual fact, deeply considered. The tea bowl symbolises this moment of mindfulness and human connection. It embodies the idea of *wabi sabi* – an appreciation of the humble, imperfect and pared down.

Machiko is interested in memory and tradition: the way her work can connect us with the makers and users of ceramics in the historical past. A tea bowl is perhaps the quintessential Japanese vessel, but this bowl seems to transcend national culture. Perhaps its universal quality owes something to the years Machiko spent studying in West Africa and South America. The glitziness of this tea bowl might be a nod to the ostentatious gilded tea house and gold tea utensils commissioned by the warlord Toyotomi Hideyoshi in the late sixteenth century, despite the *wabi sabi* ideals of the tea masters of his day. Machiko's tea bowl is golden and glowing but suggestive of humility. Managing to balance smooth beauty with rugged simplicity, it offers us something different to what we might have expected.

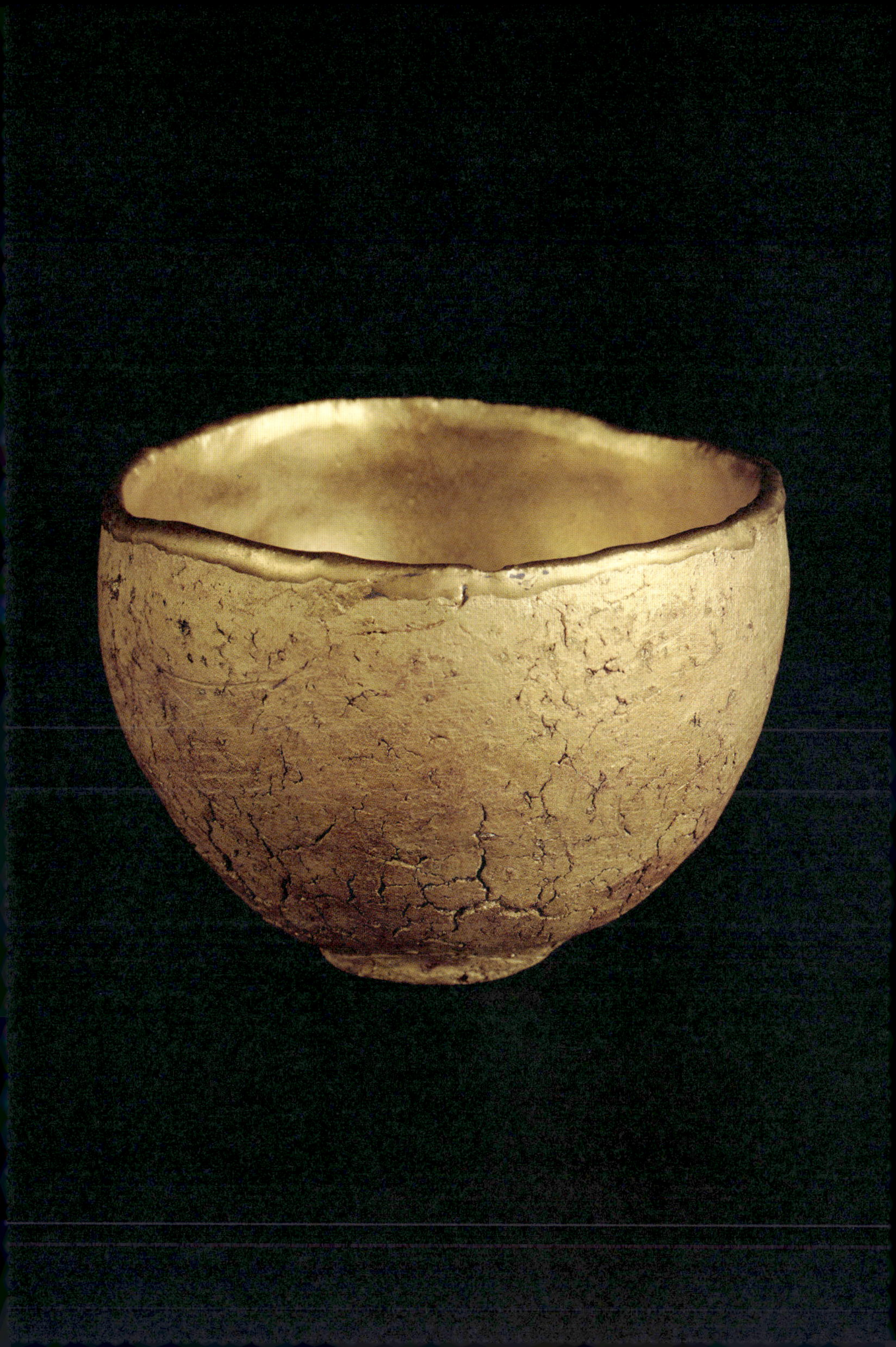

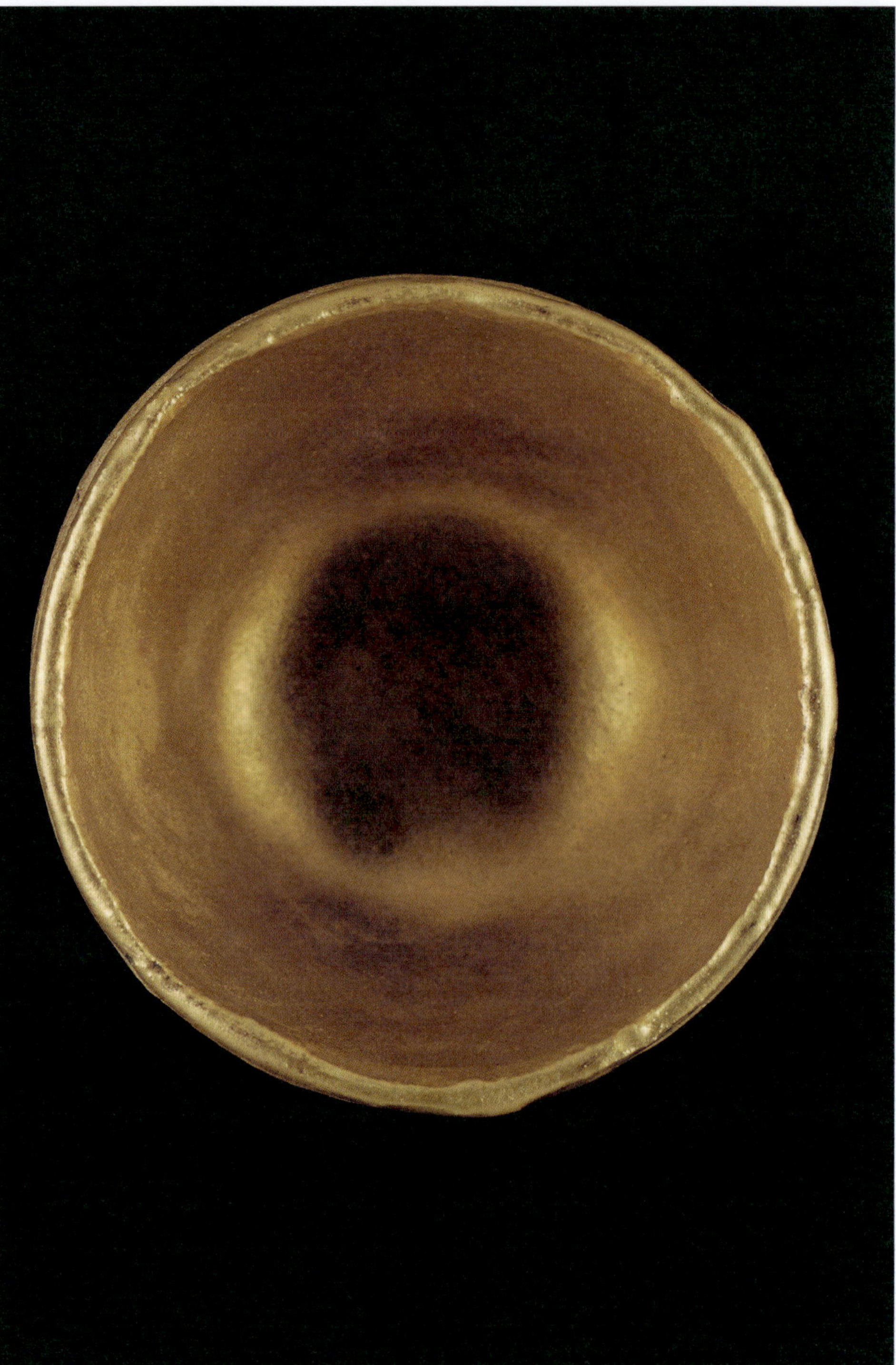

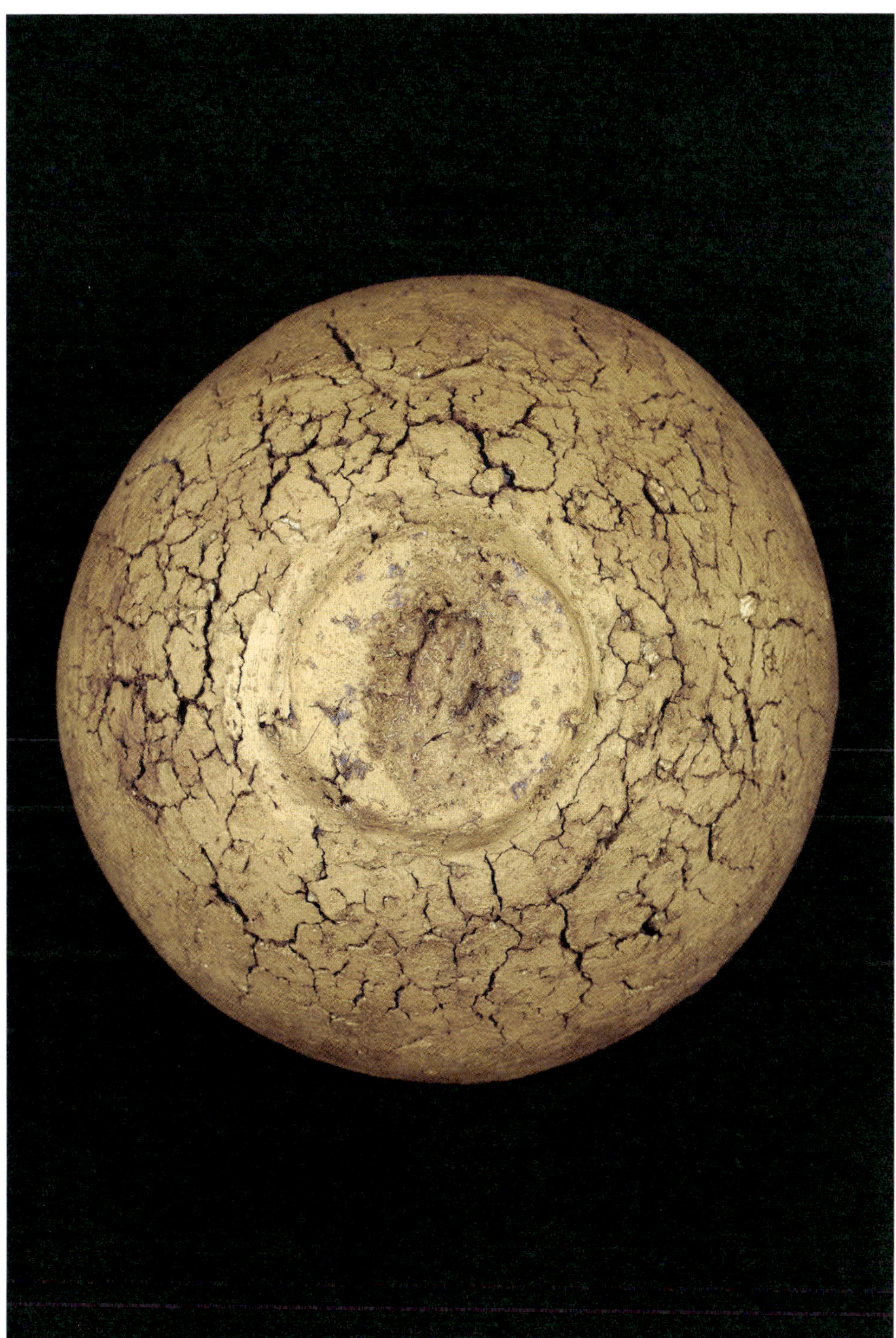

Index

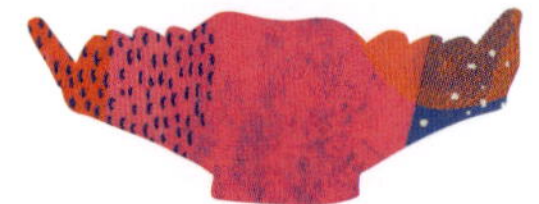

(Page numbers in bold
refer to main subject text
with photographs; italics
to all other captions/
photographs)

A

Abbasid Caliphate (750–
1517 CE) 188
Akrotiri, Thera 62
Albion Mews studio 250
Aldermaston Pottery 254,
256, *256*
Alexander, Charles *173*
Alhambra Palace 254
Althorp House 202–5
Ambushed Octopus 70
Andokides Potter 86
Anne of Brittany 160, 162
Annual Register 222
Antiphon Painter *84*
appliqué 96, *240*
Arcimboldo, Giuseppe
164
Aretino, Pietro 164
Art Nouveau 226, 246–7
Arts and Crafts 236
Augustus the Strong,
210–12, 214

B

B-ware 16, 22–5
Beazley, John 114
Belgium 186, 218, 222
Bengal 100, *195*
Bichrome Ware 76–9

Birdie, *see* osprey, seated
Birmingham, UK *224*
black-and-red ware 66–9
black-figure pottery 60,
82–3, 86–9, 92–5
black-glazed ware 128–9,
blue-and-white pottery
118, *119*, 120, 170,
180–1, 220, 290–1
bluish-white ware 148–51
boars and pigs 158–9,
218–19
Bosanquet, Robert Carr 48
Böttger, Johann Friedrich
210
bottle kilns *14*, *15*
British Museum 46, 106,
196,
Bronze Age 48, 52, 54,
60, 62, 64, 70, 74, 116,
250
Burghers, Michael 216
burial rituals 26, 40, 52,
68–9, 72, 74, 80, 152

C

C-ware 20
Caiger-Smith, Alan 254
calligraphy 126, 136,
144, 148, 190
Camberwell College of
Arts 250
Cambridge School of Art
258
chanoyu (the way of tea)
260

Charles II 200
Charles (Karolus) VIII
160, 162
Charlotte, Queen 218
Chelsea porcelain 218–
19, 222
China:
   black ware jar with
      white stripes (960–
      1279 CE) **128–9**
   blue and white vase
      with figures and a
      poem (1639 CE)
      190–1
   dish with purple splash
      (China: 960–1279
      CE) **130–3**
   earthenware figure of
      a camel (618–907
      CE) **112–15**
   five great wares 134
   jar with heavily
      moulded decoration
      in Central Asian
      style (500–600 CE)
      110–11
   Kintsugi bulb bowl
      with purple and blue
      glazes (Song/Yuan
      dynasty, C13th–
      14th) *16*, *17*, **140–3**
   The Oxford Plate
      (*c.*1745 CE) **216–17**
   three perfections 190
   white ware dish
      with incised lotus

Rei, Lucie 250
Renaissance 154, 164,
    166–8, 173, 186–8, 254
Restoration 200
Royal College of Art,
    London 234, 250, 258
Ru ware 134
Rudolf II 164

S
Safavid Empire 180
Sakahoko (sumo wrestler)
    244
Samanid/Samanian
    Empire 126
Sarah, Duchess of
    Marlborough 204–5
Sassanian/Sassanid
    Empire 110
Satsuma style 226, 230–3,
    242–4, 246
Schuylenburgh, Hendrik
    van *195*
secret colour Yue 110
sgraffito 120, *120*, 128
Shijō school 226
Shino ware 174–5
shoki-Imari ware 182, 185
Sibthorp, Humphrey 216
Silk Road 112
Song dynasty (960–1279
    CE) *16*, 128, 130–2,
    134, 140, 148, 190
Sotheby's 212
Spain 118, 144, 189, 192,
    220, 236, 254
Sprimont, Nicholas 218
Suwa Sozan 246
Syria: *albarello*, or storage
    jar (1301–1400 CE)
    154–7

T
Ta Pitaria 64
Tang dynasty (618–907
    CE) 110, 112

Tapestry Workshop 80
thermoluminescence
    dating 68, 112
Toft, Thomas 200
Tokugawa Yoshimune,
    Shōgun 138
Tomb of Rekhmire 62
Toyotomi Hideyoshi 222
Tradescant collection 176
Tutankhamun (*c*.1332–
    1323 BCE) 46

U
Ulmschneider, Katharina
    114
United Kingdom (UK)
    216
    asymmetrical "Betu"
    Series I (2009 CE)
    258–9
    boar's head tureen
    (1755–60 CE)
    218–19
    lustred bowl (2005
    CE) **254–7**
    "Peruvian face"
    bridge-spouted vessel
    (*c*.1880 CE) **234–5**
    pot (*c*.1975 CE) **250–3**
    Schools of Design 234
    teapot and lid (*c*.1772
    CE) **222–3**
    tiles with red lustre
    decoration (1882–88
    CE) **236–7**
    jug with grotesque
    decoration (1625–50
    CE) **186–9**
    Staffordshire
    slip-decorated
    earthenware Toft
    dish (*c*.1600 CE)
    200–1
University for the
    Creative Arts 258
Urbini, Francesco 164

urushi-e (hand) painting
Usui Collection 202
Uzbekistan: bowl with
    epigraphic decoration
    (*c*.10th–11th centuries)
    126–7

V
Vetancurt, Agustín de 220
votive objects **48–51**, 62,
    100

W
wabi sabi 140–2, 174,
    260
Wedgwood, Josiah *15*,
    254
West Magazines of
    Knossos 64
West Surrey College of
    Art and Design 258
white gold porcelain 214
white ware 118, 134–5,
    148–51, 206, 254
Willems, Joseph 218
Worcester Porcelain
    Factory 222

Y
Yabu Meizan 226
yellow-glazed ware 122
Yongle, Emperor 190
Yuan dynasty (1271–
    1368) *16*, 140, 180, 190
Yue kilns 110
Yūshin Rintei *228*

Z
Zwinger Museum 210

Accession credits

1. Decorated pottery jar
    Grave 1644, Naqada, Egypt
    Flinders Petrie Excavations, 1895
    AN1895.482
2. Black-topped pottery beaker
    AN1895.1220
    AN1895.795
    AN1910.692
3. "Skidding goat" pottery vessel
    AN1971.980
4. Bowl
    AN1981.986
5. Polychrome Nal funerary pottery
    AN1945.4
    AN1945.5
    AN1945.6
6. Model chariot with four-wheeler
    battle car
    AN1925.291
7. Terracotta figure of a bull or ox
    EACh.1 & 4
8. Indus Valley pots
    EAX.7288 & EAX.7424
9. Pottery lion
    Egyptian Research Account
    Excavations, 1897–99
    AN1896–1908 E.189
10. Terracotta votive figurines
    AN1896-1908.AE.990
    AN1896-1908.AE.1010
    AN1924.32
11. Red polished stag-shaped figurine
    with three legs
    AN1888.624
12. *Kernos* with two rows of
    containers and central bowl
    AN1925.677
13. Sumerian king list
    AN1923.444
14. Two-handled spouted cup
    AN1930.645
15. Rhyton
    AN1896-1908 AE.780
16. Pithos
    AN1896-1908.AE.1126
17. Megalithic small pots with finial lids
    EAX.2397
    EAX.2396
    EAX.2393
18. Jar with octopus motif
    AN1911.608
19. Vessel in the form of a zebu
    AN1964.347
20. Terracotta horse rhyton
    AN1971.858
21. Barrel-shaped jug
    Bought in Larnaca by subscription,
    1885
    AN1885.366
22. Athenian amphora
    AN1916.55
23. The Shoemaker vase
    AN1896-1908 G.247
    Secondary objects:
    AN1896-1908.G.267
    AN1896-1908.G.287
24. Attic black-figure pottery
    stemmed cup
    AN1974.344
25. Attic red-figure pottery head vases
    AN1920.106
    AN1946.85
26. Boeotian black-figure *skyphos* depicting
    Odysseus at sea and with Circe
    AN1896-1908.G.249
    Secondary object
    WA1863.5528
27. Terracotta female figurines
    EA1958.3–11
28. Plaque with *yakshi* (nature spirit)
    or mother goddess
    EAX.201

29. Reconstructed Dead Sea scroll jar
AN1951.477

30. "Magic Bowl" with Aramaic script
AN1931.176

31. Jar with heavily moulded
decoration in Central Asian style
EA1956.964

32. Earthenware figure of a camel
EA2012.189

33. Storage jar
EA2005.85

34. Two tin-glazed bowls
EA1978.2137
EA1978.2141

35. Bowl with animals and plants
EA2005.42

36. Bowl with epigraphic decoration
EA1978.1758

37. Black ware jar with white stripes
Purchased with assistance from the
friends of the Ashmolean Museum
EA1998.219

38. Dish with purple splash
EA1956.1344

39. White ware dish with incised lotus
decoration
EA1989.193

40. Jug with epigraphic decoration
EAX.3110

41. Nabeshima porcelain cup
Jeffrey Story and Walter Cook
Bequest
EA1985.51

42. *Kintsugi* bulb bowl
EAX.1549

43. Bowl with seated figures by a
stream
EA1956.33

44. White ware vase with floral
decoration
EA1956.1399

45. Spouted bowl with two queens
flanking a pillar
WA1954.11

46. Albarello, or storage jar
EA1978.1683

47. Piggy bank
EA1997.5

48. Plat de la Passion Dish
WA2017.1

49. Dish with a composite head of
penises
WA2003.136

50. Childbirth bowl and cover
WA1888.CDEF.C456

51. Porcelain ewer
WA1899.CDEF.C298
Secondary object:
WA1899.3

52. Shino style serving dish for the
Japanese tea ceremony
Purchased with the assistance of the
Story Fund, 1988
EA1988.12

53. Palissy ware dish
AN1685.B.586
AN1685.B.584
AN1685.B.585

54. Pouring vessel, or *kendi*, in the
form of an elephant
EA1978.1700

55. Dish in the form of a peach
EA1987.10

56. Jug with grotesque decoration
WA2008.65

57. Blue and white vase with figures
and a poem
EA1978.1276

58. Kraak-style plate with Dutch East
India Company monogram
Presented by M. C. and J. W.
Christie-Miller
EA1976.53

59. Misshapen baluster jar with flowers
Kakiemon workshops
EA2000.19

60. Staffordshire slip-decorated
earthenware Toft dish
AN1949.343

61. Five-piece lacquered Imari porcelain
garniture
EA2012.1a–e

62. **Vase with grape vine**
Presented by Mr and Mrs K.R.
Malcolm in memory of their son
John
EA1974.16

63. **Birdcage vase**
EA1992.114

64. **Osprey, seated**
WA2008.61

65. **The Oxford Plate**
EA1985.10

66. **Boar's head tureen**
WA2007.1

67. **Six tiles with Chinese figures**
WA1956.67.1–6

68. **Teapot and lid**
WA1957.24.1.83

69. **Vase with a design of a bird**
Purchased with the assistance of the
Story Fund
EA1993.12

70. **Satsuma-style vase with lotus plants
and ducks in high relief**
Purchased with the assistance of the
Friends of the Ashmolean Museum
and the Story Fund
EA2008.65

71. **"Peruvian face" bridge-spouted
vessel**
Gift of Barrie and Deedee Wigmore,
2020
WA2020.20

72. **Tiles with red lustre decoration**
WA1974.273.1–12

73. **Baluster vase with flattened
shoulders**
EA1956.682
Secondary object:
EA1995.87

74. **Satsuma cup with chrysanthemums
and key pattern border**
Presented by Sir Herbert Ingram, 1956
EA1956.681

75. **Art nouveau-style vase with a
design of chrysanthemums**
Purchased with the assistance of the
Story Fund
EA1997.41

76. **Vase with winter landscape**
Purchased with the assistance of the
Story Fund, 2000
EA2000.47

77. **Pot**
Courtesy of the artist's estate
WA2005.90

78. **Lustred bowl**
Courtesy of the artist's estate
WA2005.167

79. **Asymmetrical "Betu" Series I**
Courtesy of the artist
WA2010.20

80. **Tea bowl**
Courtesy of the artist
EA2020.183

Credits